Prosody:
Theory and Experiment

Text, Speech and Language Technology

VOLUME 14

The titles published in this series are listed at the end of this volume.

Prosody: Theory and Experiment

Studies Presented to Gösta Bruce

Edited by

Merle Horne
University of Lund, Sweden

KLUWER ACADEMIC PUBLISHERS
DORDRECHT / BOSTON / LONDON

Library of Congress Cataloging-in-Publication Data

Prosody, theory and experiment : studies presented to Gösta Bruce / edited by Merle Horne.
p. cm. -- (Text, speech, and information technology ; v. 14)
Includes index.
ISBN 0-7923-6579-8 (alk. paper)
1. Prosodic analysis (Linguistics) I. Bruce, Gösta, 1947- II. Horne, Merle. III. Series.

P224 .P77 2000
414'.6--dc21

00-060904

ISBN 0-7923-6579-8

Published by Kluwer Academic Publishers,
P.O. Box 17, 3300 AA Dordrecht, The Netherlands.

Sold and distributed in North, Central and South America
by Kluwer Academic Publishers,
101 Philip Drive, Norwell, MA 02061, U.S.A.

In all other countries, sold and distributed
by Kluwer Academic Publishers,
P.O. Box 322, 3300 AH Dordrecht, The Netherlands.

Printed on acid-free paper

Printed in the Netherlands

CONTENTS

INTRODUCTION

The study of prosody is perhaps the area of speech research which has undergone the most noticeable development during the past ten to fifteen years. As an indication of this, one can note, for example, that at the latest International Conference on Spoken Language Processing in Philadelphia (October 1996), there were more sessions devoted to prosody than to any other area. Not only that, but within other sessions, in particular those dealing with dialogue, several of the presentations dealt specifically with prosodic aspects of dialogue research. Even at the latest Eurospeech meeting in Rhodes (September 1997), prosody, together with speech recognition (where several contributions dealt with how prosodic cues can be exploited to improve recognition processes) were the most frequent session topics, despite the fact that there was a separate ESCA satellite workshop on intonation in conjunction with the main Eurospeech meeting which included over 80 contributions. This focus on prosodic research is partly due to the fact that developments in speech technology have made it possible to examine the acoustic parameters associated with prosodic phenomena (in particular fundamental frequency and duration) to an extent which has not been possible in other domains of speech research. It is also due to the fact that significant theoretical advances in linguistics and phonetics have been made during this time which have made it possible to obtain a better understanding of how prosodic parameters function in expressing different kinds of meaning in the languages of the world.

One of the researchers who has made a significant contribution in shaping the methodology and goals of prosody research is Gösta Bruce. His studies of Swedish intonation in sentence perspective have involved both theoretical modelling and experimental phonetic studies testing the theoretical claims. Perhaps the most influential of his contributions was the compositional analysis of Swedish intonational contours into word accents, sentence accent (associated with focus) and terminal juncture (boundary tones) which realize different combinations of two phonological level tones H and L. This insight into higher-level patterning of prosodic patterns led other researchers to apply the same methods to other languages. Janet Pierrehumbert shows in her contribution to this volume "Tonal Elements and their Alignment", how her model of English intonation was influenced substantially by Bruce's model of Swedish accent and intonation. In it, English intonation is also modelled using two underlying tones. The different pitch accents, phrase accents and boundary tones, defined in terms of H and L tones, constitute basic elements in a grammar for generating intonation contours.

Another contribution made by Bruce, as Pierrehumbert shows in her chapter consisted in showing how different tonal properties can be assigned to

M. Horne (ed.), Prosody: Theory and Experiment, 1–10.

different levels of prosodic structure, i.e. how the different tonal components could be assigned to word accents, phrasal (focal) accents and boundary tones. A further hallmark of Bruce's work, as Pierrehumbert points out, was that he contributed to our understanding of how tonal representations are associated with the segmental string, for example, the difference between the two word accents in Swedish is explained in terms of a timing difference in the association of the starred tonal-component of the word accent representations HL* (Accent I) and H*L (Accent II) with the stressed vowel. All these basic ideas were taken up by Pierrehumbert in her analysis of English, a language which differs prosodically in many respects from Swedish. For example, unlike Swedish, English pitch accents are not properties of words but rather post-lexical reflexes of pragmatic aspects of an utterance. English is also characterized by a larger inventory of pitch accents. In her contribution, Pierrehumbert discusses the main features of her model as well as modifications (e.g. reduction in the number of pitch accents, the treatment of pitch range choice and downstep) which it has experienced since its debut in 1980. These have resulted mainly from experimental work in collaboration with Mark Liberman and Mary Beckman. As Pierrehumbert shows, the empirical research has also raised questions related to the issues of tonal alignment as well as locality and lookahead in tonal implementation. A better understanding of the pragmatic meanings associated with the pitch accents has also been obtained in recent years due to work with Julia Hirschberg and practical studies of intonational variation in large data-bases have been made possible due to the development of the ToBI (Tones and Break Indices) transcription system based on Pierrehumbert's model.

In his chapter, "Bruce, Pierrehumbert, and the Elements of Intonational Phonology", D.R. Ladd takes up a number of theoretical problems surrounding the nature of tones and tonal association that arise from a comparison of Bruce's ideas and related autosegmental treatments. For example, whereas Bruce's H and L tones correspond to concrete "local maxima and minima in the contour" - turning points - Pierrehumbert's tones do not necessarily correspond to turning points, nor do turning points always reflect the phonetic realization of a tone. That is to say, the tones in her analysis of English are in some cases more abstract than Bruce's and resemble more the underlying tones in tone languages. Ladd raises the question as to how abstract should analyses of tone be in languages like English, which is assumed to have only post-lexical or intonational uses of tones. As opposed to languages with lexical tone, where one can observe alternations in the shape of morphemes and words in different contexts, languages with only post-lexical (intonational) uses of tone often do not allow stringent control over the identification of tones. Ladd also takes up questions regarding the status of 'starred' tones. These have been used by Pierrehumbert in representing English bitonal accents and also retrospectively by Bruce in the representation of

Swedish word accents. As Ladd shows, it is not clear, however, what the defining characteristic of the star is. Furthermore, he provides evidence that shows that it is not always straightforward how the alignment between a starred tone and the segmental string should take place. Ladd further takes up the distinction between association and alignment, where association is seen as an abstract phonological ''belonging together'' that makes no precise predictions about temporal coordination; alignment, on the other hand, ''is specified independently of the identity of the tonal string''. Association is ''digitial'', whereas alignment is ''analogue''.

In their contribution ''Levels of Representation and Levels of Analysis for the Description of Intonational Systems'', Daniel Hirst, Albert Di Cristo and Robert Espesser discuss a number of issues related to the phonetics/phonology interface as well as the relationship of the phonological component with the higher-level syntactic and semantic components. In particular those issues that are relevant for deciding what kinds of information should be incorporated in the phonetic ('surface phonological') transcription of prosodic parameters in cross-linguistic studies are addressed. They distinguish between 'functional representations' that provide information required for the syntactic and semantic interpretation of an utterance's prosody and 'formal representations' which provide information related to the pronunciation of an utternace. Since there is not a one-to-one correspondance between these two types of information in languages of the world, the possibility of making cross-linguistic studies relies on the ability to separate these two forms of representation in a universal theory of language structure. Language-specific parameters then should specify the mapping between form and functional categories. One specific problem discussed is how to derive an optimal symbolic phonetic representation for a F0 curve on the basis of what we know about related physiological and perceptual factors. The design and evaluation of the MOMEL algorithm developed at the University of Aix-en-Provence for automatic modelling of and synthesis of F0 curves in a number of languages is discussed in detail. The phonetic transcription system (INTSINT) designed to allow collection and classification of intonational data in languages at preliminary levels of analyses where pitch accent inventories are not known, is also taken up and discussed. Also discussed are possible ways in which the system could be modified to integrate into the transcription discourse information on variation in overall range and register within intonational units. Finally, a sketch of the way in which cross-linguistic differences between intonation systems can be formalized in terms of phonological representations relating prosodic constituents and tones is presented using data from English and French.

''Perception of Prosodic Prominence'' in particular perceptual judgements associated with variation in accentual patternings is the topic of Jacques Terken and Dik Hermes' chapter. They discuss different typologies of prominence that have been proposed since the time of Trager and Smith and conclude that there is phonetic evidence for at least four types of prominence categories: reduced, full, stressed and accented. Further they take up the issue of frequency scales used in the study of speech intonation and address the question as to which scale best expresses equivalences in the pitch ranges of men and women and which scale best expresses excursion sizes of pitch movements related to the perception of prominence (as opposed to 'liveliness'). They conclude that experimental evidence points to the optimality of the ERB-rate scale for representing excursion equivalences associated with accentual prominences and discuss the possible perceptual processes which may account for this. Variations in the strength of the prominence category 'accent' are known to be related both to the local phonetic characteristics of the pitch movement (e.g. size, slope, timing) as well as to more global characteristics such as declination. Terken and Hermes discuss models for accounting for the variation in accent strength of both single and multiple accents in an utterance. Two models, a High-Level (HL) model and a Pitch Level Difference (PLD) model are used to explain subjects' judgements of strength relations between different pitch movements. One observed stable effect of the perception experiments involving single accents is that falls induce greater strength than the rise and the rise-fall for the same excursion sizes. In the modelling of the relative strength relations between two or more accents within the same utterance, the situation is complicated due to effects from both declination and speaker-specific pitch range, as well as the mutual effect of the prominence of accents on each other's perceived prominence. A survey is made of experimental results made by a number of researchers on this intricate problem. The need for a reliable way of transcribing prominence relations between accents both within and across different pitch registers is also discussed and the authors conclude by pointing to the need for more work in the development of theories of prominence and in relating them to issues in language processing.

Carlos Gussenhoven discusses an interesting case of interaction between tone and intonation at the ends of phrases in his chapter entitled "The Lexical Tone Contrast of Roermond Dutch in Optimality Theory". He shoes how constraint-based Optimality Theory allows one to better understand a number of generalizations related to the phonological behaviour of the two tonal word accents and the intonation (focus and boundary) tones in the Roermond dialect. These include: the neutralization of the lexical tone contrast in nonfocussed nonfinal position, the spreading of the final boundary tones into the phrase, the assimilation of the lexical H-tone to L after a focal L-tone in the same syllable and, most interestingly, infixation of the phrasal boundary tones before the lexical tone. In particular, it is shown that an explanation for the final generalization concerning the 'tonal reversal' phenomenon in Roermond cannot be naturally provided by a derivational rule-based description since it leads to "an insoluable ordering paradox". By using the concept of alignment constraints within Optimality theory, however, the hierarchical ranking of these constraints can be used to determine which tone will be given preference at the end of a phrase. Gussenhoven also takes up the distinction between the concepts 'association' and 'alignment' and uses it to account for the two different patterns of leftward tone spreading in Roermond: whereas "alignment locates tones with reference to the edges of (phonological and morphological) constitutents, association creates a structural connection with a tone-bearing unit". All tones are aligned according to this view, but only some are associated. Association does not occur when there is no free tone bearing unit (TBU) available which is seen to be the case in Roermond when the more freely-timed case of tone spreading is compared with the more accurately timed case where a TBU is available.

Timing of tonal events relative to critical points in the segmental representation of a text is now understood to be an important parameter in relating meaning contrasts with sound structure. There is, however, another aspect of timing that is relevant to the issue of tune-text association and that is the timing of articulatory gestures in the production of the segments to which intonational events such as pitch accents and boundary tones are anchored. This is the topic of Mary Beckman and Bretonnel Cohen's chapter "Modelling the Articulatory Dynamics of Two Levels of Stress Contrast". In this contribution, differences in segmental timing control associated with "word stress" and "sentence stress" are studied in an experiment involving articulatory kinematics and the results are discussed in light of three different models of articulatory timing: a 'truncation' model, a 'rescaling' model and a 'hybrid' model. Jaw movement data involving durations, amplitudes and peak velocities for jaw opening and closing movements for accented, unaccented heavy and reduced CVC ([pɑp]/[pəp]) sequences uttered at 3 different speech tempi are seen to indicate the same relational effects of stress at both levels of prosodic structure. Movement durations were longest for the vowels in accented position and shortest for the reduced vowel, with accent producing a much smaller effect than syllable weight. Movement amplitudes and peak velocities show the same pattern as duration: accented syllables have the largest and fastest movements, while the reduced vowel showed the smallest and slowest movements. Here also, accent had a considerably smaller effect than syllable weight. However, the relatively more prominent sequences in each stress contrast pair had smaller velocities than expected for their larger amplitudes. These observations are explained by a hybrid model which changes both between- and within-gesture parameters by combining the between-gesture shortening mechanism of the truncation model as well as the gestural settling time and displacement changes of the rescaling model. It is suggested that the lengthening observed in accented sequences as opposed to unaccented sequences is a considerably more subtle version of the same effect as the length difference between a lexically specified full [ɑ] and a reduced [ə]. On the other hand, the durational effects observed in the two types of rhythmic prominence/stress examined in the study are different from the type of duration associated with phrase-final lengthening. Although a phrase-final vowel is longer than an identical vowel in a preceding syllable, the movements into and out of the vowel are not faster than the corresponding movements in a non-final CV-sequence.

In her comprehensive survey article, ''Phrase-Level Phonology in Speech Production Planning: Evidence for the Role of Prosodic Structure'', Stefanie Shattuck-Hufnagel discusses experimental evidence that has accumulated during recent years for the assumption that elements of the prosodic hierarchy play a role in speech production planning. Three lines of evidence from studies conducted by Shattuck-Hufnagel and her group are reviewed that support this hypothesis. These results are further related to results from other research groups. One set of studies have provided empirical evidence which suggests that ''stress shift'' is most insightfully interpreted as a prosodic phrase-level phenomenon rather than a movement of lexical stress within the word conditioned solely by stress 'clash'. The view of stress shift as a defining feature of a well-formed intonational phrase, i.e. one that is characterized by the early placement of the first pitch accent, has received strong support from a considerable number of experimental findings. A second set of investigations at several laboratories concerns the patterning of glottalization. The distribution of glottalization of word-initial vowels, for example, would appear to be sensitive to the presence and size of of phrase-level prosodic constituent boundaries as well as to the presence of prosodic prominences. Glottalization has been found to be more frequent at the beginning of intonational phrases; furthermore, glottalization for reduced vowels has been seen to be significantly greater for Full Intonational Phrases than for Intermediate Intonational Phrases. Initial consonant strengthening and pre-boundary lengthening are other processes that appear to be sensitive to prosodic structure. The third line of evidence suggests the importance of phrase-level rhythm in speech production planning. Preliminary findings from on-going research involving error elicitation in various prosodic contexts lend support to the notion of a phrase-level rhythm that builds on the rhythm defined by stronger vs. weaker vowels or syllables. Prosodic structure is thus hypothesized to function as a kind of ''performance framework into which the abstractly specified elements of the words and phrases of the morphosyntactic representation must be fitted in order to receive specifications for their duration, amplitude, F0 etc. that are related to phrase-level phonology and phonetics''. A number of themes for future research are also suggested.

''The Interaction of Constraints on Prosodic Phrasing'' is the title of Elisabeth Selkirk's chapter where a theoretical discussion of constraints related to prosodic phrase pattening is persued within the framework of Optimality

Theory. Selkirk demonstrates first of all using data from a number of Bantu languages, how Optimality theoretic constraint rankings allow a typology of crosslinguistic differences in prosodic phrasing in which varying patterns of prosodic phrasing observed in different languages can be understood as the consequence of different rankings of the same constraints. In the second part of her contribution Selkirk takes up a theoretical discussion of the constraints that interact to produce different prosodic phrasing patterns in English. In particular, it is shown how both focus-based and morphosyntactic-based constraints interact with each other as well as with more specific phonological constraints to explain preferred output representations. Cohesional and demarcative strategies are seen to compete in determining the interface between information (focus) structure, morphosyntactic structure and prosodic structure and these are ranked in different ways with respect to the phonological constraints which regulate the size of a major phrase/intermediate phrase as well as the number of accents in a minor/accentual phrase. A faithfulness constraint that regulates the relation between the input and output accentual representation is further proposed and is assumed to rank highest in the hierarchy of constraints. The initial hypotheses Selkirk makes regarding the number and interaction of constraints in the grammar of English provide the testing ground for further generalizations on phrasing and for experimental testing against more data.

As a result of the theoretical advances made in the understanding of prosodic structure patterning and in the experimental methods used to extract acoustic information on the phonetic correlates of this structure, considerable advances have been made in the development of algorithms for the automatic recognition of prosodic constituents in spoken language. This is the subject matter of Mari Ostendorf's chapter ''Prosodic Boundary Detection''. In her contribution, she surveys a number of the algorithms that have been proposed as well as their applications. Differences in theoretical modelling (e.g. those between superposition models vs. linear-sequence models) as well as problems in computational modelling of the acoustic parameters associated with prosodic constituent boundaries (e.g. multiple time-scales and overlapping of prosodic cues with information structure) have provided a basis for the development of several different types of phrase detection algorithms. However, of the computational models that can potentially be constructed, it is only those which can be automatically trained and include some notion of uncertainty that Ostendorf considers to be of interest at this point when we still do not have a complete understanding of how acoustic cues interact in signalling prosodic structure. These include probabilistic, neural network and template-matching methods, but not rule-based approaches. As Ostendorf also demonstrates, choice of a particular recognition algorithm depends to a great deal on the particular application intended for it. These include automatic prosodic labelling, where the goal is to make an explicit decision about where prosodic phrase boundaries are, constraining speech recognition and language processing search procedures by integrating language cues with phrase boundary detection, and prosodic scoring of linguistic hypotheses with regard to the consistency between the acoustic correlates of prosodic phrasing and the expected phrasing associated with the hypothesized word sequence (aids in resolving ambiguity of multiple valid interpretations). Ostendorf concludes by pointing at future directions in prosodic phrase detection research. Improvements in current recognition algorithms can be expected to occur as improvements in computational resources are made available and as advances are made in attaining a better understanding of the various cues associated with prosodic structure. Moreover, a better understanding of the patterning of prosody in spontaneous speech will aid in increasing the usefulness of prosodic phrase detection. The relationship of phrase structure and discourse structure and the associated interacting acoustic cues is another area for future research which will in its turn allow further applications using prosodic information such as human-computer dialog processing and speech translation.

Nick Campbell's contribution "Timing in Speech: a Multi-Level Process" concentrates on the issue of durational modelling for speech technology applications. Both speech synthesis and speech recognition are dependent on having access to information on timing characteristics in different contexts and speaking styles. In this respect, Campbell concentrates on showing how a measure of a segment's *lengthening* is more useful information than a measure of its duration. Campbell first presents a survey of theoretical and experimental developments in timing research since the work of Dennis Klatt and goes on to demonstrate how more recent work shows that the variability in durations is more insightfully accounted for by relating this variation to levels of prosodic structure higher than the segment. Moreover, Campbell maintains that the syllable is a mediator between higher (linguistic/semantic) and lower (physiological) levels of timing control. A syllable-level timing framework applying notions of 'elasticity' (stretching or squashing of segmental durations) and 'accomodation' of this elasticity in a syllable-sized frame is shown to make possible the prediction of segmental durations to a very high degree. Data from English and Japanese are used to illustrate the 'two-tier' model of speech timing combining physiologically-motivated inherent durational characteristics and a cognitively motivated prosodic structure at the level of the syllable and above. Data are primarily from read speech, but variability in timing observed in the comparison of different speech styles, e.g. between read and spontaneous speech is also discussed. Extension of the model to account for segmental distributions associated with the higher prosodic constituents (foot, phrase and utterance) is discussed as well as problems associated with identifyng the different kinds of lengthening that must be assumed. Data from spontaneous dialogue is also used to illustrate how vowel duration ratios can be used to identify speech-act type.

Differences between spontaneous speech and read speech constitute the central subject matter of Julia Hirschberg's chapter ''A Corpus-Based Approach to the Study of Speaking Style''. Acoustic/prosodic data including speech disfluencies, speech rate and pausal duration, intonation contour and pitch range from three American English corpus studies are discussed as regards information related to speech style-differences. In one study, it was observed for example, that the speaking rate was faster for read speech than spontaneous speech. Further, the boundary F0 contour-type on declaratives, yes-no questions and wh-questions was also seen to vary in different speaking styles. Declaratives in spontaneous speech are more often characterized by final rises than in read speech, while yes-no questions are more often characterized by a final falling intonation in spontaneous speech than in read speech. Speech disfluencies constitute a phenomenon which seems to be limited to spontaneous speech, and their associated phonetic features formed the content of another study. It is demonstrated that one type of disfluency, 'self-repair', is characterized by cues such as word fragments, glottalization and incomplete articulatory gestures, *inter alia.* Silent pauses associated with disfluency intervals are also seen to be shorter than fluent pauses and disfluent utterances studied had different intonational phrase boundary location and realization than read speech. However, speech disfluencies are, in general, seen to be unreliable cues for speaking style generation due to their low frequency of occurrence. A further study examined prosodic correlates of discourse structure and determined that both spontaneous and read speech were characterized by prosodic cues to discourse structure. Significant differences were observed between the two speaking styles as regards certain absolute measures of global parameters related to overall range and timing e.g. maximum and average F0 for each intermediate phrase, speaking rate, and duration of a preceding pause. However, there did not appear to be any significant differences as regards other relative values associated with the amount of change in prosodic parameters between phrases, e.g. amount of F0 and decibel change from a previous phrase or in the choice of nuclear pitch accent or phrase final contour. This suggests that although more global features such as overall range and timing may vary greatly over speaking-styles, the more local features associated with in-between-phrase relationships are relatively constant.

The studies presented here were initially sollicited by the editor in order to compile a festschrift tribute to Gösta Bruce on his 50th birthday. Since the contributing authors are all internationally recognized experts in different areas of the field of prosody, the content of the papers provides readers with a current and comprehensive picture of the major areas of research within the field. The contributions not only provide a survey of major developments in the field during the past 15-20 years but also present on-going work as well as point to areas where future research is needed. The chapters deal with topics in experimental phonetics, theoretical linguistics and language technology applications and combine to illustrate how researchers with different areas of expertise have interacted in exchanging knowledge on the structure and interaction of prosodic phenomena. This fruitful and productive cooperation between theoreticians and experimenters both within linguistics/phonetics and speech technology has resulted in making prosody one of the most highly developed subbranches of speech research.

Merle Horne
Lund, Sweden
June, 1998

JANET PIERREHUMBERT

TONAL ELEMENTS AND THEIR ALIGNMENT

1. INTRODUCTION

In English, many different melodies are possible on any given word or phrase. Even a monosyllabic word, such as *Anne* can be produced with many qualitatively different melodic patterns, as illustrated in Figure 1. This situation provides a contrast with languages such as Mandarin, in which the tonal pattern is an intrinsic part of the lexical representation. In English, the choice of the melody is not entailed by the choice of words, but rather functions independently to convey pragmatic information. Specifically, it conveys information about how the utterance is related to the discourse and to the mutual beliefs which interlocutors build up during the course of the

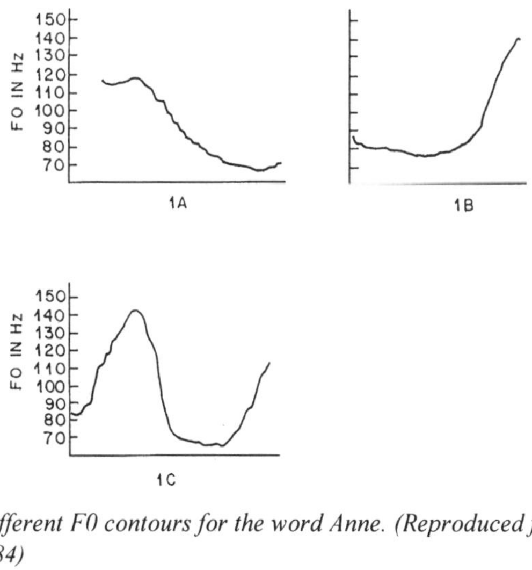

Figure 1. Three different F0 contours for the word Anne. (Reproduced from Liberman and Pierrehumbert 1984)

discourse, as discussed in Ward and Hirschberg (1985) and Pierrehumbert and Hirschberg (1990).

This situation has been recognized from the earliest work on English intonation and it has fostered attempts to phonologically abstract the melody line from the words. Like all phonological abstractions, these efforts have the

M. Horne (ed.), Prosody: Theory and Experiment, 11–36.

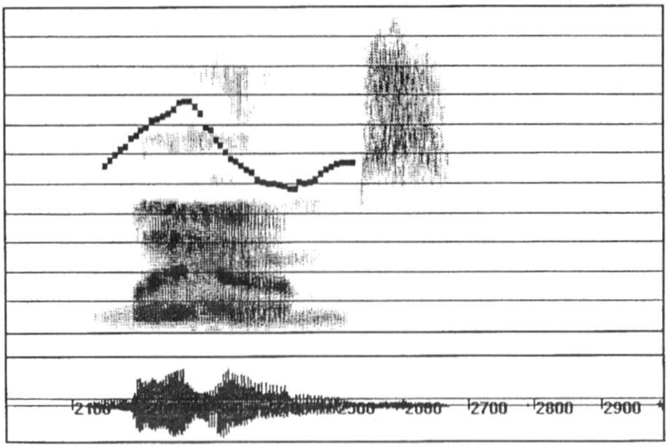

Figure 2a. A declarative pattern with a continuation rise (H L H%) on the word* limb. *The alignment of the contour with the segments can be seen by referring to the spectrogram on which the F0 contour is superimposed.*

goal of mapping out the space of contrasts in a way which treats as the same tokens which are linguistically comparable, and which treats as different tokens which contrast. The problem has not proved to be an easy one. One difficulty is illustrated in Figure 2, which displays F0 contours for a single pattern (a declarative pattern with continuation) on three different words, *limb*, *limo*, *limousine*. Each of the words was produced as the medial element in a list of three items. Although the general character of the pattern is obviously the same (a rise, a fall, and then a smaller rise) the equivalence is not captured in a syllable-by-syllable transcription of F0 levels or changes. The entire pattern is expressed on the first and only syllable of *limb*, with the result that the F0 peak is only part way into the vowel. For *limo*, the first syllable is entirely rising with the peak falling towards the end of the vowel. The F0 on first syllable of *limousine* is also entirely rising, and the rise indeed continues well into the /m/ which serves as onset for the second syllable. Thus, the patterns are only rendered equivalent by a representation which distinguishes the contour itself from the way that the contour is aligned with the syllables.

A second difficulty arises from the character of intonational meaning. Because intonation patterns are not referential (that is, they do not denote objects in the world), their meanings are notoriously slippery. Admittedly, the meanings of many words and morphemes (such as pragmatic particles) are every bit as slippery. The result, however, is that shortcuts to determining phonemic contrast are not available. There is no equivalent in intonational studies to showing someone a picture and asking *"Is this a 'pat' or a 'bat'?"*. To establish the equivalence of some patterns and the contrastiveness of

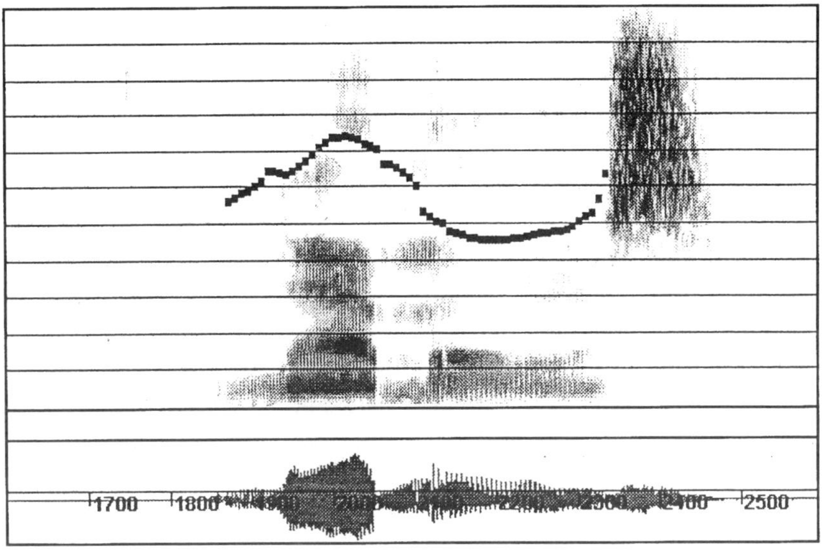

Figure 2b. A declarative pattern with a continuation rise on the word limo.

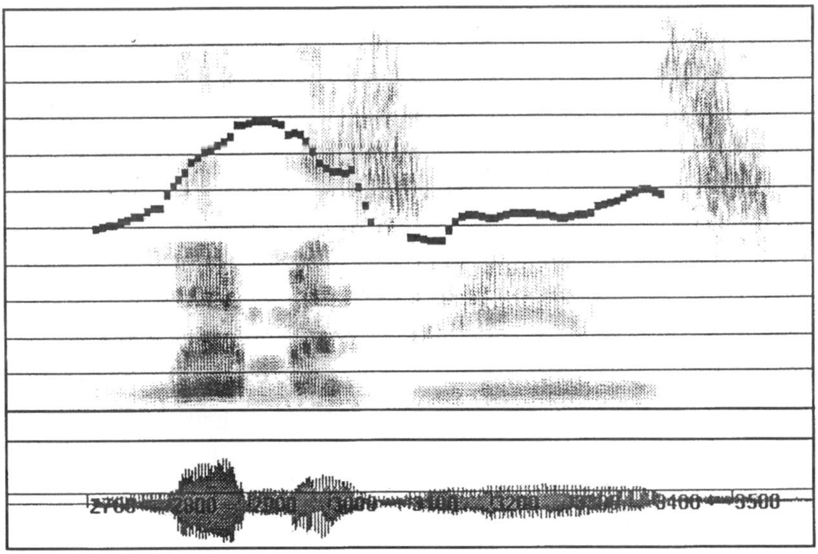

Figure 2c. A declarative pattern with a continuation rise on the word limousine.

others, exacting experiments are required. Even now, only a few of the experiments that would be relevant have actually been carried out.

A last difficulty is that it has often proved difficult to separate conventionalized and quasi-categorical aspects of the intonation pattern from stylistic variation. Indeed some scholars question whether such a separation is possible.

In the early literature, there are two competing approaches to abstracting the intonational pattern from the words. One, due to Trager and Smith (1951) and Pike (1945), decomposes the melody line in terms of tone levels, positing four phonemically distinct levels (L, LM, HM, H). The other, associated with Bolinger (1951, 1958) as well as with most phoneticians of the British school, decomposes melodies in terms of F0 changes or trajectories. Both approaches had advocates up through recent times; for example, Liberman (1975) uses four tone levels as basic units of description, whereas Ladd (1979) uses tone changes. Both approaches have intrinsic advantages and drawbacks. Two-tone models have become a recent standard because they integrate insights from the two approaches while avoiding most of the drawbacks.

An important drawback of four-tone models was already pointed out in Bolinger (1951). It is obvious that the four tones of these models do not denote absolute F0 levels (even within the speech of a single individual), but rather relative position within the pitch range; overall pitch range varies with the speaker's voice level, emotional state, and choice of style. That is, tones are relativized to an F0 space in a way which is reminiscent of vowels in the formant space (see Ladefoged and Broadbent 1957), but which is even more extensive and pervasive. As Bolinger pointed out, the sparsity of tones relative to the rate at which overall pitch range is manipulated leads to pervasive ambiguity in analysis in four-tone models. There is no principled way to distinguish a L LM L pattern produced in a large pitch range from a L H L pattern produced in a small pitch range, and linguists should not, therefore, imagine that listeners actually do so. This criticism is extremely cogent. Theories which enforce spurious distinctions (or distinctions which could not in practice be available) are indeed generally suspect. Two-tone models of intonation substantially reduce the ambiguity of analysis by reducing the inventory of different tonal strings and by making explicit provision for the role played by pitch range in scaling the phonetic outcome.

A further problem with four-tone models, discussed in Pierrehumbert (1980), relates to the existence of stepping patterns, as illustrated in Figure 3. Reserving L for the low termination at the end of the whole pattern, the three steps in this pattern would be described as H, HM, LM. Since only four tones are available in this theory, the implication is that stepping patterns can have only three steps (plus the terminal fall). This implication is false; phrases containing four or even five steps are attested, and the only real limitation appears to be the length of the intonation phrase. But if we countenanced six or

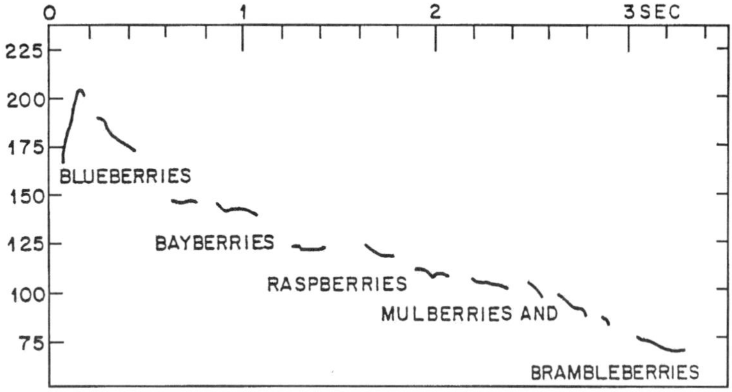

Figure 3. An F0 contour with a series of downsteps (Reproduced from Liberman and Pierrehumbert 1984).

seven different tones in the system, it would have far more tonal distinctions than were needed for any other purpose, and the problem raised by Bolinger would be further aggravated.

The solution to this problem proposed in Pierrehumbert (1980) is to reduce the tonal inventory to H and L and to posit a process of downstep which successively lowers the realization of the H tones. That is, the pattern shown in Figure 3 has the same tonal component repeatedly, with each one stepped down compared to the one before. This solution is inspired by the downstep found in many African tone languages. As pointed out in Anderson (1978), the potentially infinite number of steps in a downstepped sequence can be described as an abstract exponential. Liberman and Pierrehumbert (1984) build on this insight to develop a model of downstep in English in which the process literally is an exponential decay under the correct representation of F0 scaling.

F0-change models also have characteristic weaknesses. The first weakness concerns the issue raised in Figure 2, that of the alignment of the pattern to the segmental material. In order to describe the alignment under different conditions of stress and phrasing, it proves necessary to refer to the endpoints or extrema of the F0 changes. For example, the termination of the pattern illustrated in Figure 2 has a local maximum aligned exactly to the end of the phrase. Any reference to the endpoints of the changes amounts to decomposing the changes into successive F0 levels. Then, too, English has some qualitative distinctions in the positioning of a change in the pitch range. The most uncontroversial are probably the distinction between the "high rise" found in some questions and the "low rise" found in declaratives with continuation, and the distinction between melodies which fall to the bottom of the range and those which fall only part way. Distinguishing these patterns phonologically

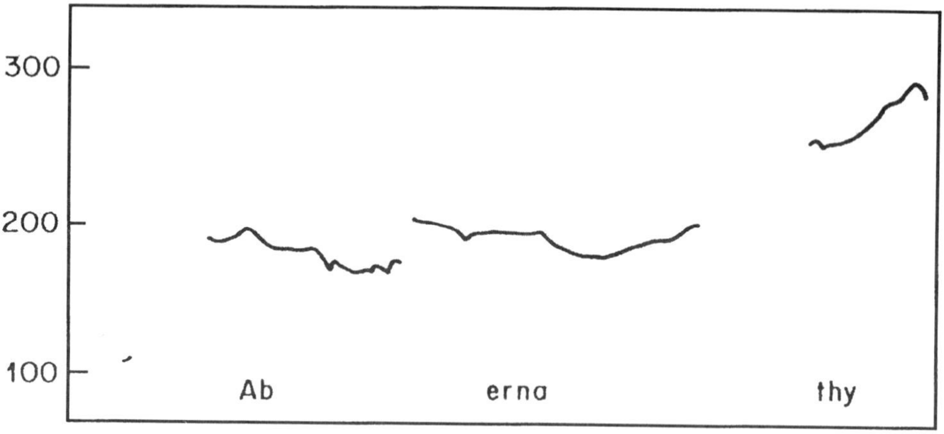

Figure 4. A low rising F0 contour (L L H%) on the word* Abernathy, *illustrating the fact that a pitch accent does not necessarily entail a pitch inflection on the accented syllable.*

requires some way to allude to position in the range, not just to the direction of change. Lastly, as argued in Pierrehumbert (1980), pitch accents can in certain circumstances emerge as F0 levels rather than F0 changes. For example, the nuclear stress on the utterance *Abernathy* whose F0 contour is displayed in Figure 4 has a relatively low F0 value on the syllable /æb/, which is continued directly into a low rising terminal configuration. Despite the impression of prominence on this syllable, there is no movement associated with it as such.

2. TWO-TONE MODELS OF ENGLISH INTONATION

A comprehensive model of English intonation using two tones (L and H) was proposed in Pierrehumbert (1980). It provided a grammar of English melodies and sketched an algorithm for mapping outputs of this grammar into F0 contours. The tonal phonology was revised in Beckman and Pierrehumbert (1986) and the F0 scaling algorithm was revised in Liberman and Pierrehumbert (1984). An intonation synthesis algorithm based on this model is presented in Anderson et al. (1984). This entire body of work builds substantially on Bruce's (1977) model of Swedish accent and intonation. Before presenting it, I would accordingly like to review the contributions of Bruce's work.

2.1 Bruce (1977)

Bruce's (1977) Ph.D. dissertation *Swedish Word Accents in Sentence Perspective* was a unified treatment of accent and intonation in the Stockholm dialect of Swedish. By examining the F0 contours of utterances which systematically varied accent, stress, and phrasing, Bruce arrived both at a theoretically novel treatment of Swedish melody, and at a method for synthesizing F0 contours. The treatment of the melodic line decomposes it into H and L tones, at a relevant level of abstraction. Some tones originate from the words (via the well-known Accent I – Accent II distinction amongst Swedish words) and others originate from the phrase. The synthesis scheme reconstructs the F0 contour by mapping tones onto F0 targets and interpolating between the targets. When tones are crowded together, some tonal targets are not fully realized; the priority amongst the targets is determined by their phonological status.

The work is distinguished first by its unified treatment of phonology and phonetics. Rather than taking the phonology as given, the work uses phonetic data to clarify phonological issues. Overall, the work is a lesson in the fact that phonology and phonetics cannot be studied separately, but only together.

A second distinctive feature of this work is its abstract view of the basic tonal elements. Much early work on prosody and intonation (such as Fry 1958) takes citation forms of words as basic. Insofar as the intonation of continuous speech was treated at all, it was in terms of concatenation and reduction of word patterns which could have been found in isolation. Bruce, in contrast, adopted the working hypothesis that the "basic" word patterns were abstract patterns whose character would be revealed by examining the full range of variation found when words are produced in different contexts. From his survey of contextual variants, the prenuclear rather than the isolation version of each accent emerges as the most basic form, provided that abundant segmental material gives it a vehicle for full expression. The phonctic form that accents acquire in isolation arises through the interaction of the accent itself with tonal correlates of phrasing; hence the isolation form is more complicated than the prenuclear form. In addition, the postnuclear forms of the accents are phonetically modified by downstepping, while all accents are subject to undershooting in situations of tonal crowding. Because of these effects, no particular tone corresponds to an invariant level or F0 configuration. The abstract use of L tones is particularly striking, because some L tones emerge as valleys in the F0 contour, others as elbows (in downstepped contexts), and others are barely expressed at all.

Bruce's approach also made it possible for him to discover the common denominator of the Swedish Accent I and Accent II patterns – namely, the H L found in both cases – and to show that the critical difference between the two relates to their alignment. For Accent I, the HL is aligned earlier with respect to the segmental material, and for Accent II, the HL is aligned later. The contrast is displayed in Figure 5, reproduced from Figure 28 of Bruce (1977). Perception experiments using synthetic speech with controlled

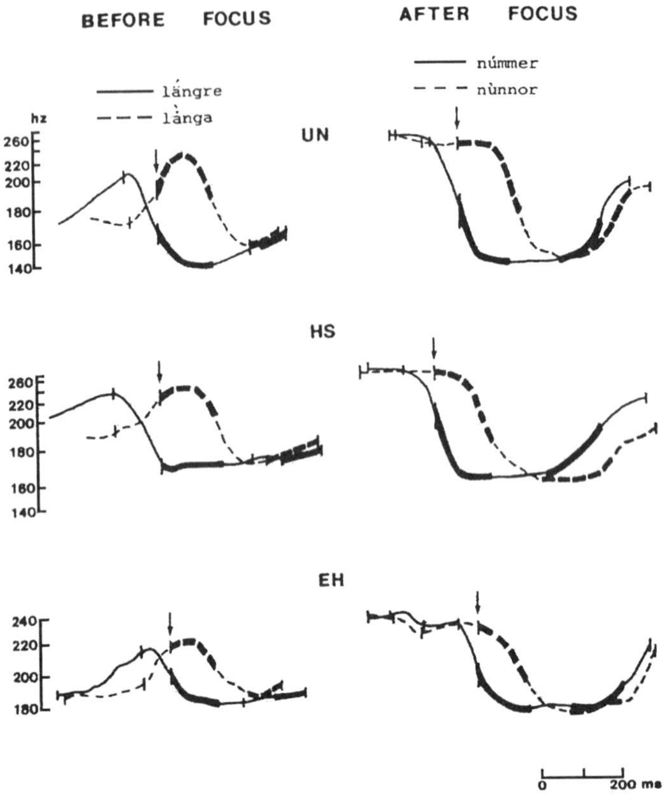

Figure 5. Figure 28 of Bruce (1977), illustrating how Swedish Accent I and Accent II differ in their alignment characteristics. Bruce's figure caption is:"The word accent distinction in non-focal position. F0-contours of accent I- and accent II-words in the second and third positions for three speakers. The line-up point is at the CV boundary of the stressed syllable."

variation of the F0 contours further substantiate his point that relative alignment is the heart of the Accent I/Accent II distinction.

The actual F0 contour on any word emerges from Bruce's model through the interaction of the word accent with tonal features assigned at the phrase level. Specifically, the phrase contributes a H tone which is phonetically manifested towards the end of the word carrying phrasal stress, and the initial and terminal junctures also contribute tonal targets. The citation form is then reconstructed as the form produced in a specific prosodic context – when the word is both phrase-final and bears the main stress of the phrase. The importance of this point cannot be overemphasized. In effect, there is no such thing as an intonation pattern without a prosodic context. The nuclear position and the phrase-final position are both particular contexts, and as such leave their traces in the intonation pattern.

In Bruce's model, some parts of the F0 contour are more phonologically critical than others. The critical parts are those at which paradigmatic or syntagmatic information is concentrated. The differential importance of different parts of the contour is captured by a modified target-interpolation model for the phonetics. Phonological tones are mapped onto F0 targets, and the targets are connected together to make a continuous F0 contour which includes F0 values for sonorant regions which are not tonally critical. Bruce acknowledges the fact that in situations of tonal crowding, phenomena of undershooting and readjustment may create departures from the capabilities of a pure target-interpolation model. These deviations from the model should not disguise the fundamental insight, which is that some parts of the F0 contour are more critical than others, and that tonal elements are semi-localized in their phonetic manifestations.

Bruce's treatment of tonal alignment provides an important antecedent to the present understanding of licensing. Bruce's work appeared just after the first works in autosegmental and metrical phonology (Leben 1973, Goldsmith 1976, Liberman 1975). Autosegmental phonology was originally motivated by regularities in the tone patterns of African tone languages. To describe these patterns, the theory attributed unaligned melodic strings to the underlying representations of morphemes or words. The surface alignment of the tones was derived by rules for associating the elements of the strings (namely tones) to tone-bearing units. A typical rule mapped the tones left-to-right in one-to-one association with tone bearing units, with some provisions (such as spreading, crowding, and deletion) for situations in which the length of the tonal string and the number of tone-bearing units differ. Liberman's metrical treatment of English intonation differed from autosegmental accounts of tone languages in attributing melodies to phrases rather than to words, a consequence of the nonlexical status of the English melodic line. It also differed in its specific proposal for aligning the tones: the alignment algorithm appealed to hierarchical structure in order to align the tones to the metrically strongest elements in the phrase. However, the general outlines of the two proposals are similar. In both, the entire tonal sequence for a meaningful phonological construct ends up aligned to specific tone-bearing units of that construct through an across-the-board process.

Bruce's treatment of Swedish accent and intonation fits in this mold by distinguishing the morphological source of a tone from the phonological spot where it shows up on the surface. By the morphological source, I mean the domain in which a choice of tone is contrastive. In intonation languages – in which tones are relatively sparse – the underlying domain is typically bigger than the critical phonetic region for the tone. For example, English has the edge of each phrase marked with a tone; in this case the underlying domain is the intonation phrase, but the surface alignment is to the last syllable or the boundary. An important difference between Bruce's treatment and those just mentioned is that there is more than one source of tones; the eventual melody arises from the interplay of word tones and phrasal tones. Furthermore, tones from each respective source have their own characteristic timing and scaling behavior. This more complicated picture finds an analogy in the present understanding of licensing. Compare, for example, the treatment of syllable-level phonotactics in Goldsmith (1990) and Coleman (1992). These accounts differ in detail, but in both different nodes in the syllable carry different featural properties. The properties at each level have a characteristic temporal scope, and the surface form of the syllable arises from the interplay of the properties at the various levels.

2.2 Pierrehumbert (1980)

Pierrehumbert's (1980) model of English intonation (later revised in Beckman and Pierrehumbert 1986) adopted many of the main insights of Bruce's model of Swedish. Specifically, it described even very complex F0 contours in terms using just two basic tone levels (H and L). It proposed bitonal pitch accents, phonologically located on metrically prominent syllables. Early-aligned accents are phonologically distinguished from late-aligned accents. Relative alignment is indicated notationally by a *; L*+H has L on the stressed syllable with a trailing H, whereas L+H* contrasts by having a H on the stressed syllable with a leading L. The model also distinguished pitch accents from boundary tones on the basis of their characteristic timing behavior. Making this distinction allows the inventory of pitch accents to be the same in prenuclear and in nuclear position. The relatively complex F0 contours found on phrase-final nuclear syllables arise not from a special nuclear inventory, but rather from the crowding of the pitch accent and the boundary tones onto a single syllable. Lastly, a fully explicit but non-trivial phonetic implementation component maps phonological tones onto context-dependent F0 targets. Interpolation and smoothing between targets is responsible for the continuous F0 contour observed. All of these basic features also characterize Bruce's model for Swedish.

English contrasts with Swedish in that the pitch accents are not underlying properties of words. Instead, they are independent pragmatic morphemes which are co-produced with words. The fact that the pitch accents land on the metrically prominent elements of words may be attributed to the general process of entrainment in motor control – this is the process whereby your two hands become synchronized if you pat your head and rub your stomach at the same time.

A further contrast between English and Swedish is that English has more different kinds of pitch accents. Pierrehumbert (1980) proposed seven different pitch accents: H*, L*, L+H*, L*+H, H+L*, H*+L, H*+H. Beckman and Pierrehumbert (1986) reduced the inventory to six by eliminating the H*+H as a categorically contrastive element. English also has more different boundary treatments than Swedish. In the Beckman and Pierrehumbert model, there are two levels of intonational phrasing, the intermediate phrase and the full intonation phrase. Each of these has a boundary tone, either L or H, although the timing behavior of these tones is rather different. The intermediate phrase boundary tone tends to spread over the entire region from the nuclear accent to the end of the phrase, whereas the intonational boundary tone is more localized right at the phrasal edge. The cross-product of the two choices creates four different post-nuclear configurations for intermediate phrase boundaries which are also intonational phrase boundaries. An optional phrase-initial boundary tone for the intonation phrase is also posited.

Given this inventory of elements, a full grammar of possible patterns is shown in Figure 6. This grammar is graphed as a finite-state network; any path through the network yields a well-formed phrasal melody. The grammar is constructed as if an intermediate phrase could have an indefinitely large number of pitch accents; this is obviously an idealization. In practice, most phrases have one, two, or three pitch accents, and it is extremely unusual to find a phrase with as many as five accents. The responsibility for describing this limitation need not fall on the intonational grammar per se, however; provided that the accents are phonologically constrained to align with metrically strong syllables, then the number of strong syllables in real live phrases naturally limits the number of accents. The grammar is also constructed on the assumption that the pitch accents and boundary tones are found in all possible combinations. As noted above, all combinations of pitch accent with boundary tones are indeed found. However, combinations of different accents within the same phrase are not found to the extent expected, see below.

In the model, tonal sequences are mapped onto F0 targets by locally context sensitive implementation rules. A regular process of interpolation contexts the targets, as discussed above. The realization rules resemble the transformational rules of generative phonology by virtue of applying when their structural description is met, with a structural description being a fragment of a complete tonal analysis. For example, the rule applying to a H% after H applies to any H% in this local context, regardless of what pitch accents may occur in the greater context. They differ in that the output of the rules are not phonological descriptions, but rather parametric phonetic values. In addition, the rules are not ordered in a derivation. Instead, they apply in a "running window" over the phonological description, mimicking the process whereby speakers transform their abstract intentions for an utterance into actual phonetic outcomes with particular physical characteristics.

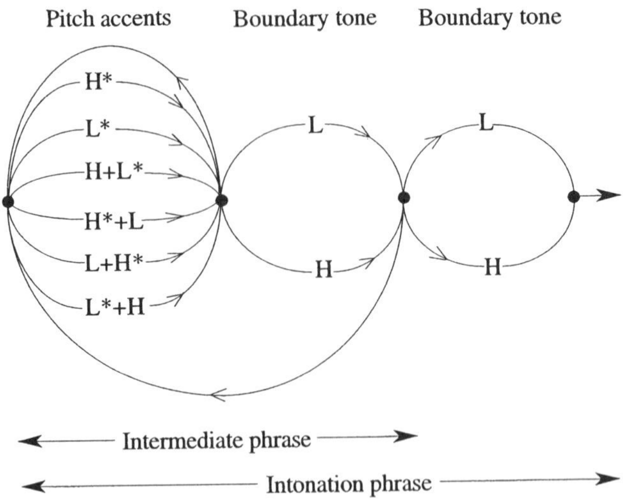

Figure 6. The grammar of English intonation patterns, according to Beckman and Pierrehumbert (1986).

There are three main contextual dependencies in the model. First, the entire space of F0 targets depends on the speaker's choice of pitch range. In the original Pierrehumbert (1980) model, a choice of pitch range was equated with the choice of first peak value. Experimental results in Liberman and Pierrehumbert (1984) and Pierrehumbert and Beckman (1988) led to a revision; choice of pitch range is now understood as selection of an abstract phrase-level parameter which in effect determines the graph paper on which the tones for the phrase are graphed.

The second major context dependency in the model is downstep. In many African tone languages, the second H in a H L H sequence is lower than the first. If the alternating pattern continues (H L H L H...), then each H is lower than the one before, so that a descending staircase results; see Figure 3 above. An analogous effect is found in Japanese, but triggered only by the H+L accent (and not by H and L tones from other sources, even in alternation). Pierrehumbert (1980) proposed that English has a downstep rule affecting H+L H and H L+H sequences. Sequences displaying these particular accent configurations then staircase downwards, in contrast to the bumpy or slightly descending outcome for a sequence of plain H* accents; see Figure 7. Beckman and Pierrehumbert (1986) revised the proposal about the trigger, holding that downstep in English is triggered by two-tone accents. However, the core insight is retained: specifically, the insight that positing a downstep rule for English makes it possible to analyze the many observed F0 target levels for H tones as manifestations of an abstract two-tone system. This proposal is particularly important in analyzing so-called "calling" contours, in

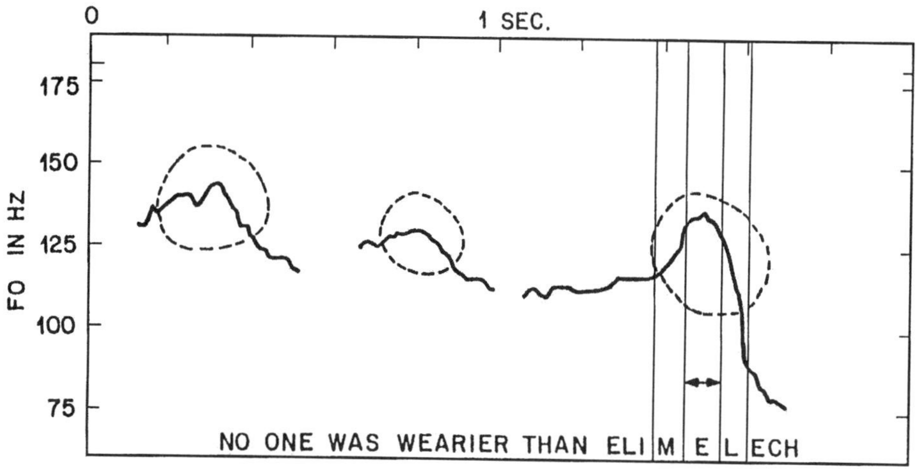

Figure 7. An F0 contour with three H accents, with corresponding F0 peaks circled.*

which the F0 falls after a nuclear peak, but levels out in the middle of the pitch range instead of falling all the way to the bottom. Previously analyzed using an otherwise unsupported M (mid) phrasal tone, these patterns can now be understood as having a downstepped H phrasal tone.

American English is also claimed in this model to have an upstep rule, which applies only to intonation phrase boundary tones (whether L% or H%) after H. This rule is responsible for the fact that F0 contours with a H phrase accent either sustain the same level, or else rise even further at the end (in the canonical yes/no question pattern). Unlike other languages, such as Hungarian, an F0 pattern which rises to a H phrase accent and then falls down again to the boundary does not occur. Some dialects of British English also have the upstep rule, but it appears that others do not. In the dialects which lack the upstep rule, the high-rising questions do not occur, but the rising-falling post-nuclear configuration does occur. As a result, the contrast between a relatively high (or H) and a relatively low (or L) termination after H is still found, and the two-tone decomposition is thus supported.

A complete set of schemata of nuclear/postnuclear configurations is found in Figure 8, reproduced from Pierrehumbert and Hirschberg (1990). Pierrehumbert and Hirschberg also sketch a compositional account of the meanings of the contours. The core concept in this model is the relationship of each phrase to the mutual beliefs as they are built up by interlocutors during a discourse. The H* accent is used to mark focused information which is to be added to the mutual beliefs; the L* accent marks information which is salient but which is for some reason not proposed as an addition. (For example, it may be already present in the mutual beliefs, or it may be doubtful or false.) The L+H accents mark information which is selected from a small domain of

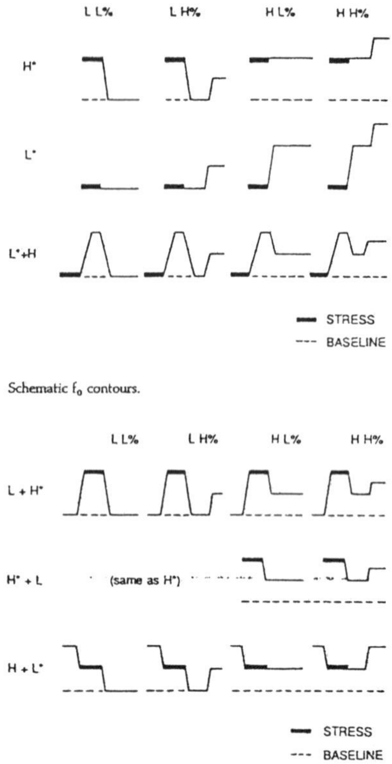

Figure 8. The full inventory of nuclear accents in combination with phrase-final tones, as schematized in Pierrehumbert and Hirschberg (1990).

alternatives, with the L+H* marking an "add" and the L*+H marking a "non-add". The H+L accents represent an instruction to identify a relationship of the information to information which is already mutually believed. The boundary tones differ from the pitch accents in this model in marking the discourse status of the phrase as a whole rather than the status of individual focused elements in the phrase. For a broadly related but competing proposal about the morphemes of the intonational system, see also Gussenhoven and Rietveld (1991).

2.3 ToBI

ToBI (an acronym standing for "Tone and Break Indices") is a standard for transcription of English intonation which was developed and tested by an international group of researchers in the field (see Silverman et al. 1992). The purpose of the standard is to further scientific study of intonation and technological development by permitting researchers at different laboratories to interpret each other's data and to pool resources in developing on-line databases of prosodically transcribed speech. The immediate antecedents of the standard are Pierrehumbert (1980) and Beckman and Pierrehumbert (1986) (for the decomposition of the melody into L and H tones, organized into bitonal accents and phrasal tones); Ladd (1983) (for the treatment of downstep) and Wightman et al. (1992) for the treatment of juncture. An experiment on inter-transcriber reliability was carried out to validate the system, with 26 independent transcribers analyzing 34 varied utterances. As reported in Pitrelli et al. (1994), the protocols permitted very good reliability, as assessed by the number of transcriber pairs agreeing on the labeling of each word. There was 88% agreement on the presence of tonal elements, 81% agreement on the exact label of tonal elements, and 92% agreement to within one level on the assignment of break indices. This level of reliability is much higher than for previous systems of intonation transcription and will permit the development of shared intonational corpora.

ToBI provides for four parallel channels of transcription. One is the orthographic or phonetic transcription of the words. The second is the melody line, which follows the general outline of Beckman and Pierrehumbert (1986) by providing for monotonal and bitonal accents plus two levels of phrasal tones. Transcribers can also mark the F0 maximum in each phrase, providing a crude but replicable index of the current pitch range. The third channel carries indices describing the strength of the juncture between each two lexical items, ranging from 0 (indicating that cliticization has turned two lexical items into a single prosodic word) to 4 (indicating a maximal, or fully-marked, intonational phrase boundary). The last provides for comments of any type.

Although the standard has obviously been influenced by current theory to a significant extent, it is theory-neutral in several important respects. First, there has been and continues to be controversy about whether downstep in English is predictable from the type and grouping of pitch accents, or whether it is an independent dimension of choice. In ToBI, downsteps are explicitly transcribed (in the style of Ladd 1983) in the hope that researchers will eventually gather enough data to settle this issue. Secondly, Beckman and Pierrehumbert (1986) advanced the hypothesis that non-accentual tones are tightly linked to levels in a hierarchical prosodic structure, with the intermediate phrase contributing one tone (L or H) and the intonation phrase contributing another. For the ToBI reliability trial, Pitrelli developed a transcription parser which enforces this regularity. However, the transcription standard itself provides no impediments to recording junctures which are not synchronized in any particular way to the melodic stream. Anyone with a different theory of how tones and junctures are related could write a different transcription checker to verify the relationship between the channels in the light of their own theory. Lastly, the inventors of ToBI do not claim that the melody and break index channels exhaust the information relevant to the understanding of intonation. These two channels represent information about which there is a broad consensus in the speech community. The fourth channel can be used to record any further observations about the intonation or prosody that may be of interest. Observations about expressive use of voice quality would be an obvious example.

After ToBI was developed, there arose considerable interest in how it might be applied to other languages. It is important to understand that ToBI is not directly applicable to other languages (or even to some dialects of English) because it is at the level of abstraction of a broad phonemic transcription, or rationalized spelling system, such as those of Korean and Finnish. Just as a broad phonemic transcription for any language must be guided by the phoneme inventory of that language (as revealed by the lexical contrasts), a ToBI-style transcription of the prosody and intonation of any language must be guided by an inventory of its prosodic and intonational patterns.

As such inventories are made, we find many recurring themes in the dimensions of phonetic contrast which are employed in prosodic and intonational systems. Recurrent dimensions include relative F0 height, relative duration, relative alignment, relative force of articulation, and so forth. In a similar way, inventories of vowel systems in the languages of the world reveal the existence of recurrent dimensions such as front/back and rounding. However, equating any particular vowel in one language with a particular vowel in another is highly problematic. For example, although the French high front unrounded vowel is broadly analogous to the English one, its exact degree of height, frontness, and spreading is different, as is its pattern of variation in context. Similarly, we do not expect to see tonal elements literally equated across languages. Instead, the expectation is that broad patterns of contrast and tonal realization might be echoed from one system to the next.

3. EXPERIMENTAL VALIDATION

Some broad features of the two-tone models have been validated by subsequent work; others have been called into question, and others have not been tested experimentally at all.

First, consider the claim that the accent inventory is the same in prenuclear and in nuclear position, with the more complex configurations found in nuclear position being attributable to extra tones originating at the phrasal level. Partial support in favor of this claim is found in experiments by Steele (1986) and by Silverman and Pierrehumbert (1990), who explored the timing of the F0 peak for H* accents in various positions. Previous work (notably, Silverman 1987) had reported that the timing was early in the stressed syllable for nuclear accents and late for prenuclear accents, suggesting that the claimed unity of the

H* accent in these two positions might be illusory. Steele (1986) and Silverman and Pierrehumbert (1990) fleshed out the picture by examining nuclear accents separated from the phrase boundary by varying amounts of material, as well as prenuclear accents separated by varying amounts of material from the nuclear accents. These experiments demonstrated that the two extremes of timing just mentioned actually fall on a continuous gradient. This point was already illustrated by the variable placement of the F0 peaks in Figure 2 relative to the segmental material. Factors contributing to the full gradient include the variable strength of boundaries, the variable sonority and length of the segments, and the variable amount of tonal crowding found in different contexts. With these results, the phonological unity of the prenuclear/nuclear H* accent is supported.

The existence of contours with mixed accent types provided a second line of evidence for the claim that the phonological inventory of accents is the same in nuclear and in prenuclear position. For example, H* and L* accents occur in all (four) possible combinations and orders in two-accent phrases; insofar as all proposed accents occur in all combinations and orders, this provides evidence that each one is an independent phonological unit. The intonational grammar in Beckman and Pierrehumbert (1986) generates 36 accentual combinations for phrases with two accents, and 216 combinations for phrases with three accents, even disregarding the contribution of phrasal tones. Although various combinations of accents are found, nothing like the full set generated by the grammar has ever been documented. For three accent phrases, the typical pattern is either to use the same accent type in all three positions, or else to use one type of accent in both prenuclear positions, and a different type in nuclear position. To understand this observation, it is worthwhile to bear in mind how sparsely languages in general sample the cross-product of the available units. Because of phonotactic constraints, most combinations of phonemes do not represent possible words and most phonotactically possible words do not happen to be real words. Syntactic and semantic constraints have the consequence that most combinations of real words do not constitute potentially observable sentences. Similarly, we need to work out what factors cause gaps in the set of intonation patterns observed. Are there phonotactic factors that are not yet understood? Are phrasal intonation patterns lexicalized as single units, with some being accidentally missing? Are some potential patterns missing because the meanings of the component parts cannot be coherently combined?

A second important feature of the Pierrehumbert (1980) is that it treated the relative alignment of two-tone accents to the segmental material as phonologically contrastive. The two L+H accents differed in which tone controlled the alignment of the whole accent via its affiliation with the stressed syllable. In the L*+H, the L* is phonologically aligned with the stressed syllable and the H falls soon after the L*; in the L+H*, the H* is aligned with the stressed syllable and the L falls soon before it. The key concept – that

English has two distinct relative alignments for the same scooped F0 configuration – was validated by the experiment described in Pierrehumbert and Steele (1989). Pierrehumbert and Steele used LPC analysis and resynthesis to create a set of 15 versions of the phrase *Only a millionaire*, differing only in the relative alignment of the L+H portion of a L+H L H% intonation pattern with the word *millionaire*. These were played 30 times in blocked randomized order to subjects, who were asked to imitate what they heard. The peak alignment in their imitations was measured. Despite the fact that the stimuli had evenly graded peak alignments, two preferred peak alignments were observed in the productions. A related set of experiments by Kohler (1987a, 1987b) also established that relative alignment is contrastive in German. The other relative alignment distinction that Pierrehumbert (1980) proposed for English (H*+L versus H+L*) has never been tested experimentally.

In contrast to relative alignment, phrasal pitch range is claimed by Pierrehumbert (1980) to be a thoroughly gradient reflex of style and discourse structure. One of the pitch scaling experiments discussed in Liberman and Pierrehumbert (1984) provides a partial test of this claim. In the experiment, subjects produced instances of the sentence *Anna came with Manny* with two different focus structures in ten different overall pitch ranges. The desired pitch range was signaled by a number from one to ten written on a note card underneath the sentence. For every subject, the resulting productions showed a smooth gradient of F0 peak values, with no preferred values. In contrast to the Pierrehumbert and Steele experiment, gradient instructions produced gradient results. Thus, phrasal pitch range does not appear to be categorical. More informal confirmation of this point may be found in Hirschberg and Pierrehumbert (1986), who measured all phrasal pitch ranges in a monologue previously recorded for a sociolinguistic study. The phrasal pitch range was found to correspond well to the discourse structure of the monologue, with larger pitch ranges used to mark the introduction of new topics and subtopics.

In the F0 scaling model developed in Liberman and Pierrehumbert (1984), a single parameter stands as the reflex of the phrasal pitch range. It is designated as the "reference line", and represents the F0 value to which an arbitrarily long series of downsteps would assymptote. It is higher than the baseline, which is viewed as a fixed property of the speaker's voice. Related models have also been developed for Japanese (Pierrehumbert and Beckman 1988) and Spanish (Prieto et al. 1996) on the basis of data from the same experimental paradigm, in which speakers are asked to produce intonation patterns involving downsteps at different overall voice levels. In addition, work by Ladd (1993) and Terken (1993) provides further evidence that implicit parameters control the scaling of F0 contours.

The paradigm just summarized does not, however, yield the last word on H tone scaling because it tackles only one source of pitch range variation, namely overall voice level. Other choices of speech style may affect the F0 scaling in other ways. For example, when people read stories to children using a small voice for a small character, they probably modify the baseline. Further experiments on the full range of stylistic choices would be valuable. In addition, note that F0 is only one of the phonetic parameters affected by overall voice level. Raising the voice also affects the amplitude, the spectral tilt, the degree of coupling of the subglottal system, and other phonetic characteristics. Pierrehumbert (1997) describes a pilot study on the interaction of intonation pattern and overall voice level as determinants of the voice source characteristics. Further work exploring these interactions would lead us to a much fuller understanding of the phonetic correlates of tone. Such understanding would help to address concerns about the ambiguity of F0 contours (which may be less ambiguous when all properties of the signal are considered), and it would also support improvements in the quality of synthetic speech.

The algebraic approach to F0 scaling taken by Liberman and Pierrehumbert (1984) has not been notably successful in describing the scaling of L tones, particularly those near the bottom of the range. One reason may be the complicated articulatory strategies involved in active F0 lowering; see the review in Beckman and Pierrehumbert (1992). Another problem is that what counts perceptually and phonologically as a very low tone may, from a mathematical point of view, not have a well-defined F0 at all. Vocal fold adduction and/or low subglottal pressure can produce irregular movements of the vocal folds which are not periodic at all. A unified treatment of the phonetics of tone is therefore likely to require innovations in the parameterization of the phonetic outcome.

Pierrehumbert (1980) advanced a strong hypothesis about the character of the phonetic implementation rules for tones. For any given target tone, the implementation was held to depend only on the identity and prosodic position of the tone itself, and on the identity and phonetic realization of the preceding tone. The two hallmarks of this "running window" are its strict locality and its temporal asymmetry. The current outcome can depend on actual outcomes in the past, but has no access to actual outcomes for future elements. The model thus presented a very strong contrast to superpositional models of F0 realization, such as Fujisaki (1989), in which the phrasal F0 contour arises through superposition of a phrasal F0 contour with local accent-related F0 contours. A number of experimental studies have undermined the strong claims about locality made in Pierrehumbert (1980). However, the nonlocal effects that have been found are not all amenable to treatment in a strictly superpositional approach.

Detailed studies of tone scaling (such as Liberman and Pierrehumbert 1984, Pierrehumbert and Beckman 1988) indicated the need for implicit phrase-level parameters controlling pitch range. In Pierrehumbert and Beckman (1988), these are treated formally using the conceptual apparatus of attribute grammars. I have already noted that the implementation of a tone can depend on its prosodic position; in just the same way that the node labels on the hierarchical structure dominating a target tone can be examined by the implementation algorithm, the pitch range parameters carried by these nodes can also be examined. Formally, this is a very simple extension of the original proposal. It obviously tends in the direction of superposition of local and phrasal components; however, there are important differences. One difference is that each pitch range parameter is just a single number and not a time function. Another difference is that the phrasal pitch range parameters are not required to combine with the local components via superposition as such; rather, they figure as values for arguments in a function which computes actual F0 values. They can only be established by fitting an entire model to an entire data set. In principle, they can differentially affect different tones – for example, they might and probably do affect H tones and L tones differently. Straight superposition models do not have this capability, since the term 'superposition' by definition means that components are combined exactly one way, namely by addition on some appropriate scale.

Silverman and Pierrehumbert's findings on tonal alignment provide one example of the need for look-ahead to the upcoming tone in producing the current tone. The phonetic alignment of the tone is shown to be readjusted when another tone is coming up soon; see also the discussion in Bruce (1990). Another example is provided by Laniran and Clements' (1995) study of downstep in Yoruba; they report raising of H tones in anticipation of downsteps. Detailed consideration of final lowering provides further evidence of the need for look-ahead. In Liberman and Pierrehumbert's experiment on downstep, the last step in the sequence was found to be lowered (with reference to the decaying exponential otherwise traced out by the sequence of downsteps); this phenomenon was referred to as "final lowering". How does a target tone "know" that it is last? The hypothesis which fits in best with Pierrehumbert's original model is that exactly nuclear accents undergo the lowering; the question "is this tone in nuclear position" can be answered by examining the prosodic nodes dominating the tone, without the need to refer to flanking material. However, this hypothesis is not really tenable. An unpublished experiment by Pierrehumbert and Liberman found that the amount of lowering varies according to the distance of the nuclear accent from the end of the phrase. Data on Danish collected by Thorsen (1980a, 1980b) plainly show a small but regular effect on the penultimate accent as well as on the last one. Herman (1996) demonstrated the existence of a gradual final lowering effect spanning the last four syllables of the phrase in Kipare, a tone language of the Bantu family. In addition, Herman et al. (1996) have found instrumental support for the observation by Hirschberg and Pierrehumbert (1986) that final lowering depends on the identity of the upcoming boundary tone.

Pierrehumbert and Beckman (1988) effectively weakened Pierrehumbert's previous stance on locality by permitting the realization of any given tone to depend not only on its immediate context, but also on any attribute of any node dominating the tone. For example, a phrase-final H% is an attribute of the intonation phrase node, and as such would be accessible as an influence on the realization of any tone within the phrase. Myers (1996) draws on exactly this capability of the model to describe the tonal realization principles of Chichewa. In general, hierarchical structures provide a way to reencode apparent lookahead in terms of an upward search in the tree structure. However, the model still prohibits phonetic (as opposed to phonological) lookahead. For example, the implementation of a tone could not depend on whether the phonetic outcome for a future final boundary tone would be above or below 140 Hz.

Another strong hypothesis advanced by Pierrehumbert (1980) and Beckman and Pierrehumbert (1986) is that the occurrence of downstep is predictable in English. Pierrehumbert (1980) proposed a downstep rule which was highly analogous to that found in some African languages, in which the second H in a H L H configuration is downstepped relative to the first. Specifically, Pierrehumbert proposed that the second H in a H+ L H or a H L+H configuration is downstepped. According to Beckman and Pierrehumbert, any two-tone accent in English triggers downstep. However, neither of these claims has been substantiated by a large-scale study of naturally occurring speech; substantiation of the claim does require a full inventory of naturally occurring variation, because in any given experimental situation, subjects confine their behavior to a small subset of their full range of capabilities. The alternative to a rule predicting downstep is a phonologically contrastive downstep feature (typically transcribed as !). For example, instead of transcribing H*+L H* L L% for a sequence with a downstepped nuclear accent, the transcription could be H* !H* L L%, attributing the downstep to a distinctive attribute of the nuclear accent itself. This is the solution adopted in ToBI. A drawback of this solution is that the downstep feature is never contrastive in initial position; for example, in phrases with one accent, there is no contrast between H* and !H*. In short, downsteps are stepped down in relation to what came before; it makes no sense to posit a downstep if nothing came before. A defective distribution such as this provides a standard argument for a more abstract analysis. However, the Beckman-Pierrehumbert approach also encounters some problems. In particular, experience to date with ToBI suggests that a L+H accent configuration can be followed either by a downstepped accent, or by one of essentially equal height.

Lastly, the intonational grammar displayed in Figure 6 incorporates the claim that every intermediate phrase has at least one pitch accent. The case in which this claim has been most called into question is that of tags. Figure 9 shows F0 contours for a single phrase with a deaccented object versus an utterance which has two phrases (under the Beckman and Pierrehumbert (1986) analysis), with the second having a L* L H% contour. Obviously, the main difference between the two is the timing of the F0 contour. However, under the analysis, the word *Manny* has an accent in the second case but not in the first. Fans of abstract analyses see no difficulty with this claim. Others (notably Bing 1979) prefer to suggest that the tag has no accent. If the tag has no accent, then the details of the grammar in Figure 6 need to be modified. However, the general approach survives. To adjudicate between these positions, more work documenting the full range of intonational variation found in tags is needed. The analysis with the accent is more plausible if this accent contrasts with other possible choices of accent in the same position.

4. CONCLUSION

The style of intonational analysis initiated by Bruce for Swedish has now been successfully applied to English as well as to other languages. Hallmarks of this approach are a limited tonal inventory (with two tones sufficing for every language so far studied which has a pitch accent or intonation system rather than a lexical tone system); a clear attribution of tonal properties to different levels of prosodic structure; and explicit, nontrivial principles of phonetic implementation.

These hallmarks emerge from a methodology in which tonal contrasts and their phonetic manifestations are systematically examined in continuous speech. With the perspective provided by continuous speech, we see that isolation forms are not in any sense basic. Instead, they simultaneously display the complexities of initial, final, and nuclear position.

In surveying the intonational system of any language, priority must be placed on identifying the dimensions of contrast in the language. Controlled experiments can be used to identify regularities of timing and F0 scaling, and to distinguish gradient from categorical effects. Recent improvements in transcriptional tools (such as ToBI, with its associated software utilities) also provide unprecedented capability for exploring the full range of intonational variation found in expressive natural speech.

F0 has proved to be an extremely useful parameter for exploring intonation systems. One reason is that it is single-dimensional (facilitating statistical analysis of the data); another is that F0 data can be obtained and analyzed in immense quantity. Yet a third is that it can be naturalistically manipulated to construct stimuli for perception experiments. Bruce (1977) made full use of this situation. The example he set is one reason why the study of intonation is more advanced than the study of any other aspect of continuous speech.

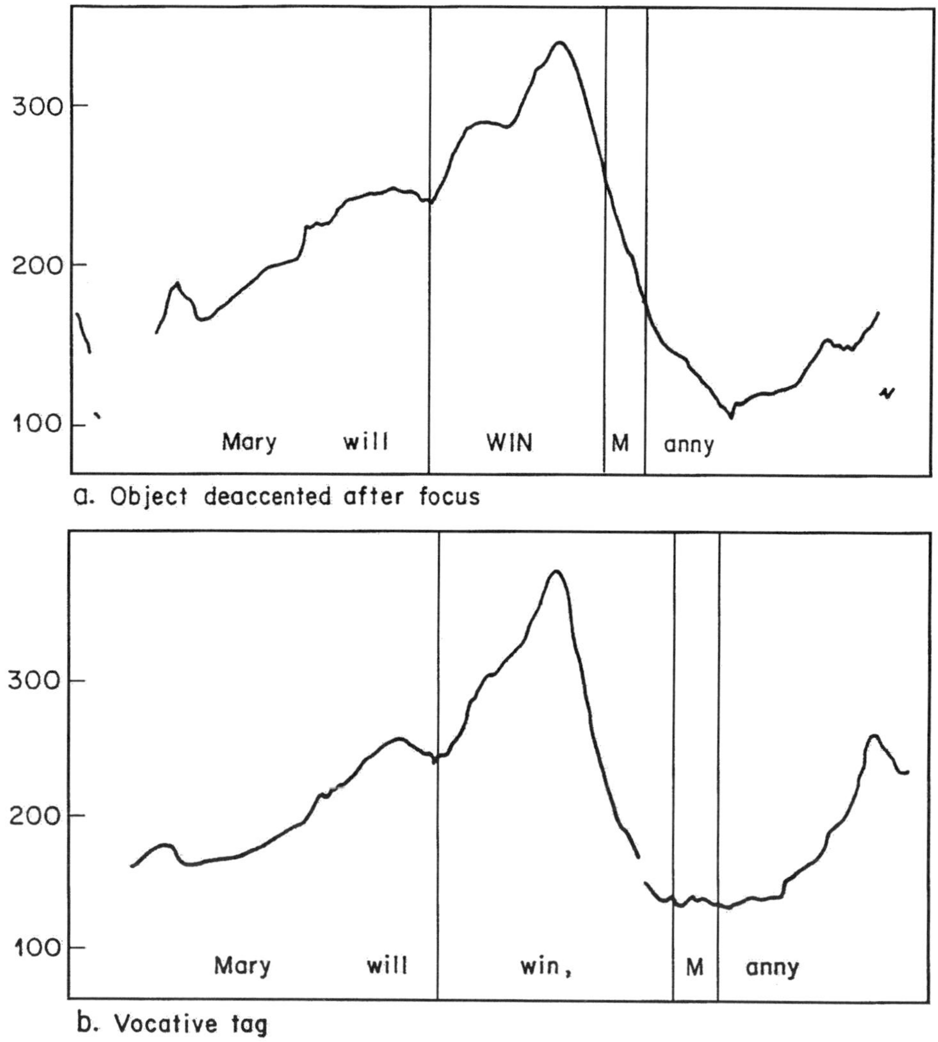

Figure 9a. A H L H% pattern produced on a single phrase in which the word* win *is under focus and* Manny *is deaccented (Reproduced from Beckman and Pierrehumbert 1986). Figure 9b. The highly similar F0 pattern (H* L | L* L H%) in which* win *carries a nuclear accent and* Manny *is a vocative tag standing as a separate intermediate phrase (Reproduced from Beckman and Pierrehumbert 1986).*

REFERENCES

Anderson, M.J., Pierrehumbert, J. and Liberman, M.Y. 1984. Synthesis by rule of English intonation patterns. *Proc. IEEE Congress on Acoustics, Speech, and Signal Processing* I, 2.8.1-2.8.4.

Anderson, S.R. 1978. Tone features. In V.A. Fromkin (ed), *Tone: A Linguistic Survey*. New York: Academic Press, 133-176.

Beckman, M.E. and Pierrehumbert, J. 1986. Intonational structure in Japanese and English. *Phonology Yearbook* 3, 15-70.

Beckman, M.E. and Pierrehumbert, J. 1992. Tactics and strategies for thinking about F0 variation. In G. Docherty and D.R. Ladd (eds), *Papers in Laboratory Phonology II: Gesture, Segment, Prosody*. Cambridge: Cambridge University Press, 387-387.

Bing, J. 1979. *Aspects of English Prosody*. Ph.D. Dissertation, Univ. of Massachusetts at Amherst.

Bolinger, D. 1951. Intonation: Levels versus configurations. *Word* 7, 199-210.

Bolinger, D. 1958. A theory of pitch accent in English. *Word* 14, 109-149.

Bruce, G. 1977. *Swedish Word Accents in Sentence Perspective*. Lund: Gleerup.

Bruce, G. 1990. Alignment and composition of tonal accents: comments on Silverman and Pierrehumbert's paper. In J. Kingston and M.E. Beckman (eds), 107-114.

Clements, G.N. and Ford, K. 1979. Kikuyu tone shift and its synchronic consequences. *Linguistic Inquiry* 10, 179-210.

Coleman, J. 1992. The phonetic interpretation of headed phonological structures containing overlapping constituents. *Phonology* 9, 1-44.

Fry, D. 1958. Experiments in the perception of stress. *Language and Speech* 1, 125-152.

Goldsmith, J. 1976. *Autosegmental Phonology*. Ph.D. dissertation, MIT. [Published in 1979 by Garland Publishing, New York.]

Goldsmith, J. 1990. *Autosegmental and Metrical Phonology*. Cambridge MA: Blackwell.

Gussenhoven, C. and Rietveld, A.C.M. 1991. An experimental evaluation of two nuclear-tone taxonomies. *Linguistics* 29, 423-449.

Herman, R. 1996. Final lowering in Kipare. *Phonology* 13, 171-196.

Herman, R, Beckman, M. and Honda, K. 1996. Subglottal pressure and Final Lowering in English. *Proc. International Congress of Spoken Language Processing*, vol. 1, 145-148.

Hirschberg, J. and Pierrehumbert, J. 1986. Intonational structuring of discourse. *Proc. 24th Meeting of the Association for Computational Linguistics*, 136-144.

Kingston, J. and Beckman, M.E. (eds). 1990. *Papers in Laboratory Phonology I: Between the Grammar and Physics of Speech*. Cambridge: Cambridge University Press.

Kohler, K.J. 1987a. Categorical pitch perception. *Proc. XIth International Congress of Phonetic Sciences* (Tallin), vol. 5, 331-333.

Kohler, K.J. 1987b. The linguistic functions of F0 peaks. *Proc. XIth International Congress of Phonetic Sciences* (Tallin), vol. 3, 149-151.

Ladd, D.R. 1979. *The Structure of Intonational Meaning*. Ph.D. dissertation, Cornell University.

Ladd, D.R. 1983. Phonological features of intonational peaks. *Language* 59, 721-759.

Ladd, D.R. 1993. On the theoretical status of 'the baseline' in modeling intonation. *Language and Speech* 36, 435-451.

Ladefoged, P. and Broadbent, D. 1957. Information conveyed by vowels. *Journal of the Acoustical Society of America* 29, 98-104.

Laniran, Y. and G.N. Clements. 1995. A long-distance dependency in Yoruba tone realization. *Proc. XIIIth International Congress of Phonetic Sciences* (Stockholm), vol. 2, 734-737.

Leben, W. 1973. *Suprasegmental Phonology*. Ph.D. dissertation, MIT.

Liberman, M.Y. 1975. *The Intonation System of English*. Ph.D. dissertation, MIT. [Published by Garland Publishing, New York].

Liberman, M.Y. and Pierrehumbert, J. 1984. Intonational invariance under changes in pitch range and length. In M. Aronoff and R. Oehrle (eds), *Language Sound Structure*. Cambridge MA: MIT Press, 157-233.

Myers, S. 1996. Boundary tones and the phonetic implementation of tone in Chichewa. *Studies in African Linguistics* 25, 29-60.

Pierrehumbert, J. 1980. *The Phonology and Phonetics of English Intonation*. Ph.D. dissertation, MIT. Reproduced by Indiana University Linguistics Club, Bloomington.

Pierrehumbert, J. 1997. Consequences of intonation for the voice source. In S. Kiritani, H. Hirose, and H. Fujisaki (eds), *Speech Production and Language*, Speech Research 13. Berlin: Mouton de Gruyter, 111-131.

Pierrehumbert, J. and Beckman, M.E. 1988. *Japanese Tone Structure*. Cambridge, Mass.: MIT Press.

Pierrehumbert, J. and Hirschberg, J. 1990. The meaning of intonational contours in the interpretation of discourse. In P. Cohen, J. Morgan, and M. Pollack (eds), *Intentions in Communication*. Cambridge, Mass.: MIT Press, 271-311.

Pierrehumbert, J. and Steele, S. 1989. Categories of tonal alignment in English, *Phonetica* 46, 181-196.

Pitrelli, J.F., Beckman, M.E. and Hirschberg, J. 1994. Evaluation of prosodic transcription labeling reliability in the ToBI framework. *Proc. International Conference on Spoken Language Processing* (Yokohama), 123-126.

Pike, K.L. 1945. *The Intonation of American English*. Ann Arbor: University of Michigan Press.

Prieto, P., Shih, C. and Nibert, H. 1996. Pitch downtrend in Spanish. *Journal of Phonetics* 24, 445-473.

Silverman, K. 1987. *The Structure and Processing of Fundamental Frequency Contours*. Ph.D. dissertation, Cambridge University.

Silverman, K. and Pierrehumbert, J. 1990. The timing of prenuclear High accents in English. In J. Kingston and M.E. Beckman (eds), 72-106.

Silverman, K., Beckman, M, Pitrelli, J., Ostendorf, M., Wightman, C., Price, P., Pierrehumbert, J. and Hirschberg, J. 1992. ToBI: A standard for labelling English prosody. *Proc. International Conference on Spoken Language Processing* (Banff), vol. 2, 867-870.

Steele, S. 1986. Nuclear accent F0 peak location: effects of rate, vowel, and number of following syllables. *Journal of the Acoustical Society of America* 80 Supplement 1, S51.

Terken, J. 1993. Baselines revisited. Reply to Ladd. *Language and Speech* 36, 453-459.

Thorsen, N. 1980a. Intonation contours and stress group patterns in declarative sentences of varying length in ASC Danish. *Annual Report of the Institute of Phonetics, University of Copenhagen* 14, 1-29.

Thorsen, N. 1980b. A study of the perception of sentence intonation – evidence from Danish. *Journal of the Acoustical Society of America* 67, 1014-1030.

Trager, G.L. and Smith, H.L. 1951. *An Outline of English Structure* 41. Norman OK: Battenburg Press.

Ward, G. and Hirschberg, J. 1985. Implicating uncertainty: the pragmatics of fall-rise. *Language* 61, 747-776.

Wightman, C., Shattuck-Hufnagel, S., Ostendorf, M. and Price, P.J. 1992. Segmental durations in the vicinity of prosodic phrase boundaries. *Journal of the Acoustical Society of America* 91, 1707-1717.

Department of Linguistics, Northwestern University, Evanston, IL, USA

D. ROBERT LADD

BRUCE, PIERREHUMBERT, AND THE ELEMENTS OF INTONATIONAL PHONOLOGY

INTRODUCTION

Over the past two decades a great deal of work on intonation has been based on the general set of assumptions that I have elsewhere (Ladd 1996) referred to as the 'autosegmental-metrical' (AM) approach. This is the theory generally associated with the work of Janet Pierrehumbert (e.g. Pierrehumbert 1980, Pierrehumbert and Beckman 1988, Pierrehumbert, this volume). The central claim of the AM view is that an intonation contour is represented phonologically as a sequence of tones associated in well-defined ways with the segmental string. The AM theory and Pierrehumbert's notational conventions have been widely adopted by investigators of other languages (e.g. Gussenhoven 1984 on Dutch, Hayes and Lahiri 1991 on Bengali, Uhmann 1991 on German, Grice 1995a on Italian), and it is unquestionably the currently dominant approach to intonational description. This dominance is now being institutionalised in the form of ToBI-style transcription systems for a variety of languages and language varieties (e.g. Silverman et al. 1992, Roach 1994).

However, in this essay – and I acknowledge that it is no more than that – I wish to raise some questions about the empirical and theoretical assumptions underlying the AM tonal analysis of intonation. In particular, I want to look at the definitions, explicit and implicit, formal and phonetic, of the tones themselves, and of the way they are associated to the segmental string. What do the Hs and Ls and *s of an AM description represent? For most adherents of the theory, there is no real question here. Intonational tones are formally identical to the tones of tone languages: they are autosegments on a tonal or laryngeal tier, and they are linked to a segmental root node or to various domain edges or domain nodes, either directly or through some sort of tonal root node. None of this seems to make much difference to everyday descriptive and empirical work based on the AM theory. Yet if we look back at the intellectual background to Pierrehumbert's original version of the theory, we can see that there are actually two distinct sources of her ideas, with subtly divergent assumptions about what tones are, how they are organised, and how they are phonetically manifested. These divergent assumptions are the basis of what I have to say here.

M. Horne (ed.), Prosody: Theory and Experiment, 37–50.

1. BACKGROUND TO PIERREHUMBERT (1980)

The most obvious source of the theory presented in Pierrehumbert's thesis (1980) is the early work on autosegmental phonology, in particular work by Leben, Goldsmith, and Liberman (e.g. Leben 1973, 1976; Goldsmith 1976, 1981; Liberman 1975). In a certain sense, of course, some of the autosegmental ideas about intonation themselves go back to Pike (1945) and more generally to the received American structuralist analysis of intonation (Wells 1945, Trager and Smith 1951). However, the theoretical issues that concerned Leben, Goldsmith, and Liberman were not ones that grew out of the intonational work of the 1940s and 1950s, but out of 1960s generative phonology – in particular, the problem of incorporating suprasegmental phenomena into the rigorously segmental phonological formalism proposed in SPE (Chomsky and Halle 1968). Their work on intonation built on the basic autosegmental idea that phonological elements occur on autonomous 'tiers' with various kinds of linkages between elements on one tier and elements on another. In particular, they explored Liberman's notion of 'tune-text association' – the idea that there is a lawful basis for the linkage between the segmental string and the autonomous tonal tier representing the intonation contour. They also strongly supported the view that the elements on the tonal tier are 'pitch levels' entirely analogous to the tones of the African tone languages, which were the central empirical phenomenon that autosegmental phonology first set out to deal with. These early autosegmental issues dominated the theoretical discussion in Pierrehumbert's thesis.

There was, however, an additional important source of inspiration for Pierrehumbert, namely Bruce's analysis of Swedish word accents (Bruce 1977). The basic question addressed by Bruce was the variable phonetic realisation of the Stockholm Swedish word accent distinction in different sentence contexts. The problem is as follows. In citation form, the distinction between the Accent I and Accent II is essentially a difference between a 'single-peaked' and a 'double-peaked' contour. In an Accent I word like *anden* 'the duck' (morphologically *and* + *en*), the pitch rises on the stressed syllable *an*- and falls again on *-den*; in an Accent II word like *anden* 'the spirit' (morphologically *ande* + *en*), the pitch falls from an early peak on the stressed syllable *an*- and rises again to a second peak on *-den* before a final fall. By contrast, when the words occur non-finally in a sentence, the difference between the two accents is an extremely fine difference of the pitch movement on the stressed syllable and bears no obvious relation to the single-peaked and double-peaked distinction. However, Bruce was able to show how the superficial variability of the word contours masked a deeper invariance.

Specifically, Bruce showed that the citation form contours consist of three distinct linguistic events, *in sequence*: a word accent contour, a separate 'phrase accent' peak, and a sentence-final low. When the citation form contours are analysed as sequences in this way, it becomes clear that the word accent distinction in the citation form contour is realised in the same way as in other sentence contexts, provided we focus our attention on the actual word accent part of the contour and are not distracted by the following phrase accent and sentence final low. In *all contexts*, the word accent contour is a fall or High – Low sequence, and what signals the distinction between Accent I and Accent II is the *alignment of the fall with the accented syllable*. In Stockholm Swedish, for Accent I, the High is aligned with the vowel *preceding* the accented vowel (if there a preceding vowel) and the Low is aligned at the beginning of the accented vowel; for Accent II, the High is aligned at the beginning of the accented vowel and the Low near the end of the accented syllable. The 'double-peaked' contour, in other words, is merely the result of the later alignment of the Accent II word accent contour, which yields a perceptible fall on the stressed syllable and a separate peak (for the phrase accent) on the following syllable. This approach to analysing European pitch accent systems is now generally accepted; for further discussion of the Swedish case and others like it, see Gussenhoven (This volume).

Bruce's work contributed two key ideas to the intonational theory presented in Pierrehumbert (1980). One of these ideas was the 'phrase accent', and the other was the idea of a basic tonal opposition between H(igh) and L(ow). Let us briefly consider each of these in turn.

The phrase accent, as we just saw, is an additional pitch peak that follows the lexically specified accent on a given word if the word occurs in particular sentence contexts (including citation forms). This idea gives crucial support to the view that the intonation contour is essentially linear – a string or sequence of what we might call linguistic events that happen at fairly well-defined points in the utterance – and that the local linguistic events can signal global functions (e.g. focus) as well as local ones (e.g. word accent). As I have argued elsewhere (Ladd 1995), linearity is what distinguishes Bruce and the autosegmentalists – and for that matter, most of the ideas of the British tradition (e.g. Crystal 1969) or of the IPO school (e.g. 't Hart et al. 1990) – from what I have called the 'overlay' approach to describing intonation (e.g. Fujisaki 1983, O'Shaughnessy and Allen 1983, Grönnum Thorsen 1985). In the overlay view, the primes of the description are contour shapes that are draped over domains of various sizes – word, phrase, paragraph – and the shapes for domains of one size are superimposed on domains of the next size. The domain spanned by a contour shape is the domain to which it applies functionally: that is, 'sentence intonation' is not just an intonational feature that affects the interpretation of the sentence, but a feature that *extends phonetically across the sentence's entire domain*. This central tenet of the overlay view is difficult to reconcile with the idea of the phrase accent, which is functionally a phrase-level or sentence-level feature but is phonetically localised at a specific point in the contour. By contrast, the phrase accent is strikingly consistent with the autosegmental idea of an autonomous string of tones.

The other idea of Bruce's that was crucial to the development of Pierrehumbert's theory is the notion that intonational phonology can be thought of in terms of a basic opposition between high (H) and low (L), and that a detailed phonetic description of intonation contours can be based on a sequence of Hs and Ls *so long as the description contains a rich enough phonetic component to convert the abstract highs and lows into actual F0 values*. In Bruce's original model (Bruce 1977, ch. 8), there are three components, one that we might call phonological and two that are phonetic. In the phonological component, 'pitch rules' locate Hs and Ls with respect to the segmental string. This part of the description is very similar to the descriptions in the early autosegmental works on intonation cited above. Then the 'F0 rules' convert the abstract Hs and Ls into actual F0 values[1], and 'join rules' connect one modelled F0 value to the next. Bruce made it explicit (1977:129) that his approach was intended to provide a basis for a quantitative model that could be used in generating synthetic speech.

The notion of a rich phonetic component capable of generating actual F0 values was of considerable interest to Pierrehumbert, both for practical applications in synthetic speech (Pierrehumbert 1981) and for theoretical reasons having to do with the place of phonetics in a grammar (e.g. Pierrehumbert and Beckman 1988, ch. 1). This aspect of the autosegmental-metrical theory has been extremely productive of empirical research (e.g. Liberman and Pierrehumbert 1984), implemented models of intonation (Anderson et al. 1984, Ladd 1987), and interchange between researchers with different ideas about the general problem of 'scaling' F0 targets (for a very thorough review see Terken and Hermes, this volume). However, in the context of her original theoretical synthesis, what mattered most to Pierrehumbert about Bruce's two-level phonology was that it provided a way to resolve the 'levels vs. configurations' controversy.

Up until about 1980, anyone who wanted to refer to pitch level in a description of intonation had to contend with Bolinger's early critique (1951) of the orthodox American structuralist analysis of intonation (Pike 1945, Wells 1945, Trager and Smith 1951, etc.). There were plenty of reasons for wanting to refer to pitch level, including both the practical motivation of phonetic explicitness and the theoretical analogy between intonation and lexical tone; these reasons were among those that led Liberman (1975) to revive and update the American structuralist analysis. But Bolinger's essentially phonological arguments against distinctive pitch levels applied as much to Liberman's revival as to the 1940's original of the American structuralist analysis – because both had four distinctive levels.

Pierrehumbert's insight, drawing on Bruce's work, was that "difficulties noted [by Bolinger and others] arise from the postulation of four phonemically different tones, and that they can be circumvented by a system which has two tones and an appropriate phonetics'' (1980:55-56). In retrospect, Bolinger's arguments were not as strong as they seemed; they rested in part on the specific proposal of four levels and on certain unrelated theoretical assumptions of the 1940s and 1950s, not just on the idea of distinctive levels itself (see Ladd 1983a for more discussion). By seeing this weakness, Pierrehumbert showed that the levels-vs.-configurations dichotomy was largely spurious, and was able to develop the autosegmental idea that the elements of intonation are level tones. Perhaps more importantly, she thereby laid the foundation for descriptions of intonation that successfully integrate the best features of previous 'levels' and 'configurations' work.

2. THREE QUESTIONS ABOUT TONES

In the wake of the success of Pierrehumbert's theory, Bruce's analysis of the Swedish word accents has generally been recast (even by Bruce himself, e.g. 1987) in more orthodox AM terms. Most obviously, the difference in alignment between Accent I and Accent II has been reinterpreted on the basis of Pierrehumbert's notion of 'starred tone' (about which more shortly): in the AM version of Bruce's Swedish analysis, the early-aligned Accent I is analysed as H+L*, and late-aligned Accent II as H*+L. With this modification, Bruce's analysis can be seen as simply one of many autosegmental analyses of intonation systems, like the works of Gussenhoven and others cited at the beginning of the paper. The only slight anomaly in this assessment is that Bruce's analysis underlies some of Pierrehumbert's ideas rather than being derivative from them.

However, I think that this seamless incorporation of Bruce's analysis into the overall AM tradition is misleading, and represents a potential loss for the field. In this second half of the paper, I would like to raise three questions about the nature of tones and tonal association that arise from a consideration of the subtle divergences between Bruce's original assumptions and the autosegmental versions of the same basic ideas. I am aware of the potential sterility of merely raising questions, and of performing detailed exegesis on 20-year-old work. I will try to avoid this sterility by demonstrating that there are empirical consequences to the way we answer questions about what tones are, and that the generally assumed (i.e. autosegmental-metrical) answers to the questions may not be right.

2.1. What is a Tone?

In his original work, Bruce drew no strong analogy between the Hs and Ls of his description and the 'tones' of lexical tone languages. He defined his Hs and

Ls on a strictly phonetic basis, as local maxima and minima in the contour – turning points, in his terminology[2]. In his early work he never actually uses the word 'tone', though he does refer to the maxima and minima as abstract 'tonal points' in the phonological part of his model (the 'pitch rules' mentioned above). That is, his work is very much a *phonetic* work in the sense that it represents an attempt to model some physical aspect of speech, in this case fundamental frequency contours. The turning points are simply observable physical properties of contours, based on no prior theoretical considerations. Turning points can be completely described in terms of two parameters: their F0, and their alignment in time with segmental landmarks in the signal. If we call them 'tones', then we have to say that Bruce's conception of 'tone' is a very concrete one.

This concrete conception of tones as turning points is not shared by Pierrehumbert (1980). In Pierrehumbert's approach, as in Bruce's, intonation contours are modelled as sequences of linguistic events made up of H and L tones, which are subject to a variety of phonetic realisation rules. But in Pierrehumbert's analysis of English, tones need not always correspond to turning points, and turning points need not always reflect the phonetic realisation of a tone. The former case is exemplified by the fact that the L in an H*+L accent is never realised as a low target or turning point, but serves only to trigger downstep on the next H tone. An example of the latter is the fact that the low turning point between two H* accents is not taken to reflect a L tone of any sort, but instead is treated as the result of a 'sagging transition'. Pierrehumbert has gone beyond Bruce's concrete and data-driven notion and has drawn an explicit connection between the tones in European intonation systems and tones in languages with lexical tone.

A priori there is no reason to reject Pierrehumbert's approach, or to prefer Bruce's. It is abundantly clear from the tonal phonology of many African languages that we must allow for the possibility that tones in the underlying phonological string may not be realised phonetically as distinct F0 targets (e.g. cases of downstep, in which the L of a H..L..H sequence has no independent realisation as a turning point, but is manifested only in the lowering of the second H relative to the first). That is, the simple equation 'tone = turning point' is too restrictive to serve as a universally valid principle of tonal realisation. At the same time, in dealing with intonation languages, the restrictiveness of Bruce's empirical approach may be valuable. In languages with lexical tone, we often have independent evidence for postulated tones that do not surface as target points, because we can observe alternations in the shape of individual morphemes and words in different phonological contexts. In languages with only post-lexical or intonational uses of tone, we have no such ready way of constraining our descriptions. Identifying tones with turning points puts severe limits on the range of possible phonological interpretations of a given contour.

By abandoning Bruce's simple equation we throw away a powerful empirical check on our theorising.

Pierrehumbert was well aware of this, and was in fact unhappy with the 'sagging transition' analysis of the low turning point between H* accents. As she puts it, "this complication in the interpolation rules is in some ways unattractive, and we have made a serious attempt to get rid of it by developing an account under which the dip in contours [with sequences of H* accents] arises from a L tone" (1980:70). However, given other analytical decisions that she had made, none of the obvious ways of representing the low turning point with a L tone were available to her. In practice, that is, Pierrehumbert's analysis of English uses the internal consistency of the description, rather than any principled theoretical considerations, as the main motivation for deviating from the 'tone = turning point' equation. In my view, this is part of the reason that the debate about some of her descriptive decisions (reviewed in her chapter in this volume) has gone on so long without a clear resolution. Clearer theoretical criteria are needed to deal with the empirical uncertainties of how to identify tones in intonational data.

2.2. What is a Starred Tone?

In Bruce's analysis of Swedish, as we saw, a pitch accent consists of two tones, H and L. When Pierrehumbert adopted the idea of bitonal accents in her analysis of English, she drew a distinction between 'starred' and 'unstarred' tones: exactly one tone of a pitch accent in English is starred. As we saw earlier, this requirement has been imposed retrospectively on the bitonal accents in Bruce's analysis of Swedish.

The notation on which this idea is based – the asterisk or star to indicate accent – appears to have been introduced by Goldsmith in his early discussions of English intonation. Goldsmith suggested that the characteristic intonation patterns of English typically involve two or more tones, one of which must be associated with the most prominent syllable of the text to which the tune is applied. For example, Goldsmith analysed the basic declarative intonation of American English as MHL, in which the H tone is associated with the most prominent syllable. Goldsmith marked both the most prominent syllable and the tone that associates with it using an asterisk, thus:

(1)

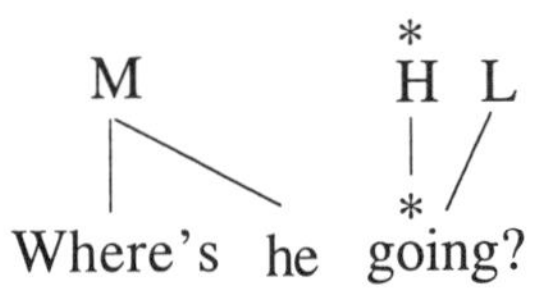

Similar ideas were explored about the same time by both Liberman (1975) and Leben (1976).

The star notation can thus readily be used to distinguish pitch accents from other tonal events (like phrase accents and boundary tones), and this is part of Pierrehumbert's use of it: for example, H* is a pitch accent and H^- is a 'phrase accent' like the one proposed in Bruce's analysis of Swedish. But Pierrehumbert's innovation was to suggest that, in a bitonal pitch accent, one of the tones is starred and the other is not. Her analysis of English involves an inventory of pitch accents, each consisting of one 'starred tone' and optionally of another tone as well. This means that we can have pairs of otherwise identical pitch accents like L*+H and L+H*, which differ only in the choice of the starred tone. This theoretical construct 'starred tone' is assumed without comment in virtually all work based on Pierrehumbert.

Yet the notion of 'starred' tone, though it makes a certain amount of intuitive sense for speakers of English, is hard to define in a way that is both explicit and general enough to cover a wide range of languages. Pierrehumbert (1980:22ff) presented the notion of starred tone in terms of relative prominence or metrical strength: in a bitonal pitch accent, the starred tone is strong relative to the unstarred tone. But she also defined the starred tone as the one that is temporally aligned with the accented vowel (cf. also Silverman and Pierrehumbert 1990), and proposed that unstarred tones, but not starred tones, can be subject to tone-spreading rules in certain circumstances. That is, the property of 'starredness' seems to involve various differences of observable phonetic and phonological behaviour in addition to whatever its defining characteristic is taken to be.

Recently, Grice (1995a, 1995b) has taken up the notion of starred tone as the metrically strong one, and proposes that pitch accents have internal metrical organisation. Obviously, in a system like Pierrehumbert's where pitch accents are restricted to a maximum of two tones, there is no advantage in replacing the notation T+T* with

(2)

However, Grice proposes that pitch accents may consist of three or even more tones, thereby providing new motivation for the idea of internal structure. On the other hand, Grice's proposal raises the new question of what it means for a tone to be 'metrically strong', and why a metrically strong tone should necessariy be aligned with an accented syllable. It also raises the question of why metrical strength, temporal alignment with the accented vowel, resistance to spreading, and so forth, should all go together as properties of a certain type of tone.

In fact, there is plenty of evidence that these properties do not always go together. Arvaniti et al. (1996) discuss the case of prenuclear accents in Modern Greek, in which a local minimum (presumably a L tone) is aligned just before the accented syllable and a local maximum (presumably a H tone) is aligned just after the accented syllable. This means that the contour typically rises throughout the entire accented syllable, and neither tonal target is aligned in time with the accented syllable. This makes it a little difficult to be certain which is the 'starred tone', or even that there is a starred tone; yet the *pair* of tones is consistently associated with accent (i.e. it is not 'unstarred' in the way that a boundary tone is) and both the L and the H are normally manifested as clear turning points (i.e. they occur at a particular place in structure rather than being subject to tone spreading).

Arvaniti et al. suggest that phonetic invariance may provide a clue here: they show that in cases of stress clash, where two prenuclear accents occur on adjacent syllables and there is no space for the full phonetic manifestation of the L and the H of both accents, the second L is 'undershot' while the scaling of the two Hs is maintained. This might argue for treating the Greek prenuclear accent as L+H*. However, Arvaniti et al. also concede that one could look at this differently and say that both the first H and the second L are affected, the latter in its scaling and the former in its temporal alignment. The very inconclusiveness of the Greek evidence leads Arvaniti et al. to state that the notion of starredness is problematical and in need of clear definition.

One promising approach may be to look at what happens when pitch range is modified. Fant and Kruckenberg (1994), while not contradicting Bruce's basic analysis of the two Swedish word accents, have in effect shown that the L in Accent II is not the same as the L in Accent I. When overall emphasis is increased on Accent I, all three turning points – accentual H and L and phrasal H – increase in F0. By contrast, when overall emphasis is increased on Accent II, the H tones are raised and the L tone is lowered. It is not implausible to suggest that the L tone in Accent II is somehow more 'central' or 'important' than that in Accent I – and to equate this 'centrality' or 'importance' with starredness. But if Accent II is H+L*, then Accent I must be something else – H*+L, perhaps, or just H*. Either of these possibilities would mean that the starred tone in Accent I is the accentual H. This in turn would mean that we cannot equate starredness with temporal alignment, since the accentual H of Accent I is aligned well before the accented syllable. Instead, starredness must be reflected in the scaling of tonal targets, in ways still to be explored.

2.3. What do we Mean by 'Association'?

More generally, the difficulties with defining tones and starredness reflect a theoretical uncertainty about what it means phonetically for a tone to be associated with a certain segment or syllable. In the early autosegmental work, as we just saw, starredness was defined as association with the accented syllable, and association was defined in terms of temporal alignment between tones and segments. The notation

(3)

(where T stands for any tone and σ for any syllable) could be interpreted phonetically in something like the following way: the phonetic reflex of T occurs during the temporal interval of σ. In the impressionistic description of languages with lexical tone, this definition appears adequate, because in many cases there is one tone per syllable and it is clear which goes with which. Moreover, many tone languages exhibit phenomena that lend themselves to being described in impressionistic autosegmental terms of tone spreading and reassociation. A good example is the allophony of H and L tones in Yoruba, in which e.g. the H toned syllable in a LH sequence is realised with a steep rise. This can be represented by the spreading of the preceding L to the underlyingly H-toned syllable, as follows:

(4)

For languages like these, then, there is rarely any need to look beyond intuitive 'belonging together' as the definition of association. And few instrumental descriptions of lexical tones are available that might shed light on whether the phonetic reflexes of tones really occur during the temporal interval of the syllable with which they are intuitively associated.

When we consider the European intonation languages, however, it becomes clear that the definition of association needs some attention. First, there are a number of cases where the intuitive belonging together of tones and syllables is not reflected phonetically, or is reflected phonetically in a variety of different ways. For example, in the case of the Greek prenuclear accents just discussed, the L and the H are clearly part of the phonetic manifestation of the accent and their alignment is determined relative to the accented syllable, yet neither is aligned in time with the accented syllable itself. This case is not unique: nuclear accents in Glasgow English are similar to the Greek prenuclear accents in that they involve a L aligned before the accented syllable and a H aligned after it (Ladd 1996: 143-147).

More importantly, the precise details of tonal alignment are a point on which otherwise similar varieties of the same language may differ. Bruce and Gårding (1978) present a prosodic typology of Swedish dialects based squarely on Bruce's analysis of Stockholm. They show that, for all the dialects that exhibit the word accent distinction, the word accent contour is a HL sequence *somewhere in the neighbourhood of the accented syllable*, and the distinction between Accent I and Accent II is invariably that the HL sequence is aligned later in Accent II than in Accent I. But the alignment of Accent I in one dialect can be more or less identical to that of Accent II in another.

The assumptions underlying this account of Swedish dialect variation are quite unlike those of the autosegmentalists. Bruce and Gårding, following Bruce's original analysis of Stockholm Swedish, treat the accentual fall *in both Accent I and Accent II* as one and the same HL pattern. They treat the detailed differences of alignment as the basis not only of the phonetic differences among dialects, but also of the phonological distinction between Accent I and Accent II in all dialects. They offer no explicitly phonological analysis of either of these kinds differences of alignment: they take the intuitive belonging together of the accentual fall and the accented syllable for granted, and assume the cross-dialect identity of the accentual Hs and Ls.

It does not seem to distort Bruce's original intention to relate the Bruce and Gårding typology to autosegmental terminology as follows. In both Accent I and Accent II, and in all dialects, the accentual contour is 'associated' with the accented syllable, and in all cases has *the same representation*. This means that there is a distinction between association and alignment, as proposed (on the basis of Bruce's work) by Ladd (1983b). Association – and by extension starredness – is an abstract phonological 'belonging together' that makes no precise predictions about temporal coordination, while alignment is a variable phonetic dimension that is specified independently of the identity of the tonal string. In the standard autosegmental approach, which takes for granted that association implies temporal alignment, no such distinction is possible; a tone can be associated with one syllable or another, but not with the midpoint of a syllable, or the end, or 15 ms. into the vowel, or 60 ms. after the immediately preceding tone. In effect, alignment is 'digital' for the autosegmentalists and 'analogue' for Bruce[3].

This is not just a difference of opinion but an empirical question. Is the alignment of tonal targets with segmentals limited to certain fixed patterns that can plausibly be represented in a 'digital' phonological notation that distinguishes only H*+L and H+L*? Or is the range of possible ways in which, say, an accentual fall can be aligned with its accented syllable so great that we must acknowledge the abstract phonological identity of all 'HL' accentual contours and treat the alignment differences as a matter of gradient phonetic detail? Even in his later, more explicitly autosegmental work (e.g. Bruce 1987), Bruce questions whether the syllable-by-syllable basis of tonal association is valid for

Swedish. The alignment data dicussed by Arvaniti et al. (1996) also provide reason to doubt that alignment is purely digital. But considerable empirical work remains to be done.

3. CONCLUSION

My purpose in this paper has been to show that certain foundational questions about the nature of tones and tonal association have been largely ignored during the ascendancy of the autosegmental-metrical theory of intonation. Plainly, the failure to deal with these questions has not made the theory any less productive: the basic idea of a linear string of H and L tones associated in a metrically-governed way with the text of an utterance has yielded many discoveries and will continue to do so. Nevertheless, the theory's very success now begins to reveal where its weak points lie. By going back to the beginnings of the theory, and in particular by reexamining Bruce's early contributions, I have tried to identify some lines of enquiry that will not only provide answers to specific empirical questions but may ultimately strengthen the theory as a whole.

ACKNOWLEDGEMENT

Section 2.2 of this chapter overlaps substantially with the paper by Arvaniti, Ladd, and Mennen (1996) referred to here. I acknowledge both the contribution of Amalia Arvaniti and Ineke Mennen, my collaborators on that paper, and the support of the UK Economic and Social Research Council, which funded the project 'Phonological and Phonetic Properties of Tonal Targets in Modern Greek Intonation' through grant no. R000-23-5614 to Edinburgh University.

NOTES

1. Actually, in Bruce's original version the F0 levels are still idealised and abstract, but more finely differentiated than simply H vs. L.
2. Bruce does not appear to have used the term 'turning point' in his thesis, but it is used extensively in Bruce and Gårding (1978).
3. The 'digital/analogue' metaphor for describing the distinction between the two views is due to Amalia Arvaniti.

REFERENCES

Anderson, M., Pierrehumbert, J. and Liberman, M. 1984. Synthesis by rule of English intonation patterns. *Proc. IEEE International Conference on Acoustics, Speech, and Signal Processing*, 2.8.2-2.8.4, New York: IEEE.

Arvaniti, A., Ladd, D.R. and Mennen, I. 1996. What is a starred tone? Evidence from Modern Greek. Paper presented at the Fifth Conference on Laboratory Phonology, Evanston IL, July 1996, submitted for publication in the proceedings.

Bolinger, D.L. 1951. Intonation: levels versus configurations. *Word* 7, 199-210.

Bruce, G. 1977. *Swedish Word Accents in Sentence Perspective*. Lund: Gleerup.

Bruce, G. 1987. How floating is focal accent? In K. Gregersen and H. Basboll (eds), *Nordic Prosody IV*. Odense: Odense University Press, 41-49.

Bruce, G. and Gårding, E. 1978. A prosodic typology for Swedish dialects. In E. Gårding, G. Bruce and R. Bannert (eds), *Nordic Prosody*. Lund: Gleerup, 219-228.

Chomsky, N. and Halle, M. 1968. *The Sound Pattern of English*. New York: Harper and Row.

Crystal, D. 1969. *Prosodic Systems and Intonation in English*. Cambridge: Cambridge University Press.

Fant, G. and Kruckenberg, A. 1994. Notes on stress and word accent in Swedish. *Speech Transmission Laboratory Quarterly Status Report*, October 15, 1994. Department of Speech Communication and Music Acoustics, Royal Institute of Technology, Stockholm.

Fujisaki, H. 1983. Dynamic characteristics of voice fundamental frequency in speech and singing. In P.F. MacNeilage (ed), *The Production of Speech*. New York: Springer, 39-55.

Goldsmith, J. 1976. *Autosegmental Phonology*. MIT PhD dissertation, published 1979 by Garland Press, New York.

Goldsmith, J. 1981. English as a tone language. In D. Goyvaerts (ed), *Phonology in the 1980's*. Ghent: Story-Scientia, *287-308*.

Grice, M. 1995a. *The Intonation of Interrogation in Palermo Italian: Implications for Intonation Theory*. Tübingen: Niemeyer.

Grice, M. 1995b. Leading tones and downstep in English. *Phonology* 12, 183-233.

Grönnum Thorsen, N. 1985. Intonation and text in standard Danish. *Journal of the Acoustical Society of America* 77, 1205-1216.

Gussenhoven, C. 1984. *On the Grammar and Semantics of Sentence Accents*. Dordrecht: Foris.

't Hart, J., Collier, R. and Cohen, A. 1990. *A Perceptual Study of Intonation: An Experimental-Phonetic Approach*. Cambridge: Cambridge University Press.

Hayes, B. and Lahiri, A. 1991. Bengali intonational phonology. *Natural Language and Linguistic Theory* 9, 47-96.

Ladd, D.R. 1983a. Levels versus configurations, revisited. In F.B. Agard, G.B. Kelley, A. Makkai, and V. B. Makkai (eds), *Essays in Honor of Charles F. Hockett*. Leiden: E.J. Brill.

Ladd, D.R. 1983b. Phonological features of intonational peaks. *Language* 59, 721-759.

Ladd, D.R. 1987. A phonological model of intonation for use in speech synthesis by rule. *Proc. Eurospeech* (Edinburgh), 21-24.

Ladd, D.R. 1995. 'Linear' and 'Overlay' descriptions: An autosegmental-metrical middle way. *Proc. XIIIth International Congress of Phonetic Sciences* (Stockholm), vol. 2, 116-123.

Ladd, D.R. 1996. *Intonational Phonology*. Cambridge: Cambridge University Press.

Leben, W. 1973. *Suprasegmental Phonology*. PhD dissertation, MIT, publ. 1978 by Garland Press, New York.

Leben, W. 1976. The tones in English intonation. *Linguistic Analysis* 2, 69-107.

Liberman, M. 1975. *The Intonational System of English*. PhD dissertation, MIT, publ. 1978 by Garland Press, New York.

Liberman, M. and Pierrehumbert, J. 1984. Intonational invariance under changes in pitch range and length. In M. Aronoff and R. Oerhle (eds), *Language Sound Structure*. Cambridge: MIT Press, 157-233.

O'Shaughnessy, D. and Allen, J. 1983. Linguistic modality effects on fundamental frequency in speech. *Journal of the Acoustical Society of America* 74, 1155-1171.

Pierrehumbert, J. 1980. *The Phonology and Phonetics of English Intonation*. PhD dissertation, MIT, publ. 1988 by Indiana University Linguistics Club.

Pierrehumbert, J. 1981. Synthesizing intonation. *Journal of the Acoustical Society of America* 70, 985-995.

Pierrehumbert, J. This volume. Tonal elements and their alignment.

Pierrehumbert, J. and Beckman, M. 1988. *Japanese Tone Structure*. Cambridge, MA: MIT Press.

Pike, K.L. 1945. *The Intonation of American English*. Ann Arbor: Univ. of Michigan Press.

Roach, P.J. 1994. Conversion between prosodic transcription systems: 'Standard British' and ToBI. *Speech Communication* 15, 91-99.

Silverman, K. and Pierrehumbert, J. 1990. The timing of prenuclear High accents in English. In J. Kingston and M. Beckman (eds), *Papers in Laboratory Phonology I*. Cambridge: Cambridge University Press, 72-106.

Silverman, K., Beckman, M., Pitrelli, J., Ostendorf, M., Wightman, C., Price, P., Pierrehumbert, J. and Hirschberg, J. 1992. ToBI: A standard for labeling English prosody. *Proc. International Conference on Spoken Language Processing* (Banff), vol. 2, 867-870.

Terken, J. and Hermes, D. This volume. The perception of prosodic prominence.

Trager, G. L. and Smith, H. L. 1951. *An Outline of English Structure*. Norman, Oklahoma: Battenburg Press; reprinted 1957 by American Council of Learned Societies, Washington.

Uhmann, S. 1991. *Fokusphonologie. Ein Analyse Deutscher Intonationskonturen im Rahmen der Nicht-Linearen Phonologie*. Tübingen: Niemeyer.

Wells, R. 1945. The pitch phonemes of English. *Language* 21, 27-40.

Dept. of Linguistics, University of Edinburgh, Edinburgh, UK

DANIEL HIRST, ALBERT DI CRISTO AND ROBERT ESPESSER

LEVELS OF REPRESENTATION AND LEVELS OF ANALYSIS FOR THE DESCRIPTION OF INTONATION SYSTEMS

1. INTRODUCTION

The linguistic description of the intonation systems of different languages, like that of any other aspect of language, can be thought of as a rather indirect process of extracting linguistic information from measurable physical data. As has long been known, there is no automatic technique for performing this operation. As Chomsky (1964) has pointed out, there is no general 'discovery procedure' we can appeal to.

The development of large corpus-based studies and the introduction of widely available automatic modelling techniques do, however, bring the hope that our knowledge of these systems may increase significantly over the next few years. One of the reasons for this is that as we increase our database we are able to formulate and test more and more empirical predictions about the data. The comparison of the predictions with the observed data in turn leads us to formulate more constrained hypotheses about the nature of the phonological representations we hope to bring to light.

At the same time, the development of studies dealing with inter-speaker and intra-speaker variability will teach us more and more about the way in which prosodic systems can vary within the same language or dialect. Similarly it can be expected that the availability of comparable data derived from descriptions of a number of different languages or dialects will enable us to separate out with more and more confidence the language-specific from the universal characteristics of prosodic systems and this will allow us once more to formulate considerable constraints on the nature of these systems.

It must be emphasised that this continuous dialogue between empirical data and linguistic theory is at each step a process of formulating hypotheses on the basis of available data and of testing these on new data. All hypotheses involved are, naturally, only provisional and are liable to be questioned at all times. The higher the degree of abstraction of a hypothesis, however, the greater the quantity of data necessary before we call it into question and look for a better hypothesis to replace it. This means that the fact that different

M. Horne (ed.), Prosody: Theory and Experiment, 51–87.

teams of researchers work with different theoretical backgrounds, far from being a handicap, is in fact a guarantee that research is not confined to what may after all turn out to be a blind alley.

In this paper we formulate a general picture of an overall phonological and phonetic description of intonation and we then make a number of specific proposals for implementing the different levels of representation. Much of this material has been presented elsewhere[1] but we are grateful to the editor for this opportunity to present an overall synthesis here.

2. LEVELS OF REPRESENTATION AND LEVELS OF ANALYSIS

It is obvious that we need to distinguish at least two levels of representation (cf. 't Hart and Collier 1975). At the most abstract linguistic level we want to be able to represent somehow the knowledge that a speaker needs to acquire when he learns a language. At the other extreme, we want to relate such a representation to the physical manifestations of this linguistic knowledge: the corresponding acoustic and physiological characteristics of utterances. Although the distinction between abstract linguistic representations and concrete physical representations is fairly uncontroversial, different approaches tend to differ in the relative importance which they attach to each of these two levels (cf. discussion in Cutler and Ladd 1983).

We can note that in these two cases the term 'representation' is not being used in strictly the same way. When we refer to linguistic representations, we assume that we are describing, however imperfectly, the way in which speakers of the language themselves represent the information in their mind. With physical representations, we are rather describing the way in which we, as scientists, choose to analyse the data. In order to distinguish these two types of representation we can use the specific terms *cognitive representation* and *analytical representation*. In many cases of representations which are intermediate between the purely physical and the purely cognitive extremes, it is an empirical question whether they should be considered cognitive or analytical. Henceforth, when we refer simply to 'representations' without any qualifying term, it can be assumed that we are being deliberately vague as to which of the two meanings is intended. It could be argued that one of the principal aims of linguistic investigation is to provide a satisfactory theory of how analytical and cognitive representations are related.

While the existence of an abstract cognitive representation is fairly widely accepted, there is little or no consensus as to its nature. For many linguists, linguistic representations are composite objects and the pronunciation of an utterance makes use of information from a number of different linguistic components: syntax and semantics as well as phonology as in the following diagram:

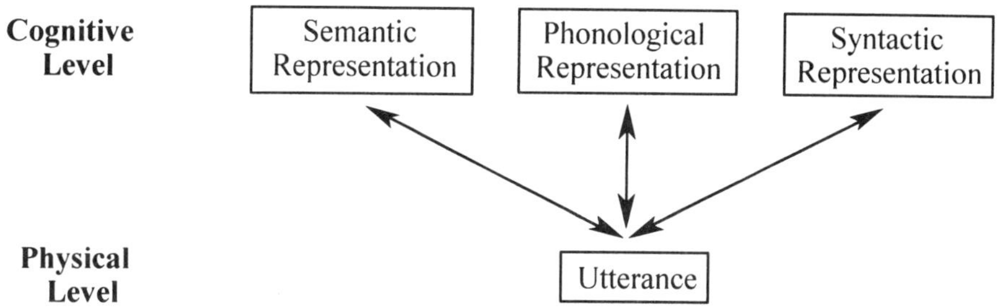

Diagram 1

For others (cf. in particular Selkirk 1978, 1984, 1986) there exists an autonomous level of phonological representation which encodes all (and only) the information concerning the pronunciation of an utterance. Schematically:

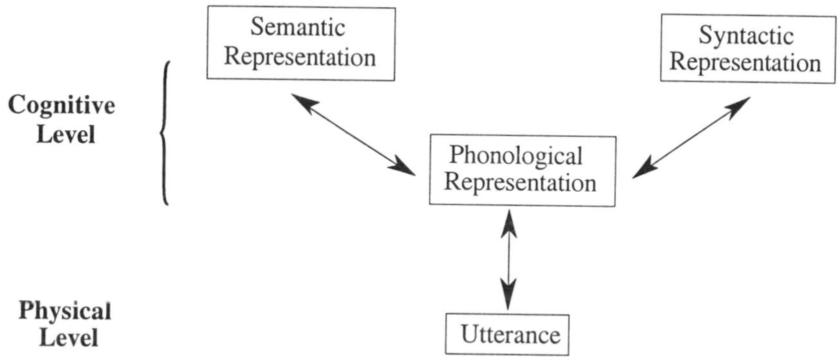

Diagram 2

This hypothesis places a very strong constraint on the nature of linguistic representations and in the absence of convincing evidence against it we shall consequently assume such an organisation. In fact many linguists would place even stronger constraints and assume an organisation such as the following:

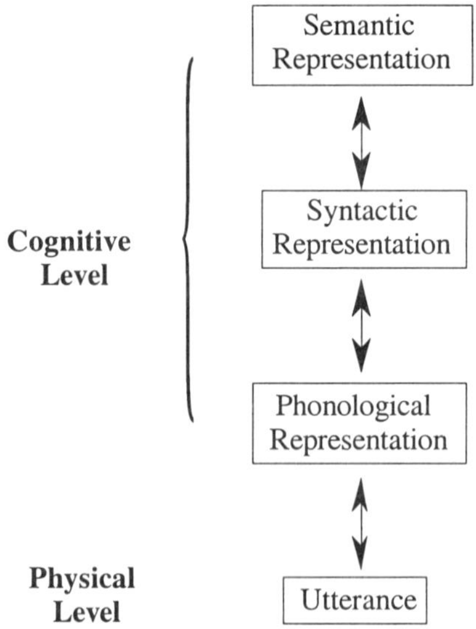

Diagram 3

where the syntactic component (including the lexicon) is assumed to communicate with both the phonological and the semantic component but where there is no direct link between semantics and phonology. Nothing crucial in the rest of this chapter depends on the difference between this framework (that of the so-called 'Extended Standard Theory' of generative grammar (Chomsky 1981)) and the one mentioned previously. A lot depends on the precise role which the theory attributes to the semantic component and its relation to pragmatic interpretation. There are a number of very important issues at stake here but a discussion of these would obviously take us far beyond the scope of this chapter. See, however, Hirst (1987, 1993).

It can be seen from the above diagrams that a phonological representation must fulfill two purposes: it must provide both the information necessary for the pronunciation of an utterance and the information necessary for its syntactic and semantic interpretation. This in fact provides us with a useful constraint which we shall refer to as the 'Interpretability Condition' and which states:

Interpretability Condition:
Representations at all intermediate levels must be interpretable at both adjacent levels: the more abstract and the more concrete.

There is nothing in what we have said so far which is specific to intonation. In the case of lexical items, for example, this framework embodies the insights of de Saussure 1916 concerning the double nature of the linguistic sign as an arbitrary association between *signifiant* and *signifié*. In the case of prosody and intonation, we can usefully make a similar distinction between *functional representations* which encode the information necessary for the syntactic and semantic interpretation of the prosody of an utterance and *formal representations* which encode the prosodic information necessary for its pronunciation. Most systems of transcription for intonation mix functional and formal characteristics and it is of course an empirical question whether these should be encoded in separate representations or not. The fact that different languages make use of different prosodic forms for encoding the same prosodic functions seems to us evidence, though, that they should be separated: the inventories of prosodic forms and prosodic functions might then both be part of universal linguistic theory while the specific mapping between forms and functions in any given language would be defined by language specific parameters.

An example of such a difference is the distinction between lexical and non-lexical use of prosody in different languages. It has often been noted that there are no specific acoustic characteristics which distinguish languages which make use of lexically distinctive stress and tone from other languages which do not. Thus, there is no obvious acoustic cue which would allow us to distinguish stressed syllables in languages with 'free' or 'distinctive' stress (English, Russian etc.) from those in languages with 'fixed' or 'non-distinctive' stress (French, Polish, Finnish etc). Presumably the (surface) phonological representation of stress is the same in both types of languages.

Similarly, in tone languages (Chinese, Yoruba etc), we can assume that at least some of the melodic characteristics of utterances are lexically determined. It has, however, become standard in recent years to formulate phonological models of intonation which derive intonation patterns in non-tone languages from phonological tones, together with appropriate rules specifying how they are aligned with the accented syllables. This suggests the fascinating possibility that phonological representations in all languages draw from a universal set of prosodic characteristics which are either lexically specified or which are introduced in conformity with language specific parameters and which are subsequently mapped onto phonetic characteristics.

Once again it is an empirical question whether phonetic representation is a distinct level or whether, as suggested by Pierrehumbert and Beckman (1988), we should think of phonetics rather as a *dynamic interpretative process* mapping phonology directly onto acoustics and physiology. Whatever the theoretical status of phonetic representation, we believe that it is a useful heuristic strategy to postulate a distinctive level of phonetic representation which can then be mapped (following the interpretability condition) onto both physics and phonology. As Pierrehumbert and Beckman themselves put it: "the division of labor between the phonology and the phonetics is an empirical question, which can only be decided by constructing complete models in which the role of both in describing the sound structure is made explicit.'' (*op. cit.* p. 5).

We can note that the term 'phonetic representation' has at times been used to cover a number of different phenomena which really need to be distinguished. Phonetics is sometimes used as a synonym for acoustics and physiology. It should be clear that we wish to distinguish these levels of analysis. Phonetics is also sometimes used as a synonym for 'surface phonology' as in the terms 'phonetic transcription' or the 'International Phonetic Alphabet'. For Trubetskoy (1949), the distinction between phonology and phonetics is one between discrete and continuous phenomena. In this sense then a 'phonetic transcription' would more appropriately be termed a 'surface phonological transcription'.

Between the underlying phonological representation and the physical representation we wish then to postulate two distinct levels: the level of surface phonology and the level of phonetics. The level of surface phonology is a level of distinctive discrete categories with which we can describe surface phenomena cross-linguistically. The level of phonetics is the level of continuously variable phenomena from which we have factored out universal constraints on the production and perception of sounds. We can illustrate these distinctions with the example of durational characteristics. Duration is often referred to as one of the three prosodic acoustic parameters, the other two being fundamental frequency and intensity. Unlike the other parameters, however, duration is not purely acoustic: it is impossible for a machine to produce something like a 'duration curve' in the same way that machines can produce intensity curves or fundamental frequency curves. In order to measure duration we need to posit boundaries which are associated with phonological categories such as phonemes or syllables. Since duration is a continuously variable relationship between phonological units and physical parameters it fits precisely the definition we have given of a phonetic characteristic. Studies of the duration of phonological units of a large number of languages might well lead us to the conclusion that only some small finite number of durational distinctions are ever distinctive on a cross-language basis. This could then lead us to set up discrete surface phonological categories and we should then hope to be able to predict the observed range of values from an even more restricted number of underlying representations.

The description of the prosody of languages can thus be seen as a continuous process of defining representations at different levels together with constraints on these representations. The more we learn about representations at any one of the different levels we have described above, the more we shall know about other levels since we project supplementary constraints via the Interpretability Condition onto adjacent levels. The picture we have built up so far can be illustrated as follows:

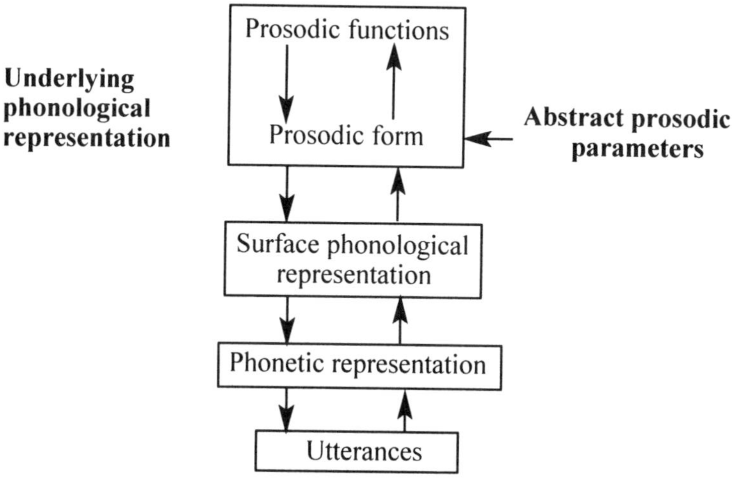

Diagram 4

Prosodic functions are mapped onto underlying prosodic forms in conformity with language specific abstract prosodic parameters. These underlying forms are mapped onto surface forms then onto phonetic representations before ending up as physical correlates of utterances.

To summarise, we argue that a satisfactory global theory of intonation will require four levels of analysis: (i) physical (acoustic, physiological) (ii) phonetic (iii) surface phonological and (iv) underlying phonological, the latter comprising both a representation of prosodic functions and prosodic forms.

In the rest of this chapter we present specific proposals for a model respecting the overall architecture we have just sketched. A number of the features of the model are yet to be developed in particular those needed for the representation of durational and intensity characteristics as well as the exact nature of the alignment between the tonal categories and the segmental categories. This should in no way be taken as reflecting a lack of interest in these aspects but rather our current ignorance of the way in which these characteristics can most appropriately be integrated into represention at different levels.

3. SPECIFIC PROPOSALS

3.1. Phonetic representation

The hypothesis behind the distinction between physical and phonetic levels of representation is that all languages obey the same physical constraints in the production and perception of variations in fundamental frequency, intensity and duration. A phonetic representation, then, is one where such universal constraints have been factored out (cf. Di Cristo 1985: 31).

Phonetic models of fundamental frequency patterns have been developed which attempt to account for physiological aspects of the production of intonation contours, as in the work of Fujisaki and colleagues (Fujisaki and Nagashima 1969; Fujisaki 1988, 1997). Other researchers, in particular those of the 'Dutch school': 't Hart et al. 1990, Terken 1991, Terken and Hermes This volume, as well as Rossi et al 1980, House 1990, d'Alessandro and Mertens 1995, etc., have concentrated on modelling more particularly the perceptual aspects of intonation patterns. If our conception of phonetics (as the level where constraints on perception and production are resolved) is correct, then neither of these approaches is giving the complete picture: both aspects need to be incorporated into a more general model.

A number of different implementations of phonological/phonetic models of intonation have been designed to derive an acoustic output (F0 curve) from a symbolic input (for an overview cf. Hirst 1992). As in all fields of speech analysis, however, it is the inverse problem which is really the most challenging. Given an F0 curve, how can we recover a symbolic representation? Even if we are able to perform such symbolic coding automatically, how should we validate the output of such a programme? One way would be to require the symbolic representation to be in such a form that it can be used as input to a synthesis system, the acoustic output of which can then be directly compared to the original F0 curve. The coding problem is thus directly related to the synthesis problem and in the rest of this section we shall reserve the term 'model' for attempts to solve both the coding problem and the synthesis problem together:

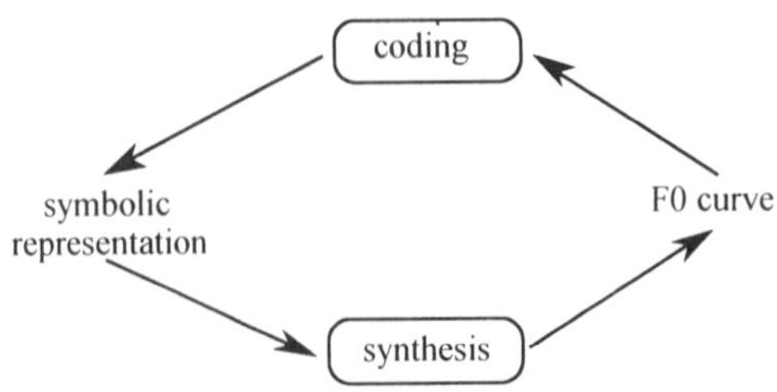

Diagram 5

It is obvious that an automatic modelling system would be highly desirable for a number of reasons. An efficient algorithm would be extremely useful for collecting data for improving both speech synthesis and automatic speech recognition. Such a tool would also of course be extremely valuable for obtaining empirical evidence for testing phonological models of intonation and examining the variability in prosodic parameters across languages, dialects and individuals.

Despite its obvious interest, until quite recently the modelling problem had received less attention from workers in the field than might have been expected. An early treatment is Scheffers (1988) who described a technique for obtaining an automatic piece-wise linear approximation of an F0 curve. More recent approaches approaches are Geoffroi (1993), Jensen et al. (1993), ten Bosch (1993), Bagshaw (1994), Taylor (1993), Morlec et al. (1995), as well as a numer of contributions to Sagisaka et al. (1997). In the rest of this section, after discussion of some of the background and assumptions behind our work, we present an algorithm which has been developed in our laboratory in Aix-en-Provence.

Common to many proposals for fundamental frequency is the idea that microsegmental characteristics of curves should be factored out (Di Cristo and Hirst 1986) so that the resulting curve is similar to that found on a sequence of entirely sonorant segments as found in sentences likes ''Molly may marry Larry''. An F0 curve, then, is modelled as the superposition of two components: a micro-prosodic component caused by the characteristics of the individual phonematic segments of the utterance and a macroprosodic component reflecting the choice of intonation pattern for the utterance.

3.1.1 THE MICROPROSODIC COMPONENT

Di Cristo and Hirst (1986) describe an experiment in which nonsense syllables ''bababa'' and ''vavava'' were pronounced by a single speaker in three different contexts:

(i) "_____, c'est un mot." (__, it's a word.)
(ii) "C'est un mot, _______." (It's a word, ___)
(ii) "C'est un mot, _______?" (Is it a word, ____ ?)

The contexts were chosen to ensure that the nonsense words were pronounced with rising, low flat and high flat pitch patterns respectively. The fundamental frequency at the centre of the consonants of the nonsense words was compared to that of the mean of the surrounding vowels. Nicolas (1989) carried out the same experiment with four subjects (two male, two female), six voiced consonants (b d g v z j) two contexts (high, low) and four repetitions. The relationship between the F0 on the consonant and on the surrounding vowels was found to be fairly linear. Although a logarithmic regression in fact gave slightly better predictions than a linear one for most speakers and most consonants, the differences between linear and logarithmic predictions were very small. We conclude then that the raw F0 curve can be factored out into a macroprosodic component onto which is superimposed a (linear or logarithmic) microprosodic component.

3.1.2. THE MACROPROSODIC COMPONENT

It follows from the above that the macroprosodic component of an F0 curve will be practically identical to the raw F0 curve observed for utterances consisting entirely of vowels and sonorant consonants, since these are known to have the smallest micromelodic effect. Observation of such curves shows that they obey two constraints: they are continuous and they are smooth. More technically, both the curve and its first derivative are everywhere continuous. The simplest function which obeys both these constraints is a quadratic spline function which we have used successfully for many years now for very close modelling of observed fundamental frequency curves and we have argued elsewhere (Hirst 1980, 1983, 1987, 1992, Hirst and Espesser 1993) that a function of this type allows us to treat a sequence of target points as an appropriate *phonetic* representation for F0 curves[2].

A spline function of degree n corresponds to a continuous sequence of polynomials of degree n, the derivatives of which, up to and including degree n-1, are everywhere continuous. Cubic splines are commonly used for interpolating values in a sequence of which only certain values are known. Quadratic splines, by contrast, interpolate monotonically between points of the curve at which which the first derivative is zero. They have the advantage that they are completely locally determined, since any given stretch of the curve between two target points is entirely determined by those points. This means that any erroneous targets will only affect the immediately adjacent sections of the curve. Quadratic splines are defined by a sequence of triples $<t,h,k>$ where t and h define the time and frequency of the target point and where k defines the spline-''knot'', that is the inflection point of the s-shaped transition between two target points. In the rest of this paper we assume for simplicity that these transitions are symmetrical[3], that is that the inflection point is always situated halfway between two adjacent targets[4].

Hans 't Hart (1991) has suggested that our attempt to synthesise F0 curves with parabolas rather than with straight lines is misguided since subjects cannot hear the difference anyway. This calls for a few comments.

Trivially, it is of course possible to approximate any complex function to an arbitrary precision by a sequence of straight lines. At the limit, one line segment per pair of F0 values will be exactly equivalent to the output of a quadratic spline function. This is not of course what 't Hart has in mind. Note however that the straight line interpolation which he compares to parabolas is not simply linear interpolation between target-points. Instead the interpolation is between horizontal plateaux so that a simple rising pattern which we would code with two target points as in:

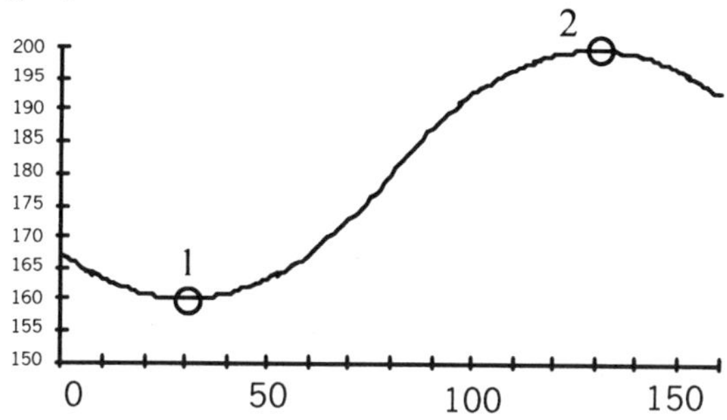

Figure 1. *Coding an F0 rise as a quadratic spline function with two target points.*

would need to be coded as a sequence of five straight line sections as in:

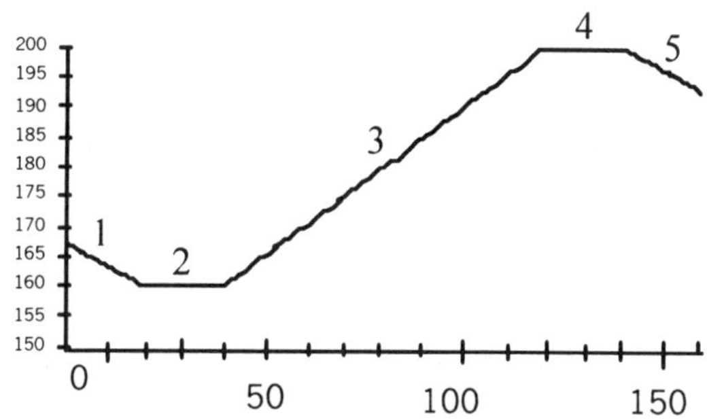

Figure 2. *Coding an F0 rise as a sequence of five straight lines.*

The economy of such a representation is not evident.

We are not entirely convinced by 't Hart's argument that straight line synthesis will always be just as good as that using quadratic splines. Since quadratic splines give a closer approximation to real F0 curves than do straight lines, it seems possible that under certain circumstances these differences will be noticeable. Even the fact that subjects are apparently unable to distinguish certain stimuli does not prove that their reactions to these stimuli will always be identical under all conditions.

We might conclude that the very fact that an automatic coding algorithm such as that which we describe below works at all is a good enough reason for using the same model for synthesis since, as we suggested above, this means that the output of the analysis feeds directly into the synthesis. It would of course be possible to adapt the output to feed into staight line synthesis system, giving an output like that of Figure 2 for example[5], but then of course the direct comparison between observed F0 and modelled F0 would no longer be possible.

Finally, we entirely agree with 't Hart's finding that listeners are much more sensitive to variations in the temporal and frequential values of target points than they are to the nature of the transitions between targets. We take this, however, far from being a weakness of our system, as a strong argument in favour of using a sequence of target points as a phonetic representation of the F0 curve.

3.1.3. MOMEL: AN ALGORITHM FOR AUTOMATIC MODELLING OF F0 CURVES

The MOMEL algorithm which we present here consists of four stages. The central part of the algorithm (stage 2) is an asymmetric version of what we call *modal regression* as defined below. This stage works on the assumption that the only effect of the microprosodic component of a fundamental frequency curve is to *lower* the values of the underlying smooth continuous macroprosodic curve. The only case where microprosodic effects actually raise the F0 curve are at the boundary between voiced and unvoiced segments. We consequently implement a preliminary stage (stage 1) to make sure that any potential raising effects have been eliminated. The second stage then applies the modal regression technique within a moving window, to provide one optimal estimate of a local fundamental frequency target centred on each value of the fundamental frequency curve. The next stage of the algorithm (stage 3) selects a partition of these target candidates. The final stage (stage 4) reduces the candidates within each segment of the partition to a single target.

The technique used in this algorithm is one which we call 'modal regression' since it bears in fact the same relation to ordinary regression as the estimation of the mode of a distribution bears to that of the mean. Both the *mean* and the *mode* of a distribution are in some sense the values which are the closest to all the items of the distribution. In the case of the *arithmetic mean* this 'closeness' is defined by calculating the square of the distance from each item of the distribution and selecting a value such that the sum of these squared distances is minimal. By contrast, the *mode* of a distribution can be defined as the value which is less than a given threshold Δ from the largest possible number of items of the distribution. While a distribution has only one mean, it may have a number of different modes, and even when there is only one mode, its value may vary in function of the value of the threshold Δ.

Ordinary regression is basically the same as the calculation of the mean except that instead of comparing the items of a distribution to a single value, the items of a series are compared to the values of a well-defined function (in the case of linear regression to a straight line, in the case of quadratic regression to a parabola, etc.), the parameters of which are selected to minimise the sum of the squared distances from the individual items. By contrast, we can define *modal regression*, as selecting the parameters of a given function such that the values of the function are less than a given distance Δ from the largest possible number of items of a series. Note that while this definition tells us what the *aim* of modal regression is, it does not provide us with a procedure for selecting the parameters to fulfill this aim. The task is further complicated by the fact that, as with the mode of a distribution, there may be more than one value for the optimal parameters of a modal regression. One method for estimating these parameters is presented below.

In the case of fundamental frequency curves, we have suggested that with the exception of some local effects linked to the onset and offset of voicing and which the first stage of the algorithm described below is designed to eliminate, all other microprosodic effects consist of a lowering of the 'underlying' macroprosodic curve which we model as a quadratic spline function. We consequently introduce the further asymmetric constraint that the quadratic spline function we wish to find is such that there are *no* values more than a distance Δ above the function and as few values as possible more than the same distance Δ below it.

The four stages:

(1) preprocessing of F0

(2) estimation of target-candidates

(3) partition of candidates

(4) reduction of candidates

are described in the rest of this section. The typical values given for each parameter of the algorithm were obtained by a process of optimisation which is described in 3.1.4 below.

(i) preprocessing of F0

All values more than a given ratio (typically 5%) higher than both their immediate neighbours are set to 0. Since unvoiced zones are coded as zero, this preprocessing has essentially the effect of eliminating one or two values (which are often dubious) at the onset of voicing, i.e. about 10 to 20ms.

(ii) estimation of target-candidates

The following steps are followed iteratively for the time value *x* of each F0 value

a. Within an analysis window of length *A* (typically 300ms) centred on *x*, values of F0 (including values for unvoiced zones) are neutralised if they are outside of a range defined by two thresholds *hzmin* and *hzmax* and are subsequently treated as missing values. The threshold *hzmin* is a constant set to 50 Hz and the adaptative threshold *hzmax* is set to the mean of the top 5% of the F0 values of the sequence multiplied by 1.3.

b. A quadratic regression is applied within the window to all non-neutralised values.

c. All values of F0 which are more than a distance Δ *below* the value of F0 estimated by the regression are neutralised (typical value of Δ fixed at 5%).

Steps **b** and **c** are reiterated until no new values are neutralised.

d. For the time value ***x*** of each F0 value a target point <*t,h*> is calculated from the regression equation:

$$\hat{y} = a + bx + cx^2, \text{ where } t = -b/(2c) \text{ and } h = a + bt + ct^2$$

If t is outside the current analysis window (i.e. if it is less than $x-(A/2)$ or greater than $x+(A/2)$) or if h is less than *hzmin* or greater than *hzmax*, then t and h are treated as missing values. Steps **b, c** and **d** are repeated for each time value x, resulting in one estimated target point <*t;h*> (or a missing value) for each original value of F0.

(iii) partitioning of target candidates

The sequence of target candidates is partitioned by means of another moving window of length R (typically 200 ms) in which the average value of the targets in the first half of the window is compared to the average value in the second half. The boundaries of the partition are then taken as those values which correspond to a local maximum for this distance and which is greater than the overall average value of the distances.

Specifically $dt(x)$ and $dh(x)$ are calculated as the absolute mean distances between the t and h values of the targets in the first half of the window and those in the second half of the window. A combined distance $d(x)$ is then obtained by weighting these distances:

$$d(x) = \frac{dt(x) * wd + dh(x) * wh}{wd + wh}$$

where:

$$wd = \frac{1}{mean(dt(x))}$$

and

$$wh = \frac{1}{mean(dh(x))}$$

The boundaries of the partition are consequently set to each value x respecting the following three conditions:

$d(x) > d(x-1)$; $d(x) > d(x+1)$; $d(x) > mean(d(x))$

(iv) reduction of candidates

Within each segment of the partition, outlying candidates (for which either $dt(x)$ or $dh(x)$ are greater than one standard deviation from the corresponding mean values for the segment) are eliminated. The mean value of the remaining targets in each segment is then calculated as the final estimate of *t* and *h* for that segment.

The following figures illustrate the application of the algorithm to a sentence taken from the corpus Eurom1-French.

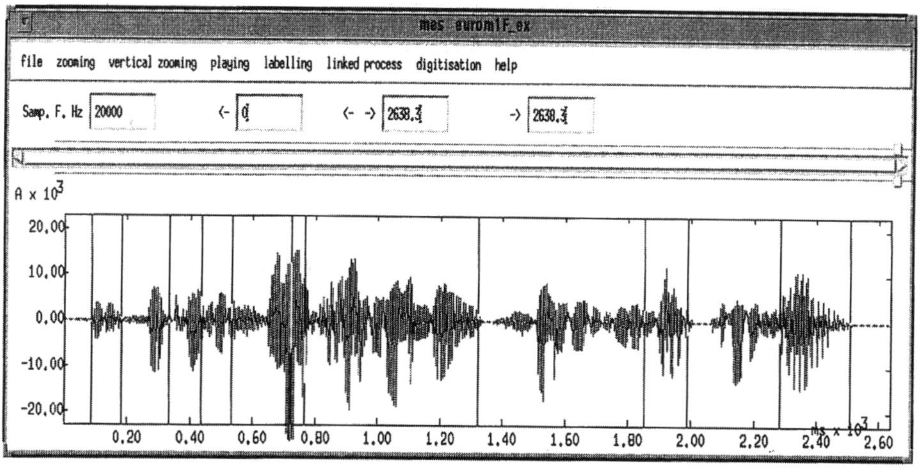

Figure 3. *Acoustic wave form for the sentence* Il faut que je sois à Grenoble samedi vers quinze heures *'I have to be in Grenoble by Saturday around 3 p.m.' taken from the French recordings of EUROM1. Vertical lines correspond to word labels time-aligned to the beginning of each word.*

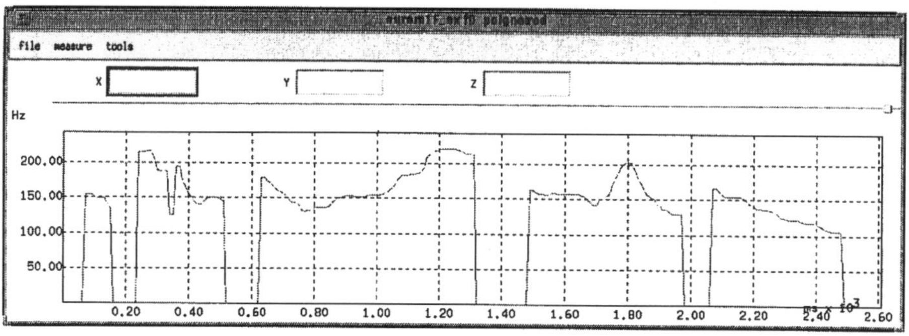

Figure 4. *Fundamental frequency for the same sentence as in Figure 3, detected using a comb function.*

In Figure 5, the different target candidates estimated by the algorithm are represented by a grey line joining the centre of the analysis window (on the abscissae) to the x,y value of the target estimated for that window.

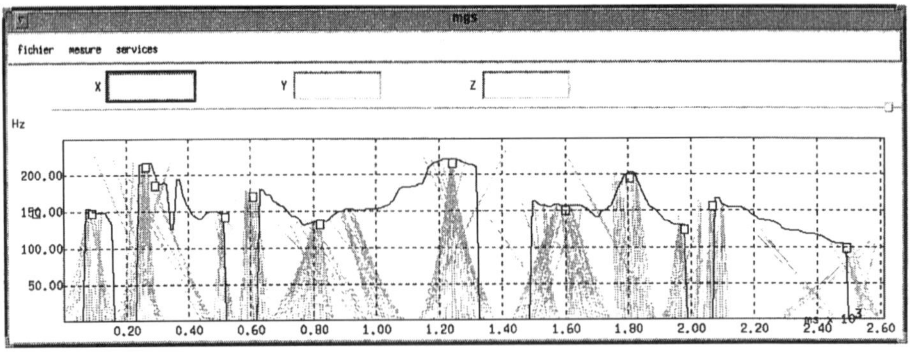

Figure 5. *Output of the automatic modelling program. Each target point candidate is visualised by a grey line connecting the centre of its analysis window on the x-axis to the target point. These candidates are then partitioned and the final target points selected are represented by the squares.*

The squares in Figure 5 correspond to the final estimates of the targets after partitioning the different candidates. The resulting modelled curve is illustrated in Figure 6 by the continuous smooth line.

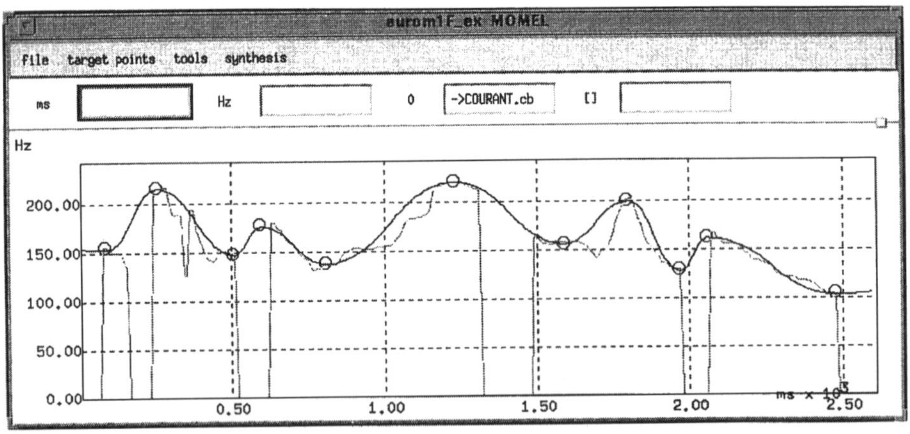

Figure 6. *The target points of Figure 5 are used to generate a quadratic spline function giving a close fit to the original.*

While the algorithm described above is still somewhat less than perfect, it does seem to constitute at least a reasonable first approximation to a working phonetic model of fundamental frequency curves, incorporating both the coding and the synthesis of such curves. Since its development, the model has been used for the analysis of fundamental frequency curves in a number of different languages including English, French, Spanish, Italian and Arabic

(Hirst et al. 1993) and is apparently fairly robust. Preliminary results obtained within the MULTEXT project (Véronis et al. 1994, Strangert and Aasa 1996, Campione et al. 1997) show that the modelling technique gives quite satisfactory results for English, French, Spanish, German and Swedish. Tests with a number of other Western and Eastern European languages are also in progress.

3.1.4. EVALUATION OF THE PHONETIC REPRESENTATION

Fundamental frequency was detected every 10 ms using a combination of three methods: a comb-function (Martin 1981, Espesser 1982), AMDF and autocorrelation. No manual corrections were made to the detected F0 values.

The algorithm described above uses 3 independant parameters :

- analysis window *[A]*
- distance threshold *[D]*
- reduction window *[R]*

In order to optimise these parameters, a small corpus was used (*corpus VNV*) consisting of two sentences, containing all the stops and fricatives (and hence all the different microprosodic configurations) of French, spoken by ten subjects (5 male, 5 female).

S1: La pipe de Jean s'est cassée en tombant de ta gabardine.
(John's pipe broke when it fell from your raincoat).
S2: La fille de Charles Sablon a voulu un petit chien en guise de cadeau.
(Charles Sablon's daughter wanted a little dog as a present).

The following criteria were adopted:

(a) subjective evaluation

The original fundamental frequency curve and the modelled curve were compared visually. The number of manifest errors consisting of either missing targets or false targets was counted. The original recordings were compared (informally) with the same recordings resynthesised using the SOLA/PSOLA technique (Roucos and Wilgus 1985, Hamon et al. 1989) in order to check the relevance of the visual analysis.

(b) objective evaluation

A mean distance was calculated between the original fundamental frequency curve (hz_i) and the modelled curve (hz'_i):

$$d = \frac{1}{n}\sum_{i=1}^{n}\left|1 - \frac{hz'_i}{hz_i}\right|$$

During the optimisation of the algorithm a good correspondance between the two types of evaluation was observed. The minimum value for F0 was found to be quite robust so that it was possible to fix the same value for all ten speakers *[hzmin]* = 50. The maximum value */hzmax/* was fixed as the mean of the top 5% of the speaker's F0 values multiplied by 1.3. For the other three parameters, the following values were found optimal for the corpus:

- *[A]* : 300 *[D]* : 5% *[R]* : 200

The algorithm was subsequently applied with its parameters fixed as above to four other corpora:

- *ATOME* : one continuous text in French read by 4 speakers (2 male, 2 female).
- *EUROM1*, English, French and German. For each language, 10 speakers (5 male and 5 female) read either 15 (English) 10 (French) or 20 (German) passages from a total of 40 different passages each consisting of 5 semantically related sentences (Chan et al. 1995)

The following table summarises the results of the analysis for the five corpora.

Corpus	Mean distance	Number of targets	Total duration
VNV	6.22 %	284	49 s
ATOME	6.28 %	654	220 s
EUROM1 French	6.00 %	6747	2190 s
EUROM1 English	5.60 %	8680	2635 s
EUROM1 German	4.66 %	13995	4419 s

Table 1. *Summary of distance measurements as defined above, number of target points detected and total duration for four different corpora.*

The mean distances recorded in Table 1 can be attributed to a number of different causes:

- erroneous detection of F0 (resulting in erroneous target points).
- errors of the algorithm itself i.e. targets either missed, added or displaced.
- microprosodic effects.

Microprosodic effects, according to preliminary estimates, probably account for no more than about one fifth of the distances recorded in Table 1 so that there is obviously still room for improvement of the algorithm.

Subjective evaluation of the application of the MOMEL algorithm to the EUROM1 passages for a number of different languages is currently in progress. Evaluators were asked to note cases where they could hear a difference between the original recording and the versions resynthesised from the target points, restricting attention as far as possible to differences in intonation. The results of this evaluation, although not amenable to statistical analysis are expected to provide us with an extremely valuable "error database" which can be used for the evaluation of future versions of the algorithm.

3.1.5 REMAINING PROBLEMS OF PHONETIC REPRESENTATION

As we mentioned above, we present in this section proposals for a phonetic representation of fundamental frequency curves. This representation is obviously far from constituting a complete phonetic model of intonation. Even in the domain of fundamental frequency, the model as presented above does not address an important issue: that of the relative scaling of the target points - in particular when comparing data from different speakers. The problem of normalisation of acoustic data is one which needs to be addressed seriously if we hope to abstract away from speaker-specific variability. Prosodic characteristics differ from other phonetic properties in that what is important is generally not the absolute value of the property but its relative value. What is perceived as loud, long or high for one speaker may in fact be less loud, long or high than what is perceived as soft, short or low for another speaker.

Recent research into the inter-speaker variability of prosodic characteristics (cf. Campbell 1992, This volume; Hermes and Van Gestel 1991, Terken and Hermes, This volume) has provided a number of scaling techniques as a means to obtain speaker independent representations of prosody. These techniques provide a useful way of factoring out universal constraints on production and perception and should consequently provide a useful step in the process of extracting a phonetic representation from acoustic data.

3.2. Surface phonological representation

The hypothesis behind the distinction between surface and deep phonological representations is that there is a level of description at which we can describe the prosody of different languages cross-linguistically using a restricted inventory of symbols much in the way that the International Phonetic Alphabet is used to describe the vowels and consonants of different languages.

Our examination of intonation patterns described for twenty different languages (Hirst and Di Cristo 1998) led us to the conclusion that there is indeed a useful level of generalisation at which we can describe intonation patterns across languages on a surface level.

3.2.1. INTSINT: AN INTERNATIONAL TRANSCRIPTION SYSTEM FOR INTONATION.

We proposed, as a first approximation, a transcription system (INTSINT) by means of which pitch patterns can be coded using a limited set of abstract tonal symbols, {T,M,B,H,S,L,U,D} (standing for: Top, Mid, Bottom, Higher, Same, Lower, Upstepped, Downstepped, respectively). Each symbol characterises one of the target points of the phonetic representation. Unlike other systems of representation such as ToBI (Silverman et al. 1992, Pierrehumbert, this volume), the use of INTSINT does not require that the inventory of pitch patterns of a given language already be known before transcriptions can be made. It can thus be used, like the International Phonetic Alphabet, as a tool for gathering data on the basis of which phonological descriptions may be elaborated.

The rationale behind the INTSINT system is that the F0 values of pitch targets are programmed in one of two ways: either as **absolute** tones {T, M, B} which are assumed to refer to the speaker's overall pitch range (at least within the current Intonation Unit), or as **relative** tones{H, S, L, U, D} assumed to refer only to the value of the preceding target point. For relative tones, a distinction is made between **non-iterative** {H, S, L} and **iterative** {U, D} tones, since in a number of descriptions it appears that iterative raising or lowering uses a smaller F0 interval than non-iterative raising or lowering. The tone S has no iterative equivalent since it would be impossible to locate intermediate S tones temporally in a sequence of such tones. For illustrative purposes we propose iconic symbols which can be used to align the tonal symbols with the orthographic transcription. The following table gives the orthographic code followed by the recommended iconic symbols (all of which are taken from the widely available font *Symbol*):

		Positive	***Neutral***	***Negative***
ABSOLUTE		T ⇑	M ⇒	B ⇓
RELATIVE	*Non-Iterative*	H ↑	S →	L ↓
	Iterative	U <	•	D >

Table 2. *Orthographic and iconic symbols for the INTSINT coding system*

3.2.2. AN AUTOMATIC CODING ALGORITHM FOR INTSINT

The following algorithm represents a first approximation to such an optimal coding:

(i) the highest and lowest target values in the utterance are coded respectively T and B.

(ii) the first target point, as well as any which follow a silent pause of more than 250ms, is coded M (unless already coded T or B).

(iii) all other target points are coded with relative tones. A target which is less than a given threshold Δ Hz from the previous target is coded S. Otherwise it is coded H, L, U or D according to its configuration with respect to the preceding and following target points as in the following diagram:

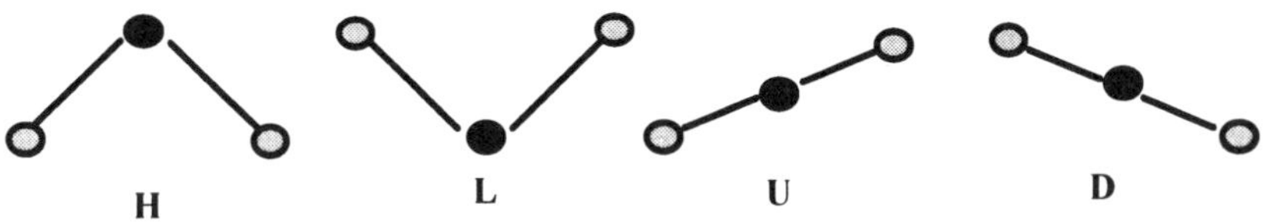

Diagram 6

Target points before a pause (>250 ms), where there is no relevant following target point, are (somewhat arbitrarily) coded as either S, H or L depending on the previous target.

(iv) the statistical value of each category of target points is then calculated: for absolute tones the mean value is taken, for relative tones a linear regression on the preceding target is calculated.

(v) Any target points originally coded H or L can be recoded as T, U, B or D if this improves the statistical model (i.e. the value predicted from the recoded tone is closer to the observed value than that predicted from the original coding).

(vi) steps (iv) and (v) are then repeated until no more points are recoded.

The fact that only H and L targets are allowed to be recoded in step (v) ensures that the algorithm converges to a (local) optimum.

The sample sentence analysed in section 3.1 above coded in this way corresponds to the following transcription:

(1) Il faut que je sois à Grenoble samedi vers quinze heures

⇒ ⇑ ↓ ↑ ↓ ⇑ ⇒ ⇑ ↓ ↑ ⇓

There are, of course, a great number of other possible statistical models which need to be tested. The simple model we have described above seems however to give quite good results for the material to which it has been applied (i.e. isolated sentences and short passages for a number of different languages) and it is now a question of collecting more empirical evidence in order to see how such a model can best be improved.

3.2.3 PRELIMINARY RESULTS FOR SURFACE PHONOLOGICAL REPRESENTATIONS

The algorithm described in §3.2.2 was applied to the EUROM1 passages for both English and French. The output of the algorithm consists of two coefficients *a* and *b* for each tone and each reading of a passage which can then be used to model the value of a given target point P_i with the linear equation:

(2) $P_i = a.P_{i-1} + b$

For 'absolute tones' (**T**, **M** and **B**) the coefficient *a* was set to 0 so that the tones were modelled as a constant (the mean for the tone). For 'relative' tones (**H**, **S**, **L**, **U**, **D**) the coefficients were derived by linear regression on the preceding value. It was observed that in over 75% of the cases the coefficient *a* was estimated as a constant with a value between 0 and 1[6]. This suggests that a more constrained model might be appropriate since when a is between 0 and 1 the linear equation (2) applied reiteratively converges to an asymptotic value[7]:

(3) $A = b/(1-a)$

so that equation (2) could be replaced by the following:

(4) $P_i = A + a.(P_{i-1} - A)$

In other words, each time a lowering or raising factor is applied, this is the equivalent of going a fixed ratio of the distance remaining between the previous target and the asymptotic value. The following table gives the estimated mean or asymptote for male and female speakers split by language:

Sex:	M		F	
Language:	English	French	English	French
T	205	200	327	377
H	160	191	287	310
U	184	232	195	336
M	162	140	261	261
S	106	78	284	343
D	53	54	143	215
L	93	111	161	206
B	92	98	139	179

Table 3. Mean (T, M, B) or asymptote (H, U, S, D, L) values calculated for male and female speakers split by language for the EUROM1 passages analysed with the MOMEL and INTSINT algorithms.

Analysis of variance showed, as might be expected, highly significant effects for both tone and sex ($p < 0.0001$). Significant interactions between tone and sex ($p=0.039$) and tone and language ($p=0.049$) were also observed but the number of speakers (5 per sex per language) was obviously far too small for us to propose any interpretation of these results. Ignoring these interactions, the following figure illustrates the means or asymptotes of the different tones expressed in ERB units (cf. Terken and Hermes, This volume) offset to the mean value of all the tones for male speakers and female speakers respectively:

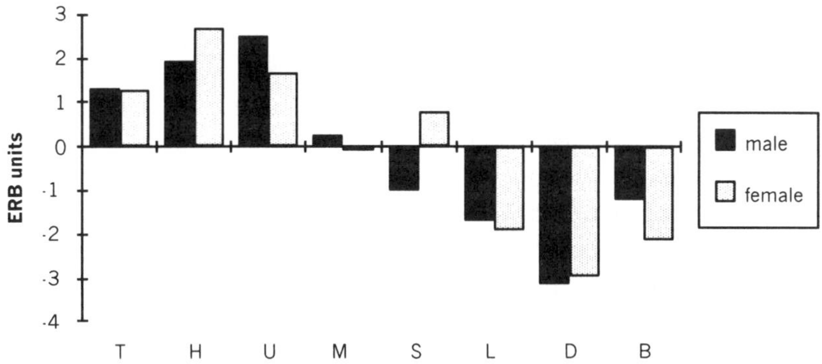

Figure 7. *Mean or asymptote values for each tone expressed in ERB units offset to the overall mean for male and female speakers, respectively.*

Hirst et al. (1991) suggested the possiblity that the same constants might be used for modelling both absolute tones and relative tones. Under this interpretation, T and H might share one target value while B and L might share another value[8]. In the case of absolute tones the target is actually reached whereas in the case of relative tones the target is simply approached. The data illustrated in Figure 7 seem compatible with this hypothesis. They are also compatible with the observation that the relationship between the two constants defining the upper and lower target values is close to that of one octave.

Making a number of simplifying assumptions, the whole system could in fact be reduced to a small number of speaker dependent variables: minimally a single parameter related to a speaker's physiological characteristics. It remains, however, to be seen how such a minimalist prosodic model can be related to the sort of inter and intra speaker variability discussed above (§1).

3.2.4. REMAINING PROBLEMS OF SURFACE PHONOLOGICAL REPRESENTATION

One obvious weakness of the model as presented above is the fact that it does not take into account the way in which 'absolute' tones T, M and B are themselves scaled throughout a more extended discourse. Nicolas (1995), Nicolas and Hirst (1995) showed that for continuous texts in French, appreciably better results were obtained if some scaling of this sort was introduced. In particular, a T tone near the beginning of an intonation unit tended to be significantly higher than average in the first Intonation Unit of higher-level pargaraph (or ''paratone'') type structures and significantly lower in the last Intonation Unit of these structures. Final B tones were also significantly lower at the end of units of this type.

Since the scaling is assumed to apply across Intonation Units, one possible solution would be to use the same set of INTSINT symbols to code not only the individual target points, but also the overall range and register of the Intonation Units somewhat in the way recommended by the IPA working group on suprasegmental transcription (Bruce 1989). Assuming that Intonation Units are delimited by square brackets and that symbols within the square brackets define target points whereas symbols in round brackets define the pitch range of the following Intonation Unit, a sequence such as example (1) above could be transcribed:

(5a) Il faut que je sois à Grenoble samedi vers quinze heures

(⇑) [⇑ ↓ ↑ ↓ ⇑] (>) [⇑ ↓ ↑ ⇓]

where the intial (⇑) indicates that the pitch range of the first intonation unit is set to an initial high (Top) value, and the symbol (>) indicates that the entire pitch range of the second Intonation Unit is scaled somewhat lower than the first. Another possibility would be for the pitch range as described for (5a) to be the default values so that the sentence could be simply transcribed:

(5b) Il faut que je sois à Grenoble samedi vers quinze heures

[⇑ ↓ ↑ ↓ ⇑] [⇑ ↓ ↑ ⇓]

where by default the pitch range of the first Intonation Unit is set to Top and that of subsequent unmarked Intonation Units is Dowstepped.

Needless to say, we have not yet developed an algorithm which will convert a representation like (4) above into one like (5a) or (5b). Dividing a signal up into Intonation Units is a problem which is far from trivial and a lot of work remains to be done in this area. Introducing global parameters on Intonation Units such as in the above example makes it possible to produce much more accurate modelling of data but at the expense of introducing a number of extra degrees of freedom with the result that it is not a simple matter to develop heuristics to choose between possible alternative codings of a given surface realisation.

3.3. Underlying phonological representation

The level of deep phonological representation can be thought of as the level of linguistically significant choices which interact with a number of language-specific abstract prosodic parameters to generate observed intonation patterns.

We made a distinction above between functional and formal representations of prosody and we mentioned that most transcription systems combine aspects of the two. Thus for example the 'tonetic stress' marking system of the 'British School' (cf. Cruttenden 1986) uses symbols such as {´; `; ^; ˇ} to indicate both the fact that a given word is highlighted, and the direction of the pitch-movement initiated on the stressed syllable of that word. Similarly the ToBI system (Silverman et al. 1992) makes use of symbols such as H* L% corresponding to a high tone associated with an accent followed by a low tone associated with a boundary. The symbols * and % encode functional aspects of the transcription whereas the symbols H and L encode formal aspects. One purely functional system of representation of intonation is ordinary punctuation: readers are surprisingly consistent in the way in which they interpret punctuated texts even though there is absolutely no formal relationship between the punctuation symbols and their corresponding intonation patterns[9].

It has long been known that intonation has "a bewildering variety" of different functions (Ladd 1993: 12). Ladd mentions "signalling focus, emphasis, phrasing distinctions, lexical distinctions, speaker attitude, and many more" (op. cit.). If we restrict our attention to the more strictly linguistic functions of intonation, following Bruce (1985) we can distinguish two major functions: a weighting function (foregrounding/backgrounding) and a grouping function (coherence/boundary signalling). Work on prosodic structure theory has suggested that a fruitful way of thinking of prominent syllables is as the head of a prosodic constituent (Selkirk 1978, 1984, 1986). Halle and Vergnaud (1987) suggest that both prominence and grouping can be reduced to a single relationship of 'government' defined as the conjunction of two 'conjugate' structures: the pure representation of heads (prominence) and the pure representation of domains (grouping) - either can be recovered from the other provided that we specify by a language-specific abstract parameter the left-headed or right-headed nature of metrical constituents (p. 16)[10].

Representations of this sort entail the prediction that in addition to the distinction accented/unaccented, there can be different positions of the constituent boundary, depending on whether the head is taken to be at the beginning or the end of the prosodic constituent. An example of such a distinction can be made between accent groups in English and French. A number of facts are consistent with the idea that accent-groups in French are "right-headed" - that is they culminate in an accented syllable rather than starting with one as in English and presumably many other Germanic languages (Wenk and Wiolland 1982; Fant et al. 1991)[11].

Assuming that this is an appropriate distinction between English and French at least would mean that phrases like:

(6a) a cup of tea
(6b) un verre de vin 'a glass of wine'

which presumably have the same syntactic structure in English and French, would be structured differently at the prosodic level as in the following:

(7a)

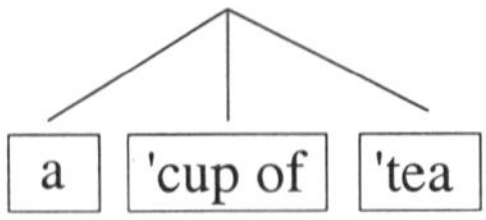

(7b)

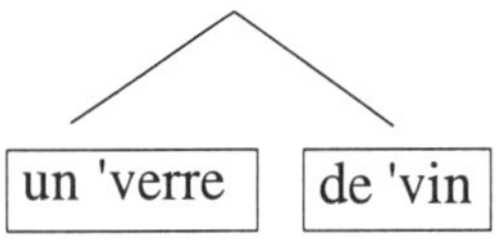

Holmes (1996) in a contrastive study of English and French monologues observed a significant difference in the distribution of pauses in the two languages: English speakers may regularly insert pauses within a syntactic phrase whereas French speakers very rarely do so. When French speakers do pause after an article or a preposition, they typically repeat the article or preposition when they resume speaking so that whereas an English speaker might produce say something like:

(8a) a... cup of … tea.

A French speaker is more likely to say:

(8b) un… un verre de… de vin

In terms of prosodic structure these observations can be re-interpreted as saying that speakers tend more to pause between accent groups rather than within them.

The left/right-headed nature of the stress-group is then an example of the sort of abstract prosodic parameter which we assume underlies the variability observed across different intonation systems. In the remainder of this section we present a sketch of the way in which we account for the differences between the basic intonation systems of English and French[12]. Since this sketch draws on characteristics of French which are not necessarily widely known we introduce this comparison by a brief account of French non-emphatic accentuation.

French is traditionally described as having systematic word-final accentuation (Halle and Vergnaud 1989). Recent studies have shown, however, that the actual accentuation of utterances in spoken French is considerably more complex than this (Hirst and Di Cristo 1984, Pasdeloup 1990, Di Cristo and Hirst 1997). In particular, pitch prominence is often given to the initial syllable of a word as well as, or instead of, to the final syllable. This initial prominence, which in the past has often been taken for some type of emphatic stress, in fact occurs quite systematically in the speech of many speakers of modern French without conferring any particular emphatic connotation to the word. Other more conservative styles of French do not give prominence to the initial syllables of words.

There have been a number of attempts to account for the ''probabilistic'' nature of French accentuation (Fónagy 1980). A fairly satisfactory first appoximation can be obtained by the following rules which can be assumed to apply from left to right:

(i) divide the utterance into intonation units
(ii) assign an accent to the final syllable of the intonation unit.
(iii) assign an accent to (the initial and) final syllable of each accentable word

where the bracketed part of (iii) only applies in the less conservative styles mentioned above. Rule (iii) obviously considerably overgenerates accents. A further principle, similar to the well-known ''stress clash'' rule in English and many other languages limits the number of syllables which are actually accented by stating:

(iv) Do not assign an accent to a syllable if a 'nearby' syllable within the same Intonation Unit is already accented.

The term 'nearby' is deliberately vague and can be made more specific in a number of ways, each of which would result in a different set of accent patterns. To simplify the discussion we shall assume here that 'nearby' is interpreted as 'adjacent' which results in a fluent and acceptable set of accent patterns. It should, however, be remembered that 'nearby' might equally well

be interpreted as 'less than x syllables away' or even as 'less than x milliseconds away' where x has some integer value.
To take a few examples:

(9a) Elle parle 'She speaks'
(9b) Elle parlait 'She was speaking'
(9c) Elle parlait français 'She was speaking French'
(9d) Elle savait très bien parler le français 'She could speak French very well'
(9e) Elle ne savait pas très bien parler le français 'She couldn't speak French very well'

The words *pas*, *très* and *bien* are all accentable in French despite the fact that they are function words. The rules given above generate the following patterns:

(10a) Elle 'parle
(10b) Elle par'lait
(10c) Elle 'parlait fran'çais
(10d) Elle 'savait 'très bien 'parler le fran'çais
(10e) Elle ne 'savait 'pas très 'bien par'ler le fran'çais

The fact that (ii) is ordered before (iii) ensures that the final syllable of each Intonation Unit will retain its accent. There is in fact independent evidence that the accent assigned to the final syllable of an Intonation Unit does not follow the same rules as the word-initial and word-final accent found elsewhere. Clitic syllables like *le*, *en*, *vous* are normally unaccentable as can be seen in (11a, 11b):

(11a) Je le tra'duis 'I translate it'
(11b) Vous vous en a'llez 'You are going away'

When these syllables occur in final position, however, they are assigned prominence as in (12 a-c).

(12a) 'Traduis-'le! 'Translate it!'
(12b) 'Allez-vous 'en! 'Go away!'
(12c) 'Où allez-'vous? 'Where are you going?'

We have accounted for unemphatic intonation patterns in English and French (Hirst 1983, Hirst and Di Cristo 1984) by assuming that High and Low tones are attached directly to the accent group (which for this reason we have preferred to call the Tonal Unit) as well as to the higher order Intonation Unit in accordance with a tonal template with the following form:

(13a) French :

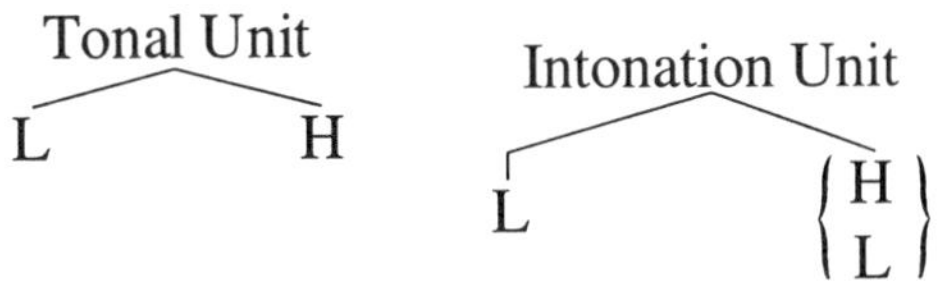

(13b) English :

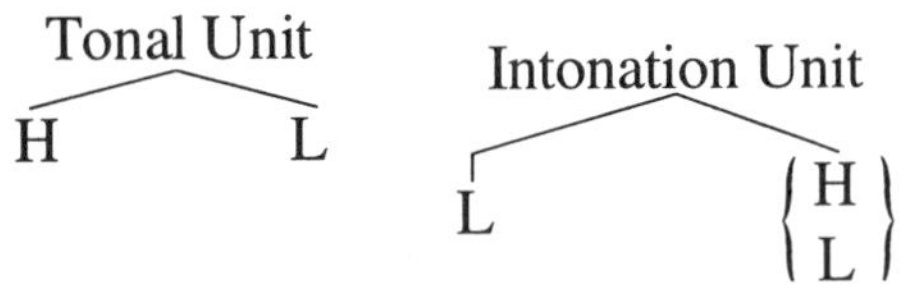

This results in a non-linear prosodic structure which is submitted to linearisation constraints, projecting the tonal segments onto a single tonal tier.

A rule of downstepping also applies under slightly different conditions in English and French. This rule, which can be roughly formulated as follows:

(14) Downstep:
$$\begin{array}{ccc} H & L & H \\ & \Downarrow & \Downarrow \\ & \varnothing & D \end{array}$$

converts a sequence HLH into a sequence H D (where D represents a downstepped (lowered) high tone[13]. The rule appears to be style/dialect dependent in English: systematic downstepping is said to be more frequent in Standard British English (RP) than in either Scottish English or American English (Hirst 1998). In French, the downstepping rule seems dependent on the syntactic mode of the sentence, applying systematically in both yes/no-questions and wh-questions but only to the final tonal unit in assertions (Hirst and Di Cristo 1984, Di Cristo 1998).

The choice of the final tone in the Intonation Unit seems to depend on pragmatic and semantic constraints in both English and French. In English and French a high final tone (final rise) is found both in continuatives and in questions and is generally held to be more common in Yes/No questions than in WH-questions in both languages.

Applied to sentence (9d) above, assuming the stress-pattern assigned in (10d), and assuming that a terminal falling pitch is chosen, templates (13a) will generate the following non-linear prosodic structure (where the symbol σ stands for 'syllable'):

(15)

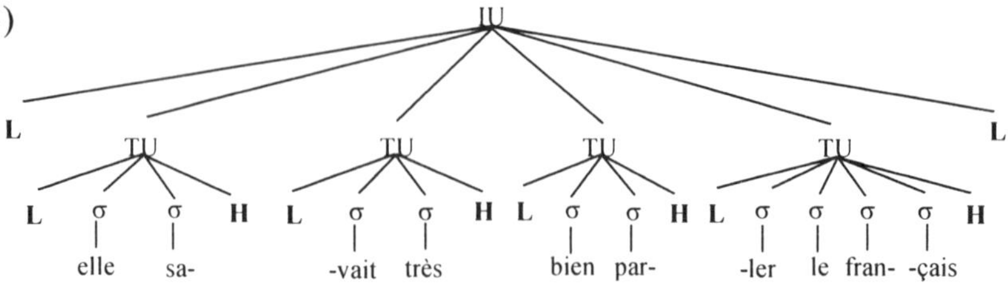

whereas the English equivalent:

(16) She could 'speak 'English 'very 'well would be assigned the structure:

(17)

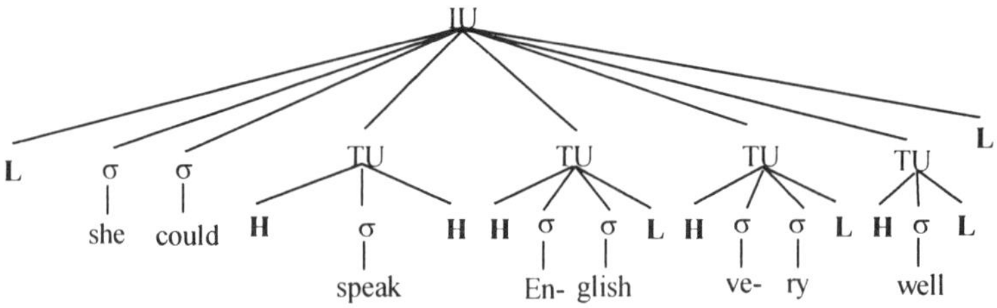

After application of Downstep and other rules of surface phonological interpretation these non-linear structures will be converted to the bi-linear sequence:

(18a)

[elle sa- -vait très bien par- -ler le français]
MT LH LH DB

(18b)

[she could speak English very well]
M T D D DB

We have argued elsewhere (Hirst 1994, Hirst and Di Cristo 1998) that representations such as these can subsequently be converted into phonetic reresentations which in turn can be used to generate synthetic fundamental frequency curves. Representations of this type have been used in two recent applications to speech synthesis for French, one based on a stochastic model (Véronis et al. 1997) and another based on a metrical model of French rhythm and intonation (Di Cristo et al. 1997) with very encouraging results.

4. CONCLUSION

The automatic generation of large quantities of symbolic data will without doubt make it possible in the next decade to develop a far more complete description of the prosody of different languages than is at present feasible. It remains to be seen how far the description of other languages, drawing both on existing work and on our own future investigations, can be accomodated within a model such as we have outlined in this chapter. It appears to us quite probable that a number of characteristics of the phonological models which are current today, (including our own), will not stand up to empirical testing by such large quantities of data. It is also quite possible that a number of characteristics of the different modules which we have outlined above will need to be improved in order to account for the observed data. It is, however, our view that a model of intonation such as that which we have sketched above, specifying explictly different levels of representation, will continue to be an extremely interesting tool for comparing the prosodic systems of different languages.

NOTES

[1] In particular, our thanks for many helpful comments and remarks to participants at a number of different meetings including the 1st International ESCA Conference on Speech Synthesis in Autrans (September 1990) the semi-plenary session on prosodic models at ICPhS XII in Aix en Provence (August 1991), the Prosody Seminar at La-Baume-les-Aix (May 1992), the ESCA Workshop on Prosody in Lund (September 1993), the National Language Institute 2nd International Symposium in Tokyo (June 1994), the AAAI-94 Workshop on the Integration of Speech and Natural Language Processing in Seattle (August 1994), the International Symposium on Prosody in Yokohama (September 1994), the IEEE/ESCA Meeting on Speech Synthesis at New York (September 1994), the COST 233 meeting in Lausanne (November 1994), the Japonic First Research Meeting at the National Language Institute in Tokyo (February

1995), the HCM Workshop on Dialog and Discourse Prosody in Stuttgart (February 1995) and the 21st Journées d'Etude sur la Parole in Avignon (June 1996), where different parts of this study were presented.

[2] These "target points" are essentially the "turning points" described by Gårding 1977, 1983, except that, unlike turning points, target points do not necessarily imply a change of direction: downstepped and upstepped targets can thus be generated directly as described below.

[3] Experimental evidence (Cavé, Hirst and Rossi 1986) suggests that the exact localisation of this inflection point is not crucial for the quality of synthesised speech. In natural speech a certain assymmetry is in fact observed with the inflection point being closer to the higher of the two targets.

[4] The "Rise/Fall/Connection" model described by Taylor (1993) is in fact formally identical to a quadratic spline function except that it also allows the possibility of linear transitions as well as curvi-linear transitions between target points with different F0 values. We have seen no convincing evidence that this option is ever in fact necessary. Note that curvi-linear transitions between successive targets with the same F0 value will be identical to linear transitions so that no special option is in fact needed to generate these with a quadratic spline function.

[5] The closest approximation to a quadratic spline function is given by using horizontal plateaux of approximately 1/3 of the duration between two target points.

[6] A similar result has now been obtained for the German recordings of EUROM1.

[7] When a is greater than 1 the reiterated function diverges; when it is less than 0 it oscillates.

[8] The case of D and U is somewhat different. Phonological arguments (Clements and Ford 1979) suggest that D is equivalent to a sequence LH from which it generally seems to be derived diachronically. If this is taken literally, we should expect the U and D targets to be closer to the middle of the pitch range than those for T or H and B or L, respectively. The data analysed here is not consistent with regard to this prediction, perhaps due to the fairly small number of examples of these tones.

[9] One attempt to provide a system of notation for English intonation based on purely formal criteria was that of Hirst (1977) who proposed a representational system based on five abstract features : [±stress; ±centre; ±emphasis; ±boundary ±terminal].

[10] More recently, Halle and Idsardi (1995) have claimed that grouping is in fact linguistically more basic than prominence which should be considered "a by-product of the grouping of the elements into constituents." (p 439)

[11] Studies of durational effects, in particular Reis (1995) for Brazilian Portuguese, Le Besnerais (1995) for Spanish, suggest at least the possibility that the left/right headedness of accent groups might be a parameter which distinguishes Germanic languages from Romance languages in general (Hirst et al. 1993; Hirst and Di Cristo 1998).

[12] More detailed accounts of the intonation systems of the two languages within the same basic framework are to be found in Di Cristo (1998) and Hirst (1998).

[13] In line with recent tendencies in phonology this downstep "rule" could be reformulated as a principle stating that under certain conditions an underlying L tone is delinked from the Tonal Unit, i.e. it no longer operates as a target but still has the effect of lowering the following H tone which is consequently interpreted as a surface D tone.

REFERENCES

d'Alessandro, C. and Mertens, P. 1995. Automatic pitch-contour stylisation using a model of tonal perception. *Computer Speech and Language* 9, 257-288.

Bagshaw, P. 1994. *Automatic Prosodic Analysis for Computer-Aided Pronunciation Teaching.* PhD thesis, University of Edinburgh.

Bosch, L. ten. 1993. On the automatic classification of pitch movements. *Proc. Eurospeech 93* (Berlin), 781-784.

Bruce, G. 1985. Structure and function of prosody. In B. Guérin and R. Carré (eds), *Franco-Swedish Seminar on Speech* (Grenoble) 2, 549-559.

Bruce, G. 1989. Report from the IPA working group on suprasegmental categories. *Working Papers* 35 (Lund University), 25-40.

Campbell, W.N. 1992. *Multi-level Timing in Speech.* PhD Thesis, University of Sussex.

Campbell, W.N. This volume. Timing in speech: a multilevel process.

Campione, E., Flachaire, E., Hirst, D.J. and Véronis, J. 1997. Stylisation and symbolic coding of F0, a quantitative model. In A. Botinis, G. Kouroupetroglou and G. Carayiannis (eds.), *Intonation: Theory, Models and Applications. Proc. ESCA Workshop.* (Athens), 71-74.

Cavé, C., Hirst, D.J. and Rossi, M. 1986. Pitch of glissandos in speech sounds. *Proc. 12th International Congress on Acoustics* (Toronto, 1986), Vol. I, A27

Chan, D., Fourcin, A., Gibbon, D., Granström, B., Huckvale, M., Kokkinakis, G., Kvale, K., Lamel, L., Lindberg, B., Moreno, A., Mouropoulos, J., Senia, F., Tracoso, I., Veld, C. and Zeiliger, J. 1995. EUROM - a spoken language resource for the EU. *Proc. Eurospeech 95* (Madrid), 867-870.

Chomsky, N. 1964. *Current Issues in Linguistic Theory.* The Hague:Mouton.

Chomsky, N. 1981. *Lectures on Government and Binding.* Dordrecht: Foris.

Clements, G.N and Ford, K.C. 1979. Kikuyu tone shift and its synchronic consequences. *Linguistic Inquiry* 10, 179-210.

Cruttenden, A. 1986. *Intonation.* Cambridge: Cambridge University Press.

Cutler, A. and Ladd, D.R. (eds). 1983. *Prosody: Models and Measurements.* Berlin:Springer.

Di Cristo, A. 1985. *De la microprosodie à l'intonosyntaxe.* Publications de l'Université de Provence, Aix en Provence.

Di Cristo, A. 1998. Intonation in French. In Hirst and Di Cristo (eds), 196-219.

Di Cristo, A., Di Cristo, P. and Véronis, J. 1997. A metrical model of rhythm and intonation for French text-to-speech synthesis. In A. Botinis, G. Kouroupetroglou and G. Carayiannis (eds), *Intonation: Theory, Models and Applications. Proc. ESCA Workshop* (Athens, September 1997), 83-86.

Di Cristo, A. and Hirst, D.J. 1986. Modelling French micromelody: analysis and synthesis. In Kohler (ed.), *Prosodic Cues for Segments* (= *Phonetica* 43), 11-30.

Di Cristo, A. and Hirst, D.J. 1997. L'accentuation non-emphatique en français: stratégies et paramètres. In J. Perrot (ed), *Polyphonie pour Ivan Fónagy*. Paris: Harmattan, 71-101.

Espesser, R. 1982. Un système de détection du voisement et de F0. *Travaux de l'Institut de Phonétique d'Aix* 8, 241-261.

Fant, G., Kruckenberg, A. and Nord, L. 1991. Language specific patterns of prosodic and segmental structures in English, Swedish and French. *Proc. XIIth International Congress of Phonetic Sciences*, 118-121.

Fónagy, I. 1980. L'accent en français: accent probabilitaire. *Studia Phonetica* 15, 123-233.

Fujisaki, H. 1988. A note on the physiological and physical basis for the phrase and accent components in the voice fundamental frequency contour. In O. Fujimura (ed), *Vocal Physiology. Voice Production, Mechanisms and Findings, 347-355.* Raven Press.

Fujisaki, H. 1997. Prosody, models and spontaneous speech. In Sagisaka et al. (eds) 27-42.

Fujisaki, H. and Nagashima, S. 1969. A model for synthesis of pitch contours of connected speech. *Annual Report Engineering Research Institute: University of Tokyo.* 28, 53-60.

Gårding, E. 1977. The importance of turning-points for the pitch patterns of Swedish accents. In L.M. Hyman (ed), *Studies in Stress and Accent (= Southern California Occasional Papers in Linguistics 4),* 27-36.

Gårding, E. 1983. A generative model of intonation. In A. Cutler and D.R. Ladd (eds), *Intonation: Models and Measurements.* Berlin: Springer, 11-26.

Geoffroi, E. 1993. A pitch contour analysis guided by prosodic event detection. In *Proc. Eurospeech95* (Berlin), 793-796.

Halle, M. and Vergnaud, J. R. 1989. *An Essay on Stress*. Cambridge Mass.: MIT Press.

Halle, M. and Idsardi, W. 1995. General properties of stress and metrical structure. In J. Goldsmith (ed), *The Handbook of Phonological Theory*. Oxford: Blackwell, 403-443.

Hamon, C., Moulines, E. and Charpentier, F. 1989. A diphone system based on time-domain modifications of speech. *Proc. IEEE Conf. on ASSP,* 238-241.

Hart, J. 't. 1991. F0 stylisation in speech: straight lines versus parabolas. *Journal of the Acoustical Society of America* 6, 3368-3370.

Hart, J. 't and Collier, R. 1975. Integrating different levels of intonation analysis. *Journal of Phonetics* 3, 235-255.

Hart, J. 't, Cohen, A. and Collier, R. 1990. *A Perceptual Study of Intonation: an Experimental-Phonetic Approach to Speech Melody.* Cambridge: Cambridge University Press.

Hermes, D.J. and Van Gestel, J.C. 1991. The frequency scale of speech intonation. *Journal of the Acoustical Society of America* 90, 97-102.

Hirst, D.J. 1977. *Intonative Features: a Syntactic Approach to English Intonation.* The Hague: Mouton.

Hirst, D.J. 1980. Un modèle de production de l'intonation. *Travaux de l'Institut de Phonétique d'Aix* 7, 297-315.

Hirst, D.J. 1983. Structures and categories in prosodic representations. In A. Cutler and D.R. Ladd (eds), *Prosody: Models and Measurements.* Berlin: Springer, 93-109.

Hirst, D.J. 1987. *La description linguistique des systèmes prosodiques: une approche cognitive.* Thèse de Doctorat d'Etat, Université de Provence.

Hirst, D.J. 1992. Prediction of prosody: an overview. In G. Bailly and C. Benoît (eds), *Talking Machines: Theories, Models and Applications.* Amsterdam: North-Holland, 199-205.

Hirst, D.J. 1993. Detaching intonation phrases from syntactic structure. *Linguistic Inquiry* 24, 781-788.

Hirst, D.J. 1994. The symbolic coding of fundamental frequency curves: from acoustics to phonology. *Proc. International Symposium on Prosody* (Yokohama), 1-5.

Hirst, D.J. 1998. Intonation in British English. In Hirst and Di Cristo (eds), 56-77.

Hirst, D.J. and Di Cristo, A. 1984. French intonation: a parametric approach. *Die Neueren Sprachen* 83, 564-569.

Hirst, D.J. and Di Cristo, A. 1998. A survey of intonation systems. In Hirst and Di Cristo (eds), 1-44.

Hirst, D.J. and Di Cristo, A. (eds), 1998. *Intonation Systems.* Cambridge: Cambridge University Press.

Hirst, D.J. and Espesser, R. 1993. Automatic modelling of fundamental frequency using a quadratic spline function. *Travaux de l'Institut de Phonétique d'Aix* 15, 71-85.

Hirst, D.J., Di Cristo, A., Le Besnerais, M., Najim, Z. and Nicolas, P. 1993. Multi-lingual modelling of intonation patterns. *Proc. ESCA Workshop on Prosody. Working Papers* (Dept. of Linguistics, Lund Univ.) 41, 204-207.

Hirst, D.J., Nicolas, P. and Espesser, R. 1991. Coding the F0 of a continuous text in French: an experimental approach. *Proc. XIIth International Congress of Phonetic Sciences* (Aix en Provence), Vol. 5, 234-237.

Holmes, V. 1996. A crosslinguistic comparison of the production of utterances in discourse. *Cognition* 54, 169-207.

House. D. 1990. *Tonal Perception of Speech.* Lund: Lund University Press.

Jensen, U., Moore, R.K., Dalsgaard, P. and Lindberg, B. 1993. Modelling of intonation contours at the sentence level using CHMMs and the 1961 O'Connor and Arnold Scheme. *Proc. Eurospeech93* (Berlin), 785-788.

Ladd, D.R. 1993. Notes on the phonology of prominence. *Proc. ESCA Workshop on Prosody. Working Papers* (Dept. of Ling., U. of Lund) 41, 10-15.

Le Besnerais, M. 1995. *Parámetros rítmicos para el estudio contrastado del francés y del español contemporáneos.* Doctoral thesis, University of Barcelona.

Martin, P. 1981. Extraction de la fréquence fondamentale par intercorrélation avec une fonction peigne. *Actes des 12e Journées d'Etudes sur la Parole* (Montréal), 221-232.

Morlec, Y., Bailly, G. and Aubergé, V. 1995. Synthesis and evalutation of intonation with a superposition model. *Proc. Eurospeech 95* (Madrid), 2043-2046.

Nicolas, P. 1989. *Amplitude des variations micromélodiques des obstruantes voisées en fonction de la hauteur de la voix.* DEA Dissertation, Université de Provence.

Nicolas, P. 1995. *La structuration prosodique du texte lu en français.* Doctoral thesis, Université de Provence.

Nicolas, P. and Hirst, D.J. 1995. Symbolic coding of higher level characteristics of fundamental frequency curves. *Proc. Eurospeech 95* (Madrid), 989-992.

Pasdeloup, V. 1990. *Modèle de règles rythmiques du français appliqué à la synthèse de la parole.* Doctoral thesis, Université de Provence.

Pierrehumbert, J. This volume. Tonal elements and their alignment.

Pierrehumbert, J. and Beckman, M. 1988. *Japanese Tone Structure.* Cambridge, Mass.: MIT Press.

Reis, C. 1995. *L'interaction entre accent intonation et rythme en portugais brésilien.* Doctoral thesis, Université de Provence.

Roucos, S. and Wilgus, A. 1985. High quality time-scale modification for speech. *Proc. ICASSP*, 493-496.

Rossi, M., Di Cristo, A., Hirst, D.J., Martin, P. and Nishinuma, Y. 1980. *L'intonation: de l'acoustique à la sémantique.* Paris, Klincksieck.

Sagisaka, Y., Campbell, N., and Higuchi, N. (eds), 1997. *Computing Prosody. Computational Models for Processing Spontaneous Speech.* New York: Springer.

Saussure, F. de. 1916. *Cours de Linguistique Générale.* Paris: Payot.

Scheffers, M.T.M. 1988. Automatic stylization of F0-contours. *Proc. 7th FASE symposium: Speech '88* (Edinburgh), 981-987.

Selkirk, E. 1978. *On Prosodic Structure and its Relation to Syntactic Structure.* Bloomington: Indiana University Press.

Selkirk, E. 1984. *Phonology and Syntax: the Relation between Sound and Structure.* Cambridge, Mass.: MIT Press.

Selkirk, E. 1986. On derived domains in sentence phonology. *Phonology Yearbook* 3, 371-405.

Silverman, K., Beckman, M., Pitrelli, J., Ostendorf, M., Wightman, C., Price, P., Pierrehumbert, J., and Hirschberg, J. 1992. ToBI: a standard for labelling English prosody. *Proc. International Conference on Spoken Language Processing* (Banff), vol. 2, 867-870.

Strangert, E. and Aasa, A. 1996. Evaluation of Swedish prosody within the MULTEXT-SW project. *TMH-QPSR* 2/1996 *(Speech, Music and Hearing - Quarterly Progress and Status Report)*, KTH, Stockholm, Sweden, 37-40.

Taylor, P. 1993. Automatic recognition of intonation from F0 contours using the Rise/Fall/Connection model. *Proc. Eurospeech 93* (Berlin), 789-792.

Terken, J. 1991. Fundamental frequency and perceived prominence of accented syllables. *Journal of the Acoustical Society of America* 89, 1768-1776.

Terken, J. and Hermes, D. This volume. The perception of prosodic prominence.

Trubetzkoy, N. 1949. *Grundzüge der Phonologie.* (French translation by J. Cantineau 1957) *Principes de phonologie.* Paris: Klincksieck.

Véronis, J., Di Cristo, P., Courtois, F., and Lagrue, B. 1997. A stochastic model of intonation for French text-to-speech synthesis. *Proc. Eurospeech97* (Rhodes), 2643-2646.

Véronis, J., Hirst, D.J., Espesser, R. and Ide, N. 1994. NL and speech in MULTEXT. *Proc. AAAI-94 Workshop of the Integration of Speech and Natural Language Processing.* (Seattle; August 1994).

Wenk, B. and Wiolland, F. 1982. Is French really syllable-timed? *Journal of Phonetics* 10, 193-216.

Labaratoire Parole et Langage, Université de Provence, Aix-en-Provence, France

JACQUES TERKEN AND DIK HERMES

THE PERCEPTION OF PROSODIC PROMINENCE

1. INTRODUCTION

We say that a linguistic entity is prosodically prominent when it stands out from its environment by virtue of its prosodic characteristics. That is, we define prominence as a property of a linguistic entity relative to an entity or a set of entities in its environment. Although the definition is cast in relative terms, it includes monosyllabic utterances, because they stand out from silence. In the acoustic domain, the primary prosodic properties bringing about these relative differences are amplitude, duration and 'F0' (we use F0 as a shorthand form for the inverse of the quasi-periodicity of the speech signal). The corresponding perceptual properties are loudness, duration or length, and pitch. Also, particular aspects of timbre come into play, such as those relating to vowel reduction, spectral slope or tilt, etc., and properties relating to voice quality, e.g. creak.

The general definition of prominence in relative terms comes in handy because it provides a good starting point for understanding the phenomenon known under the name of 'stress shift' (among others Horne 1990; Shattuck-Hufnagel 1995). This refers to the situation where speakers produce ***Chi**nese* ***lan**tern*, with prominence on ***Chi***- and ***lan***- even though the most prominent syllable of *Chi-nese*, when spoken in isolation, would be *-nese*. By stressing *Chi-* instead of *-nese* the speaker achieves the presence of a less prominent syllable in between more prominent syllables, creating a background from which the prominent elements may stand out.

The intuition that elements of the speech stream vary with respect to prominence has a long history. In ancient Greek poetry the metrum was characterized in terms of a distinction between strong, i.e. prominent, and weak, i.e. non-prominent syllables. A general intuition, however, is that prominence is not only a binary property: it is generally felt that the simple distinction between prominent and non-prominent elements does not do full justice to the observed phenomena. Instead, prominence seems to be better expressed as a scalar property or a number of ordered nominal categories. One may question, then, whether prominence should be treated as a one-dimensional continuum or in terms of different types or categories, and if so, how many different categories. This will be the topic of section 2, in which we

M. Horne (ed.), Prosody: Theory and Experiment, 89–127.

will conclude that at least four types of syllables can be distinguished: syllables containing reduced vowels, syllables containing full vowels, stressed syllables, and accented syllables. In later sections we will focus more specifically on accented syllables, and summarize insights concerning psycho-acoustic aspects of accent perception (section 3), and concerning the way pitch variation affects variation in the perceived prominence of accented syllables, i.e. variation in 'accent strength' (section 4). Finally, we will look at the relation between theoretical treatments of prominence and the practice of transcribing prominence differences in speech databases (section 5). Ideally, the conclusions reached in sections 1-4 should provide the theoretical framework for the definition of linguistically motivated transcription conventions. We will conclude that there is still a considerable distance between theory and practice.

In the previous paragraph it has been implicitly assumed that syllables are the appropriate level of representation for describing prominence variation, and this will serve as point of departure for our further presentation. In addition, we need to warn the reader that our presentation will be heavily biased towards the Germanic languages. Although prominence variation seems to be an almost universal property of speech, the means to do so vary strongly across languages, and our presentation will focus strongly on correlates of prominence variation in the Germanic languages.

2. CATEGORIAL ASPECTS OF PROMINENCE

Several typologies for prominence have been proposed, from different backgrounds, and representing different assumptions about the phonetic characteristics (e.g. Bruce 1977; Liberman and Prince 1977; Beckman 1986; Cruttenden 1986; 't Hart, Collier and Cohen 1990). In particular, the term 'stress' has been used in different ways. To give an idea, novice readers have to become familiar with terms such as prominence, emphasis, stress (primary, secondary stress; lexical stress; neutral vs. emphatic or contrastive stress, etc.), accent (word accent, phrase accent, sentence accent) and to learn about their linguistic function and their phonetic realization. As a result, there has been confusion about the meaning of terms, and debate about their adequacy. In order to get a clear view, we need to look at the theoretical treatments and at the available empirical evidence.

2.1 Linguistic treatments

With respect to prominence, linguists have primarily tried to describe and account for their intuitions about prominence patterns in example sentences. The structuralist approach has been very influential. A good example of early structuralist treatments is the work of Trager and Smith (1951). They distinguish four different prominence categories (primary-secondary-tertiary-weak), which represent relative strengths of vowels in utterances. The different levels are supposed to be consistent with loudness variations, so that the term 'stress' is used. Stress levels are indicated by diacritic marks (´ ˆ ` and ˘) denoting phonemic entities,which means that the different levels are supposed to stand in meaningful opposition in the language under consideration. Primary, tertiary and weak are used to represent loudness differences within words: /élĕvàtŏr/, /hètĕrogénĕŏus, /ópĕràtŏr/ (if no mark is specified, weak stress is implied as a default). Secondary stress is used to represent relative differences between primary word stresses at the phrase level, including compounds, to capture the intuition that not all the syllables bearing primary word stress in a phrase or a compound have the same level of stress, but that the weaker ones are still stronger than the syllables with tertiary stress: /élĕvàtŏr/+ /ópĕràtŏr/ → /élĕvàtŏr ópĕràtŏr/ /bláck+bìrd/ vs. /bláck+bôard/. In fact, the distinction between primary and secondary stress captures the intuition that one entity in the phrase is more prominent than all others. This assumption has since then been adopted in all structuralist treatments of prominence.

In addition to stress levels, Trager and Smith discern a number of pitch levels, which represent relative pitch height in syllables with static tone (i.e. where there is no obvious pitch change inside the syllable): [^{1}], [^{2}], [^{3}], [^{4}] representing successively higher pitch levels. Furthermore, there are four diacritic marks to indicate variations within any level. Finally, there are three symbols for representing dynamic tone: ["] means terminal sustainment at the level previously marked, [$^+$] terminal rise from the previously marked level, [-] terminal fall from the previously marked level. Example transcriptions are as follows:

$$/\ \overset{2}{[\wedge]}\ \text{hôw} + \overset{2}{[\vee]}\ \text{do} + \overset{2}{[\circ]}\ \text{thèy} + \overset{3}{[^-]}\ \text{stú}\ \overset{1}{[\vee]}\ \text{dy}\ [^-]\ /$$

or more simply

/ ^{2}hôw + do + thèy + ^{3}stú ^{1}dy[-]/

Although in principle stress levels, relating to loudness and duration, and pitch levels are independent, Trager and Smith indicate that there are interdependencies. In the first place, primary stress is usually associated with pitch level [^{3}]. In the second place, it is assumed that pitch variations within a pitch level vary directly with stress, so that, for instance, the pitch for pitch level [^{2}] would be higher if it is associated with secondary stress than with tertiary stress. A third case of interdependency between stress and pitch is provided by contrastive stress: it is assumed that contrastive stress can be assigned by putting primary stress and pitch level [^{3}] on the contrastive item, or, if the item to be made contrastive already has primary stress, realizing it with higher pitch, i.e. promoting pitch level [^{3}] to pitch level [^{4}].

We have summarized the work of Trager and Smith at some length, because it beautifully illustrates the strong assumptions that were made about the sensitivity of the speakers/listeners of a language with respect to prominence variation. In summary, it contains the following assumptions: (1) stress variation is related to loudness variation and can be described in terms of a four-level approach; (2) the pitch patterns in speech provide a (semi-) independent layer of organization and can be described in terms of four pitch levels, supplemented by three dynamic tones.

The representation of prominence patterns in terms of stress levels has become widely adopted by later representatives of the structuralist approach. Chomsky and Halle (1968), while adopting the same basic assumptions of Trager and Smith, proposed a cyclic-transformational account of stress patterns. In simplified terms, they assign stress patterns to words on the basis of the phonological properties of the word, and then derive the patterns for phrases in a cyclic fashion by means of a Compound Stress rule and a Nuclear Stress rule. In essence, the rules copy one of the primary stresses at a given level to the next higher level, while weakening all other stress values by one. The Compound rule assigns primary stress to the non-last primary-stressed vowel of constituents of type noun, verb, and adjective/adverb (cf. the *elevator operator* example treated above). The Nuclear Stress rule assigns primary stress to the last primary-stressed vowel of all other constituent types:

1 1		2 1
bright light	→	bright light

Thus, stress patterns for sentences are derived by applying these rules in a cyclic fashion on strings of stress markers in increasingly larger domains, which are defined with reference to the surface syntactic structure. Furthermore, it is noteworthy that Chomsky and Halle do not give a treatment of pitch patterns, which suggests that they conceive of prominence patterns in speech primarily in relation to variations in loudness and duration, and do not consider pitch patterns to be relevant to prominence; but since they are rather vague about the relation between prominence patterns and acoustic correlates, it is difficult to determine whether they indeed hold this view.

Later developments provided a more insightful account by recasting the basic ideas of Chomsky and Halle in terms of operations on a tree representation instead of a string. Liberman and Prince (1977) accounted for prominence relations at the phrase level in terms of strong-weak relations between terminal and/or non-terminal nodes in a binary-branching tree. They introduced the notion of 'designated terminal element' for a terminal element that is linked to the root node of a given subconstituent only by strong nodes, and proposed the Relative Prominence Projection Rule to define the prominence relations between terminal elements: In any constituent on which the Strong-Weak relation is defined, the designated terminal element of the strong subconstituent is metrically stronger than the designated terminal element of the weak constituent. This determines the ordering of the elements in the phrase with respect to prominence, and thus specifies which entity is most prominent within a phrase. Noteworthy, metrical prominence relations and tonal prominence were treated completely independent of each other.

Still later representatives of the structuralist tradition (Selkirk 1984; Beckman 1986; Beckman and Pierrehumbert 1986) have aimed to integrate tonal and metrical aspects. They proposed a labeled tree representation, i.e. a set of hierarchically ordered levels, where each level consists of one or more labeled constituents (foot, prosodic word, intonation phrase).

Although the different accounts are clearly distinct, they share three basic assumptions:

1. prosodic structure is non-isomorphic to syntactic structure but poses its own hierarchical constituent structure;
2. constituent types can (in part) be defined in terms of the prominence characteristics of the heads at different levels of the hierarchy; this gives rise to different types of prominence;
3. at each level of the hierarchy, the head of a constituent is more prominent than the non-head elements.

In prevailing current treatments, at least two relevant constituents are distinguished: the 'stress group' and the 'intermediate phrase'.

The 'stress group' consists of a stressed syllable (the head) and all following unstressed syllables up to the next stressed syllable. It has proven to be difficult, however, to define the feature stress in a way which is both linguistically satisfactory and observationally adequate. In most treatments, stress is defined as a phonological property of a syllable (or its nucleus), which can be assigned on the basis of the phonological and morphological properties of the syllable itself and the word containing it. But the phonetic realization of stress remains unspecified, so that it remains difficult to decide for a given phrase whether the syllables which are phonologically marked for stress are indeed stressed in the phonetic realization. Beckman (1986; cf. also Beckman and Pierrehumbert 1986) suggests a way out by defining the feature ±stress in terms of vowel quality, i.e., all full vowels are defined as stressed. This solution provides a means to explain the difference between the realizations of *-man* in *German* and *mailman* in terms of stress (*German* containing one stressed syllable and *mailman* containing two stressed syllables). But it is only partially satisfactory, as it leaves untouched relative differences in the phonetic realization of stressed syllables apart from accent.

The 'intermediate phrase' is the building block for intonation phrases, i.e. each intonation phrase is made up of one or more intermediate phrases. The segmentation of an intonation phrase into intermediate phrases is established primarily by melodic and durational boundary markers. Each intermediate phrase contains one or more accents, and the final accent (the nuclear accent) is the head of the intermediate phrase.

Thus, these statements boil down to the claims that syllables containing full vowels are more prominent than syllables containing reduced vowels, that the syllable associated with the final accent in the intermediate phrase is more prominent than all the stressed syllables in the phrase, including the ones associated with non-final accents, and that, at least for American English, no further prominence differences can or need to be defined. In particular, the prominence relation between non-nuclear accents and stressed but unaccented syllables is undefined. Likewise, no further differentiation of prominence is postulated at the level of nuclear accents: all nuclear accents are supposed to be equally prominent, and the intonation phrase, which is the next highest constituent above the level of the intermediate phrase, is defined only in terms of its boundary characteristics and not (in part) in terms of the properties of its head. It should be noted that these proposals are based on attempts to integrate metrical and melodic factors contributing to prominence into an overall theory of prosodic structure, and that alternative proposals run into considerable problems or cannot be motivated empirically.

Yet, we disagree with some of the implications, for instance the consequence that the relative prominence of non-nuclear accents and stressed but unaccented syllables is undefined. Instead, we feel that accented syllables are always more prominent than stressed but unaccented syllables. Also, it is unclear what is meant by saying that nuclear accents are more prominent than non-nuclear ones. In the first instance, one might think that it is a statement about perceived (i.e. phonetic) relative prominence. However, the experiments conducted by Curry (1980, 1981) lead to the conclusion that listeners often do not select the nuclear accent as the most prominent one. Therefore, the statement that the nuclear accent is the most prominent entity in the intermediate phrase may be more satisfactorily understood in terms of its linguistic function: different locations of the nuclear accent entail different focus structures; this does not apply to the same extent to non-nuclear accents, where the presence or absence of an accent is more a matter of speech style or rhythmic considerations. If this interpretation is correct, the statement that nuclear accents are most prominent should not be interpreted as a statement about prominence relations between stressed or accented syllables at the phonetic level. Therefore, we turn to phonetic studies in order to learn more about the phonetic aspects of prominence. This will be the topic of the next subsection.

2.2 Phonetic analyses of prominence and stress

The tools for synthesizing speech-like stimuli which became available in the 50's gave rise to experiments with manipulated stimuli, the properties of which were under the control of the experimenter. In this way, the contribution of different acoustic parameters to the perception of prominence could be investigated. Fry (1955, 1958) manipulated intensity, duration and F0 characteristics of his stimuli to create synthesized word pairs such as *s'ubject-subj'ect*. He found that duration and intensity both affected judgments of stress position, with duration as the more powerful cue. When F0 was manipulated as well, it outweighed the duration cue. Thus, these results provided evidence against the assumption that stress perception is primarily related to loudness variation.

Further doubt on the adequacy of the structuralist view of prominence was provided by Lieberman (1965). He wondered whether distinctions such as made by Trager and Smith (1951) could be transcribed at all in a reliable way, and to what extent the judgments were actually based on the linguistic information in the utterance instead of its acoustic properties, and he therefore conducted an experiment to find out. By copying the F0 characteristics and/or the intensity of eight readings (meant to represent eight different emotions) of the sentence *They have bought a new car* by two speakers each, to a synthetic vowel /a/, he produced the following stimulus types: (1) envelope amplitude and F0 preserved; (2) only F0 preserved; (3) amplitude and smoothed F0 preserved; (4) only amplitude information preserved. The 32 stimuli were transcribed by two linguists, Lisker and Stockwell, following Trager and Smith's system for labeling pitch levels. One of them also transcribed stress levels. In a post-test, they transcribed the original utterances recorded without any processing. The following main findings were obtained.

1. For the linguist who transcribed both pitch and stress patterns, only small differences were found between the stress patterns transcribed for the synthetic /a/ stimuli with and without original envelope amplitude (conditions 1 and 2); this suggests that both pitch and stress patterns were transcribed mainly on the basis of pitch information, and that loudness conveyed little information.
2. There was no consistent mapping between the actual pitch and the transcribed pitch levels for the utterances in the post-test; a more consistent mapping was obtained for the synthetic /a/-stimuli. This suggests that the transcriptions of pitch patterns were based in part on the linguistic properties of the utterances. However, better consistency could be observed between linguists' transcriptions by looking at the contour properties, e.g., two transcriptions might be different but show the same relative pattern (e.g. ...-1-3-2-... and ...-2-4-3-...). This suggests that the pitch levels for

individual syllables might actually have been assigned on the basis of overall contour properties.

3. For the linguist who transcribed stress levels, there was a clear difference between the transcriptions for the utterances in the post-test and those for the synthetic /a/'s: in the transcriptions of the /a/-stimuli, most secondary and tertiary stresses that had been transcribed for the utterances had disappeared, and only the primary and weak stresses were retained. This suggests that the secondary and tertiary stresses have no distinct physical basis but are inferred on the basis of linguistic knowledge.

Thus, it seems that the Trager and Smith type transcriptions both of pitch levels and stress reflect primarily the global pitch contour of an utterance, and that only two levels of prominence ('primary' and 'weak') can be transcribed reliably, corresponding to the distinction accented/unaccented.

The acoustic determinants of accent perception were investigated by Van Katwijk (1974). He manipulated different acoustic properties and found that the presence of an F0 change was the primary cue to induce the perception of prominence. In fact, it was a sufficient cue. These findings gave rise to the view that pitch accents should be defined in terms of accent-lending pitch changes rather than in terms of pitch changes associated with prominent (stressed) syllables, as is for instance argued in the structuralist approach, and that the presence or absence of a pitch accent is the only relevant categorial distinction for prominence.

The same view was taken by Bolinger (1958, 1961). He argued that linguists confused stress and accent because they concentrated on citation forms of pairs such as '*d'igest-dig'est*', and that, if one looked at longer utterances, it would become clear immediately that stress as related to amplitude and duration is no reliable perceptual category. Instead, stress should be treated purely as a linguistic property, specifying the location in a word where an accent will land if appropriate in a particular context. Furthermore, he argued that accent is an all-or-none affair rather than an n-valued property, although he agreed that within the class of accents there may be gradient variation, implying a quantitative change in the meaning that is conveyed. Indeed, these views anticipate later views such as those of Pierrehumbert (1980) in many respects.

The main difference between Bolinger and later structuralist views such as those of Pierrehumbert concerns the way in which the role of pitch is viewed in relation to prominence. Bolinger's view (also represented by 't Hart and Cohen 1973, 't Hart and Collier 1975; 't Hart, Collier and Cohen 1990) is that pitch change is a sufficient and necessary cue for prominence, and that stress is not realized independent of pitch accent. Pierrehumbert's view (Pierrehumbert 1980; cf. also Beckman and Pierrehumbert 1986; Pierrehumbert and Hirschberg 1990) is that syllables are prominent on the basis of their metrical properties, and that the perception of a pitch accent is dependent on the association of a pitch target with a metrically strong syllable; that is, syllables are not prominent due to the presence of a pitch change, but rather due to their duration and amplitude characteristics. In this view, the association of pitch and prominence arises mainly because the metrically strong syllables are the points where the melodic structure becomes associated with the metrical structure. Furthermore, pitch accents do not require the presence of a pitch change, but arise from the association of a pitch target with a prominent syllable; the presence of a pitch change is a mere coincidence of the tonal structure (for instance when a high target is directly followed by a low target), or due to the presence of a bi-tonal accent (consisting of two tonal targets in a fixed temporal relation).

There are two pieces of evidence relevant to this debate. The first claim, holding that a pitch change is necessary for the perception of pitch accent, is addressed by one of the conditions in Hermes and Rump (1994; see next section). They used stimuli with synthetic contours which were resynthesized from a /mamama/ utterance produced by a human speaker. In the natural utterance, the middle syllable carried a pitch accent, so that it had the characteristic properties of a strong syllable (longer duration and greater amplitude). One of the synthetic contours consisted of gradually falling declination only. Listeners reported hearing a falling pitch accent in the second syllable. Since utterances with declination only are melodically ill-formed and each utterance should contain at least one pitch accent, it appeared that listeners solved the violation of the phonological constraint by inferring a high onset and a falling pitch accent associated with the strong syllable. In other words, it appears that - under certain circumstances - even in the absence of a clear pitch change, listeners may interpret a small difference between the pitch levels of a strong syllable and the preceding weak syllable as a pitch accent in order to fulfill phonological constraints.

A study of Huss (1978) addresses the second claim, holding that stress has no realization independent of pitch accent. He asked speakers to produce phrases such as ‘the employers increase benefits’, where ‘increase’ can be a noun or a verb, in two contexts; in one context ‘increase’ was associated with contrastive emphasis, and therefore carried the nuclear accent (examples a-b below); in the other condition contrastive emphasis was on ‘employers’, so that ‘increase’ would be in post-nuclear position (examples c-d). (For the sake of convenience, the contrastive items are marked by capitals.)

a) The employers’ LOSS doesn’t benefit; the employers’ INCREASE benefits (*noun, nuclear*).
b) The employers don’t ABOLISH benefits; the employers INCREASE benefits (*verb, nuclear*).

c) Whereas formerly the WORKERS' increase in deductions used to benefit from inflation, now the EMPLOYERS' increase benefits (*noun, post-nuclear*).
d) Whereas formerly the GOVERNMENT used to increase benefits, now the EMPLOYERS increase benefits (*verb, post-nuclear*).

The acoustic analysis showed that the syllables carrying word stress showed clear differences in durations and intensities when compared with the unstressed counterparts, both in nuclear and post-nuclear position; the effects were small in the post-nuclear condition, however, and there was a strong interaction with the rhythmic structure of the carrier sentence: the phonetic realizations of the strong and weak syllables in the post-nuclear position were strongly influenced by whether the context showed a strong-weak or a weak-strong alternation (cf. Table I). A perceptual test with the target words excised from context showed that the acoustic differences were perceptually relevant, although the effect was not very strong for the words produced in post-nuclear position.

		Trochaic rhythm (s-w)		Iambic rhythm (w-s)	
		syll_1	syll_2	syll_1	syll_2
NUCLEAR	Noun	158	131	-	-
	Verb	130	143	-	-
POST-	Noun	128	124	114	121
NUCLEAR	Verb	113	113	98	112

Table 1. *Syllable durations (in ms) for noun-verb pairs (íncrease-incréase) in nuclear and post-nuclear positions, for two different rhythmic contexts: Trochaic rhythm (strong-weak), favouring first syllable; iambic rhythm (weak-strong), favouring second syllable. Data for falling pitch only (declaratives) (after Huss 1978).*

Thus, it is evident that the acoustic properties of strong syllables other than F0 (duration, amplitude) have perceptual relevance. This was further confirmed by a study of Sluijter and Van Heuven (1996), who analyzed the durational and intensity characteristics of *k'anon-kan'on* 'cannon-canon' and *n'ana-nan'a* pairs produced in carrier phrases both in focus (+accent) and outside focus (-accent). Confirming Huss' results, they found clear durational and intensity differences between the stressed syllables and their unstressed counterparts both within and outside focus (cf. Table II for the durational differences). The intensity differences were most pronounced in higher frequency bands. Perceptual tests with manipulated stimuli showed that band-specific manipulations of intensity characteristics mimicking the observed effects were more effective than overall intensity manipulations, and under certain conditions were as effective as manipulations of durational characteristics.

		kanon		nana	
	stress position	syll_1	syll_2	syll_1	syll_2
+FOCUS	initial	254	233	261	209
	final	151	278	162	289
	Difference	103	-45	99	-80
-FOCUS	initial	227	214	235	190
	final	142	262	157	260
	Difference	85	-48	78	-70

Table 2: *Durations (in ms) of initial and final syllables of (pseudo-)words as a function of stress position and focus (from Sluijter and Van Heuven 1996).*

Further evidence for the phonetic reality of stress in production and perception is provided by Fear et al. (1995). They investigated the acoustic and perceptual characteristics of vowels such as /au/ in word 4-tuples such as 'autumn-automation-automata-atomic', representing primary stressed, secondary stressed, full-unstressed, and reduced-unstressed /au/, respectively, in Fear et al.'s terminology (they do not take the presence or absence of pitch accent into consideration, but the list of stimulus sentences in their Appendix suggests that in most cases the primary stressed syllables are likely to have been produced with an accent). Their results indicate that the different vowel types are clearly differentiated with respect to F0 mean and F0 s.d., duration, intensity, and spectral characteristics (logF2-logF1). A perceptual test was conducted with stimuli in which the vowels representing the different prominence categories were spliced out and interchanged; acceptability ratings showed that whereas the primary, secondary, and full-unstressed vowels could be interchanged without much loss in acceptability, splicing them into a syllable which required a reduced vowel led to a considerable decrease in acceptability (cf. Table 3).

Required vowel type	Primary	Secondary	Full	Reduced
Replaced by Primary	4.5	4.2	3.9	2.4
Replaced by Secondary	4.2	4.3	4.2	2.6
Replaced by Full-unstressed	3.6	4.0	4.4	2.9
Replaced by Reduced-Unstressed	2.4	3.0	3.5	4.4

Table 3. *Acceptability ratings on five point scale for different types of vowels as substitutes for primary, secondary, full-unstressed, and reduced-unstressed vowels (after Fear et al. 1995).*

The findings of Huss (1978) shown in Table 1 and of Van Heuven (1987) make clear, however, that the realization of stress is strongly affected by the rhythmic structure of the utterance. This influence has also been demonstrated for the durational properties of syllables by Bruce (1987). Therefore, it is not yet clear whether the phonetic realization of stress is sufficiently robust to allow the listener to distinguish between *'increase* and *incr'ease* on the basis of acoustic properties only, if the word is unaccented.

2.3 Linguistic and phonetic treatments of prominence and stress integrated

In summary, we conclude that evidence has been provided for four different prominence categories: reduced, full, stressed, accented. Schematically, the distinctions and the primary properties by which they are distinguished can be represented as in Fig. 1. The set of all vowel tokens can be divided into reduced and full vowels, the set of full vowels divides into unstressed and stressed vowels, and the set of stressed vowels divides into unaccented and accented ones, although strictly speaking it would be more appropriate to treat stress and accent as properties of the syllable than the vowel.

It is unclear yet, whether in non-laboratory situations the phonetic realization of stress is sufficiently robust to be linguistically relevant, i.e., to distinguish between the noun and verb reading of *increase*. Instead, duration and amplitude variation related to stress might be more strongly linked to the rhythmic structuring of utterances: the assignment of phonetic stress may be to some extent independent of linguistic structure, as it is subject to other factors than purely linguistic ones. That is, linguistic stress patterns are defined on the basis of the morphological and phonological properties of words, and provide the substrate for assigning phonetic stress patterns in connected speech. The assignment of phonetic stress is dependent on other considerations besides linguistic ones.

Thus, we do not discard the contribution of amplitude and duration variation to prominence. In the first place, stressed and accented syllables are not only different in terms of F0 characteristics but also in terms of duration and amplitude, so that the prominence of accented syllables does not only result from their pitch properties but also from their duration and amplitude characteristics. In the second place, there is clear prominence variation below the level of accentedness, relating to variation in duration and amplitude, but these differences may be more relevant to conveying rhythmic structure than being linguistically relevant.

Further variation in prominence occurs within the category of accented syllables, where there are perceptually and communicatively relevant variations in 'accent strength'. This kind of intra-categorial variation will be the subject of section 4, where we will provide evidence holding that F0 variation is the primary determinant of variation in accent strength. We will

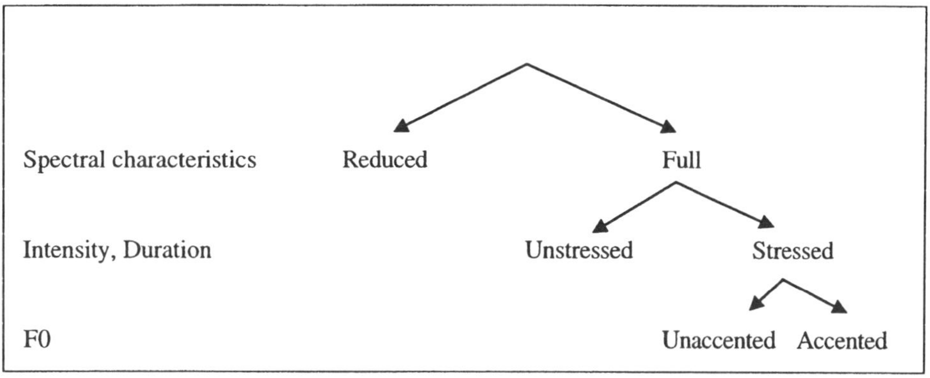

Figure 1. Schematic representation of relation between prominence categories.

first, however, consider the question of the adequacy of different frequency scales for representing speech intonation.

3. THE FREQUENCY SCALE OF SPEECH INTONATION

3.1 Introduction

This section is devoted to the appropriate frequency scale of speech intonation. Intonation of speech arises from the pitch of successive speech segments, the pitch contour, and is, therefore, a perceptual attribute. It is well known that in music, where a melody arises from the pitch of the successive tones, the physical linear frequency scale, with unit Hertz (Hz), is not appropriate. A melody is transposed to another key, not by adding a fixed amount of Hz to the original pitch frequency values of the melody, but by multiplying the pitch frequencies with a fixed number. For example, multiplying all frequencies by 2 makes the melody sound one octave higher. Since multiplication of two numbers in a linear domain is the same as adding two numbers in the logarithmic domain, this amounts to stating that, in music, two melodies run parallel when they are parallel on a logarithmic frequency scale. Because speech intonation, just as intonation in music, is a perceptual attribute, pitch contours of speech are likewise often displayed on a logarithmic frequency scale (e.g. 't Hart et al. 1990).

There is, however, a third perceptually defined frequency scale, or rather there are at least three such scales, the Mel scale (Stevens et al. 1937; Stevens and Volkman 1940), the Bark scale (Fletcher 1940; Zwicker et al. 1957; Zwicker 1961) and the equivalent-rectangular-bandwidth-(ERB)-rate scale (Patterson 1976; Glasberg and Moore 1990). Roughly speaking, the Mel scale arose from subjective magnitude estimations of pitch, the Bark scale from measurements of the so-called critical bandwidth, while the ERB-rate scale resulted from measurements of the bandwidth of the auditory filters. These frequency scales have in common that, above some 500 Hz, all three are approximately logarithmic. For lower frequencies, the Mel scale and the Bark scale are more or less linear, while the ERB-rate scale is intermediate between the linear and the logarithmic frequency scale.

Two topics will be addressed in this section. The first has to do with the pitch ranges of men and women. The question is on which frequency scale the pitch ranges of men and women are equal. The second deals with the frequency scale for sizes of pitch movements. We define the excursion size of a pitch movement as the distance between the pitch peak (i.e., the end point of a rising pitch movement or the starting point of a falling pitch movement) and the (virtual) declination line, measured perpendicular to the time axis (cf. e.g. Fig. 5). It is intuitively obvious that accents may differ with respect to their prominence, and that, for instance, in a phrase with two accents one may be clearly more prominent than the other. Furthermore, it is also clear that the prominence of such accents is related to the excursion size of the pitch movements (cf. Figs. 9-10 of Liberman and Pierrehumbert 1984): other things equal, the larger the excursion size, the larger the prominence of the accented syllable. Also, the liveliness of an utterance depends on the size of the excursions of the pitch variations in an utterance. The larger these excursions the more lively the utterance. The question is now with reference to which frequency scale the excursion sizes of pitch variations should be measured to ensure that pitch variations having equal excursion sizes in the male, female, and child's range are associated with equal perceived prominence or liveliness.

3.2 Pitch ranges of men and women compared

Various studies have been dedicated to the relative pitch ranges of men and women. Graddol (1986: 228) found that the pitch range used by most women seems to be smaller than that used by most men, when expressed in st (semitones), but larger, when expressed in Hz (Hertz). Choosing between the two, he concludes that, "whenever intervals in pitch must be compared at different frequencies, a log scale is to be preferred". Graddol did not consider the possibility that the frequency of speech intonation may best be expressed on an ERB-rate scale. The first who did were Traunmüller et al. (1989). Following Graddol (1986), they showed preference for the logarithmic frequency scale, however. In Table II of Traunmüller and Eriksson (1993: 5), an overview is presented of a number of studies in which the pitch ranges of men and women have been compared for four degrees of liveliness. It is shown that, in general, the range of women is smaller than that of men when expressed in semitones. Only for the most lively speaking condition, does the range of women appear to be larger than that of men. Traunmüller and Eriksson (1993) conclude that, on the average, the logarithmic frequency scale is most appropriate. What is not mentioned in these studies with very lively speech, is whether women have used their falsetto register. In emotional speaking conditions, the use of the falsetto register is socially accepted for women but not for men. This may explain why the range of very lively female speech is larger than that of men, even when expressed in semitones. If the most lively speaking condition is left out of consideration, it may be concluded that the range of women is smaller than that of men when expressed in semitones. This suggests that the ranges of men and women may in fact be equal on an ERB-rate scale.

3.3 Excursion sizes in high and low register compared

As to the second topic, perception research concerning the frequency scale of speech intonation has concentrated on estimations of two perceptual attributes of utterances resynthesized in different registers. The first attribute is the liveliness of an utterance (Traunmüller and Eriksson 1995) and the second is the prominence of a syllable (Rietveld and Gussenhoven 1985; Hermes and Van Gestel 1991; Rump and Hermes 1996). Rietveld and Gussenhoven (1985) set up an experiment dealing with the frequency scale for measuring excursion size in relation to accent strength. The aim of the experiment was to collect data on listeners' judgments as to what would constitute equally prominent accents in the low and high register, and to determine which measurement scale would assign equal excursion sizes to equally prominent syllables in the low and high register. Obviously, if accents in the low and high register have the same excursion size when measured on a linear scale, the excursion size of the accent in the high register is smaller than that of the accent in the low register when expressed on a logarithmic scale. So, if listeners judge accents in the low and high register to be equally prominent when they have the same excursion size on a linear scale, it is clear that the linear scale provides a better measurement scale for understanding prominence judgments than the logarithmic scale. The stimuli in the experiment conducted by Gussenhoven and Rietveld consisted of Dutch utterances with one or two accented syllables, originally spoken by a female speaker, which were were resynthesized in the original register to obtain female speech and in a lower pitch register to obtain male speech. The excursion size of the accent-lending rise-fall was varied in the low as well as in the high register. Dutch listeners then performed a pairwise comparison task for all possible combinations of excursion sizes, i.e. both within and across registers, and indicated which utterance in a pair contained the most prominent accented syllable. Comparing the actual pattern of results with the predictions that could be made from different measurement scales, Rietveld and Gussenhoven concluded that the results favoured the linear Hz scale more than the logarithmic semitone scale (the possibility of an ERB-rate scale was not yet considered), but they considered it premature to draw definite conclusions, since their stimulus set only comprised sentences recorded from a female speaker, which were then resynthesized in the low, male register without further adjustments.

Hermes and Van Gestel (1991) subsequently investigated the same issue more thoroughly. They carried out a pairwise comparison experiment with a quasi-adjustment task. The members of each pair of utterances were synthesized in different tonal registers, and each utterance contained a single accented syllable in a fixed sentential position. The stimuli were generated from a single three-syllable utterance /mamama/ rather than a real sentence, in order to avoid interference from strictly linguistic processes. The /mamama/ utterance was originally produced with an accent on the middle syllable. The modification of the pitch contour was carried out with pitch-synchronous-overlap-add (PSOLA) techniques (Moulines and Laroche 1995). Listeners were presented with a stimulus in one register, containing an accent with a certain excursion size, and were asked to select from a stimulus set in the other register the utterance containing the best matching accent. The members of a stimulus set differed only with respect to the excursion size of the accent-lending pitch change; no further adjustments were made. The rationale was the same as in the experiment by Rietveld and Gussenhoven, but more data were collected in order to be able to relate the judgments to a linear, a logarithmic, and an ERB measurement scale.

The results were in support of the ERB scale. If the excursion sizes of the accents in the low and high register were equal when expressed in ERB, the accents were judged to be equally prominent; if the excursion sizes were equal when expressed in Hz, the accent in the high register was judged less prominent; and if the excursion sizes were equal when expressed on a logarithmic scale, the accent in the high register was judged as more prominent. Thus, the results of Hermes and Van Gestel (1991) supported the ERB scale as the best scale for predicting prominence.

The conclusions drawn by Hermes and Van Gestel (1991) were criticised by Traunmüller and Eriksson (1993, 1995). They used liveliness as the perceptual attribute for their experiments instead of prominence. In their abstract they concluded that "as long as the stimuli appeared to have been produced in the modal register [of men, women, and children] listeners judged F0 intervals to be equivalent if they were equal in semitones" (1995: 1905). They argued that the results in favour of the ERB-rate scale by Hermes and Van Gestel were a consequence of the use of a falsetto voice in the high register.

Before discussing this in detail we have to outline some differences between the percept of prominence and the percept of liveliness. In the Introduction, a syllable has been defined to be prosodically prominent when it stands out from its environment by virtue of its prosodic characteristics. The concept arose from the finding that native speakers of languages like Dutch, German and English are generally well capable of indicating in an utterance whether one syllable is more prominent than another. Thus, a syllable will be more prominent than an adjacent syllable when, everything else being equal, it is louder or its duration is longer. In this respect, prominence is a local attribute of the syllable. This implicates that the prominence of a syllable should not change dramatically when the complete utterance is louder, nor should it depend on speech tempo. Liveliness, on the other hand, is a global property of the utterance. This means that factors affecting liveliness do not necessarily affect syllabic prominence. For instance, as was shown by Traunmüller and Eriksson, liveliness judgments are extremely sensitive to speaking rate. They find that "a 1% increase in liveliness can be achieved by increasing speaking rate [i.e., by decreasing the overall duration] with 0.9%" (1995: 1914). Since this does not affect the relative properties of syllables, it will not affect their prominence. But a decrease in the duration of a single syllable, which makes it shorter relative to the neighbouring syllables, will produce a decrease in the prominence of that syllable. Likewise, with respect to pitch, raising the overall pitch contour of an utterance will increase the liveliness of that utterance, but it will not affect the prominence of the accented syllables (provided that this raising is done on the appropriate scale). It may be argued, therefore, that global changes to an utterance, like increasing the speaking rate, the pitch values of the whole pitch contour, the loudness and the average formant positions, will influence the listener's percept of liveliness, but not their perception of syllabic prominence. If these changes influence the prominence of syllables at all, the effect will be small, and may be attributed to the listener confusing prominence and liveliness.

It cannot be excluded, however, that global properties indeed affected the adjustments by Hermes and Van Gestel's subjects. The end frequencies of the lower declination lines of the stimuli in the low and high register were always 75 Hz and 180 Hz, respectively, and listeners may have judged the accent strength by relating the pitch peak associated with the accent to the end frequency of the declination line, which is assumed to provide a good estimate of the floor of the pitch range in the normal speaking mode of an individual speaker. This explanation was tested in experiments reported in Gussenhoven et al. (1997), in which listeners rated the strength of accents in utterances with varying end frequencies. The variation in end frequency was produced by manipulating the slope of the segment of the pitch contour between the accent and the end of the utterance. It was found that this variation did not affect prominence judgments. If it can be assumed that this type of variation indeed affects listeners' judgments of the floor of the speaker's pitch range, these findings argue against an explanation of Hermes and Van Gestel's results in terms of global properties, and in favor of an explanation that relates perceived accent strength to excursion size.

3.4 Discussion

It appears that two accented syllables in identical positions within utterances resynthesized in different registers are equally prominent when the excursion sizes of the accent-lending pitch movements are equal on the ERB-rate scale. There are two fundamentally different perceptual processes which may underlie this phenomenon. First, it can be conceived of as a consequence of the internal organisation of the human auditory system. In this case, the perceptual mechanism underlying prominence perception derives its pitch information from an internal array on which pitch is represented along a tonotopic axis. The contribution of an accent-lending pitch movement is proportional to the number of ERBs crossed along this tonotopic array. In this model, the perceived prominence does not depend on the natural pitch range of the speaker. In the alternative model, listeners judge the prominence lent by pitch movements relative to the natural pitch range of the speaker. The smaller the pitch range, the smaller the excursion sizes necessary to realise the same prominence as when the pitch range is larger. In this view, the outcome in favour of the ERB-rate scale in the experiments of Hermes and Van Gestel (1991) and Rump and Hermes (1996) is just a consequence of the circumstance that the pitch ranges of men and women are about equal on the ERB-rate scale. So far, the data needed to decide between these alternatives are lacking.

For speech synthesis and voice conversion, the consequence of these results is that, in order to make syllables in male, female and child speech just as prominent, the excursion sizes of the accent-lending pitch movements should be equal on an ERB-rate scale. Also, in order to make utterances spoken by men, women and children sound equally lively, the pitch contours should be parallel on an ERB-rate scale.

4. THE CONTRIBUTION OF PITCH TO THE PROMINENCE OF SYLLABLES

4.1 Introduction

In section 2, we have argued that prosodic prominence may best be captured in terms of a small number of prominence categories. Furthermore, we have observed that within the category of accents there is perceptually and communicatively relevant variation of prominence. We will use the term 'accent strength' to refer to this intra-categorial variation, but alternatingly use the term 'prominence' in this more restricted sense to avoid undue repetition. Variation of accent strength relates primarily to the phonetic characteristics of the pitch movement triggering the perception of accentedness, most notably its size. This section is devoted to this relation. Roughly speaking, increasing the excursion size of an accent-lending pitch movement increases the prominence of the accented syllable. But increasing the excursion size of a pitch movement also increases the slope of a pitch movement, and the changes in the pitch contour at the onset and offset of the movement become more abrupt. Furthermore, it is not yet clear whether the timing of a pitch movement or its duration affect the prominence of a syllable, and whether rises, falls and rise-falls lend equal prominence to a syllable. Finally, more global, phrase-level trends such as declination also affect the size of pitch accents, and listeners take these global factors into consideration in judging the prominence of accents, and the question is how to account for these effects.

4.2 Models for accent strength of single accents in an utterance

The first model to quantify the contribution of an accent-lending pitch movement to accent strength was presented by Hermes and Rump (1994). The model, presented in Fig. 2a, was based on the results of adjustment experiments as illustrated in Fig. 3. Subjects were asked to adjust the excursion size of an accent-lending pitch movement in a comparison stimulus in such a way that the prominence of the accented syllable was equal to the prominence of the accented syllable in a test stimulus. Test and comparison stimulus had different kinds of pitch movement, e.g., the test stimulus could have a rise while the comparison stimulus could have a fall as shown in Fig. 3a. This experiment was carried out both with test and comparison stimulus resynthesized within the same register, and with test and comparison stimulus resynthesized within different registers (the latter situation is shown in Fig. 3). Whether test and comparison stimuli were in the same register or not, did not make much difference, except that the standard deviations of the results were larger when test and comparison stimuli were in different registers. So, apparently, the task is more difficult in the different-register condition.

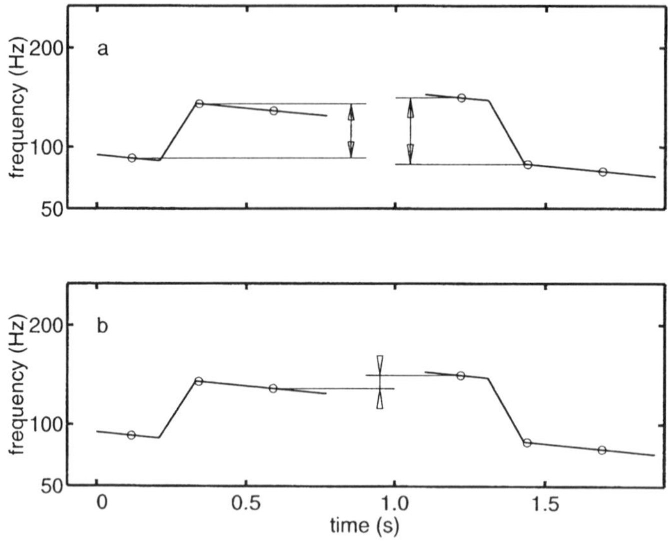

Figure 2. *Two models for the contribution of pitch to accent strength. The circles indicate the perceptual anchor points, e.g. pitch levels in the vowels, which determine the accent strength listeners perceive. In (a) the pitch-level difference between two successive positions of the pitch contour determines the accent strength. In (b) the relative accent strength for the two accents depends on the relation between levels at the upper declination line of the two stimuli.*

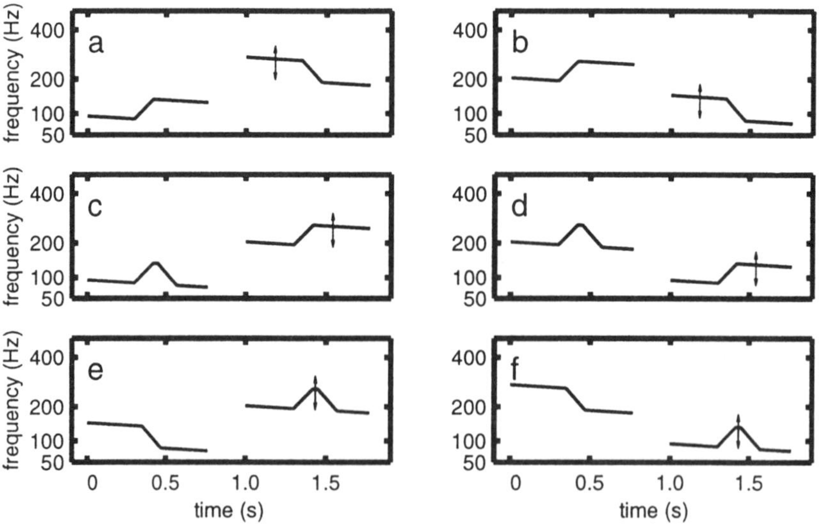

Figure 3. *Experimental set-up to investigate whether the prominence lent by rises, falls and rise-falls is different.*

It appeared that, when lending equal prominence, the rise and the rise-fall systematically had larger excursion sizes than the fall. In other words, excursion sizes being equal, the fall lends more prominence to a syllable than a rise and a rise-fall. At first, this was thought to be an effect of timing. In the resynthesized stimuli, the rise and the rise-fall started 70 ms before the vowel onset, while the fall started 20 ms before the vowel onset. By repeating the same experiment for different timings of the pitch movements, this explanation could be ruled out, however, as the timing of the fall did not influence the results in any significant way (the timing of the rise did have an effect, but this was explained as being due to a phonological difference between early and late rises in Dutch, cf. 't Hart, Collier and Cohen (1990).

A quantitative model was proposed to explain why falls lend more prominence than rises and rise-falls (cf. Fig. 2a). It assumes that the prominence lent by a pitch movement depends on the difference between two successive pitch levels in the pitch contour, and not on the excursion size of the movement (excursion size is defined in the same way as before, i.e. the distance between the pitch peak - either the end of the rise or the onset of the fall - and its projection onto the virtual declination line, perpendicular to the time axis; cf. Fig. 4). In Fig. 2a the pitch levels have been chosen at points before and immediately after the pitch movements, and the pitch level differences are indicated by the arrows. As can be seen, the pitch level difference for the fall is larger than that for the rise; the excursion sizes for the fall and the rise are equal. This is a consequence of the presence of declination: in the calculation of the pitch level difference the effect of declination is not taken into consideration, whereas it is in the calculation of excursion sizes. A simple calculation shows that the difference in length of the arrows equals 2dT, where d is the declination in ERB/s, while T is the time difference between the two anchor points in s. This model, which will be referred to as the pitch-level-difference (PLD) model, was further tested by repeating the experiment with different declinations, and the outcome was that the model could very well explain the results obtained by Hermes and Rump (1994). The first experiment comparing the accent strengths induced by different types of pitch movements was carried out both for test and comparison stimuli in the same register and for test and comparison stimuli in different registers. As mentioned, the results were the same, except that the standard deviations were larger in the conditions involving different registers. In order to obtain results with standard deviations as small as possible, the subsequent experiments were carried out with test and comparison stimuli in the same registers. Thus, the lower declination line in the stimuli with the rise, the fall and the rise-fall had the same (real or virtual) endpoints.

On the basis of reports by some subjects, an alternative model was developed, which after some simple calculation appeared to predict exactly the same results as the original model if the lower declination lines are the same and the durations of the syllables are the same. This model, referred as the high-level (HL) model, is presented in Fig. 2b. The strategy of the subjects, here, is not to compare differences in pitch levels within one utterance, but to equate the pitch level in the last syllable of the stimulus with the rise and the pitch level of the first syllable of the stimulus with the fall. At first, this strategy was thought to be less probable for three reasons. The first is that, according to this strategy, pitch levels in two unaccented syllables are matched, so that the pitch level in the accented syllable does not play a role. Intuitively, this was thought to be unlikely in prominence-matching tasks. Second, although the situation in Fig. 2b might create a favourable context for the HL strategy, since the two high-pitched syllables succeed each other without an intermediate syllable, pairs were also presented in the reverse order, with essentially the same results. The third reason why the HL strategy was thought to be unlikely was that the same results were obtained with test and comparison stimuli in different registers, where this strategy cannot play any role.

In order to test this alternative explanation, the same experiments were repeated, but with small differences between the baselines of the test and the comparison stimulus. In this situation the two models predict different results. It was found that the results were intermediate between what was predicted by the two models. In some cases, the results were closer to the HL model than to the PLD model (Rump and Hermes 1996).

It is concluded from these experiments that the PLD model still holds in the situation where test and comparison stimulus are in different registers. In the same-register conditions, however, subjects may use the HL cue, which has a strong influence on the results, and which in the case of identical baselines reduces the standard deviations of the results.

4.2.1 STABLE EFFECTS

One important finding is that subjects appear quite able to compare the accent strengths induced by different types of accent-lending pitch movements. Only for late rises do the subjects appear to have considerable difficulty in carrying out the task. This might indicate that early rises, rise-falls and falls have some common linguistic function, namely accenting a syllable. In this respect, it is remarkable to observe that there are subjects which, though quite capable of indicating in an utterance which syllables are accented, cannot tell whether the accent is lent by a rise, a fall or a rise-fall. The prominence of the syllables involved is the perceptual attribute which they perceive, and which kind of pitch movement this is lent by is not relevant for the listener.

Another stable result of these experiments is that the fall induces greater accent strength than the rise and the rise-fall for the same excursion sizes. With test and comparison stimulus in different registers, there is currently no other model than the PLD model which explains this. Due to declination, pitch level differences between points before and after a pitch movement are larger for a fall than for a rise. Note that the PLD model implicates that listeners do not compensate for declination from one syllable to the next.

If stimuli are resynthesized in the same register, subjects appear to use another cue in comparing the accent strengths induced by different types of pitch movements. The only model which may explain the results for identical lower declination lines is the HL model. What is not attractive in this model is that the baseline does not play any role at all and that only pitch levels in unaccented syllables come into play. This may be less awkward than it seems at first sight, however, as variation in the height of pitch levels at the upper declination line appears to be more salient than that of pitch levels at the lower declination line. Also, in principle it should not matter which syllable in a sequence of syllables on the upper declination line is taken as an estimate of the high pitch level, since all syllables in the sequence instantiate the high pitch level. Furthermore, in the experiments reported here the pitch contour preceding the pitch movements comprised only one syllable, and it appears that, if there is only one syllable preceding the accented syllable, a single unaccented syllable contains insufficient information to extract a reliable estimate of the corresponding pitch level (cf. Gussenhoven et al. 1997).

4.2.2 DISCUSSION

The fact that listeners can compare the accent strengths induced by early rises, rise-falls and falls, suggests that in some sense the linguistic function of these movements is similar. Actually, these experiments were inspired by the fact that, in Dutch intonation, a succession of a rise and fall can in most instances be replaced by two rise-falls without affecting the accent structure in any significant way. The accent strength induced by a late rise is essentially different in nature. This might be due to the fact that, in contrast with the other accent-lending pitch movements, where high pitch levels contribute most to prominence, the accent strength of the late rise may mostly be determined by the low pitch level at the onset of the pitch movement. Here it has to be mentioned that in the experiments in which the temporal position of the accent-lending pitch movement was varied, only the rise and the fall were varied. The rise-fall was only tested in one temporal position. Changing the temporal position of the rise-fall too much, leads to an unnatural intonation. In this situation, subjects are likely to take resort to ad-hoc strategies which are not based on their common speech processing facilities. Rise-falls with very different temporal positions were therefore not used in these experiments. The use of these stimuli will probably significantly increase the difficulty of the task, if the subjects are able to perform the task at all.

What is stable in the results is that the fall and the rise induce the same accent strength when the excursion size of the fall is somewhat smaller than that of the rise. This can most simply be explained by a model in which the accent strength is proportional to the distance between two pitch levels in the contour separated in time. The fact that the fall induces greater accent strength than a rise of equal excursion size is then simply due to declination. Where these pitch levels are located in the utterance remains a question to be investigated.

4.3 Relative accent strength of multiple accents in an utterance: The role of declination

The experiments summarized so far in section 4 all addressed the question as to which factors govern the perceived prominence for accents that constitute the only accent in a phrase or utterance. Briefly, the results so far suggest that the strength of a single accent in a phrase is related to the difference between two successive pitch levels when measured in ERBs. In addition, a model was proposed and tested as to how this pitch level difference (PLD) should be measured.

Since many phrases contain more than one accent, a logical next question is whether the relative strength of several accents in a phrase is directly related to their PLDs. However, the PLD hypothesis is relatively recent, and most research so far has used the older notion 'excursion size' as a measure for the magnitude of the pitch change (cf. Fig. 4); as before, the excursion size is defined as the distance between the peak of an accent and its projection onto a virtual baseline, which can be thought of as a regression line for the pitch minima in the contour. In most utterances where this line can be determined, it has a negative slope, which has been attributed to a decrease in subglottal pressure and the relaxation of laryngeal muscles, and which has become known as 'F0 declination'.

All experiments to be discussed here employed rise-fall pitch accents, so that the excursion sizes and the PLDs differ by a small and constant amount. Therefore, the conclusions we will draw will not be strongly affected by whether we approach the matter in terms of PLDs or excursion sizes, and we feel justified to reformulate the question in terms of the relation between accent strength and excursion size.

The first relevant study was conducted by Pierrehumbert (1979), who investigated the relation between the perceived relative height of two accent peaks in a phrase and their F0 characteristics. She asked listeners to judge which of two peaks in an utterance had higher pitch. A clear tendency could be observed for listeners to overestimate the pitch of the second peak: for the second peak to be heard as having the same pitch as the first one, it actually

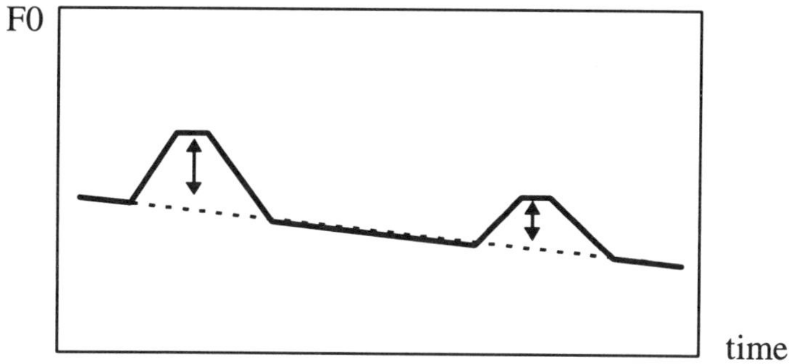

Figure 4. Schematic representation of pitch contour with two rise-fall accents. The dashed line represents the virtual baseline, relative to which the excursion sizes of the accents are measured (represented by the arrows).

should have a lower F0 peak. Pierrehumbert explained this finding as a tendency of the listeners to compensate for expected declination.

Later research showed that this compensation for declination does not only affect the pitch of F0 peaks, but also their accent strength (Gussenhoven and Rietveld 1988). They investigated the supposed compensation for declination in more detail, by measuring the frequency difference that gave rise to the perception of equal accent strength for two different temporal intervals. It was found that the difference in frequency that was required to give the impression of equal accent strength indeed varied with the temporal interval between the two peaks. However, the actual difference was not linearly related to the temporal interval, but there was a constant additional frequency difference. This led Gussenhoven and Rietveld to conclude that listeners also compensated for final lowering, i.e., a lowering of the final peak in addition to the lowering that is expected on the basis of declination only as observed by Liberman and Pierrehumbert (1984).

The conclusion that the accent strength associated with the last peak in the utterance is also influenced by the tendency of the listener to compensate for final lowering was challenged by Terken (1989), who argued that the constant difference was observed only when collapsing across conditions that should not be collapsed. This argument was supported by the results of Terken (1994) who showed that the results for utterances containing two peaks generalized to utterances containing three peaks, for which the relative accent strength of the first and second peak had to be judged.

In the experiments mentioned so far the baseline was kept constant and only the distance between the peaks and the baseline was varied, in order to assess the difference in frequency that is compensated for with different heights of the first peak and different temporal intervals. It was pointed out by Terken

(1991), however, that the slope of the baseline may vary as well, and that variation in the slope of the baseline may affect the frequency difference between the peaks that is compensated for in deciding whether two peaks are equally prominent. In order to test this hypothesis, he compared the frequency difference between two peaks needed to give equal prominence under two conditions, and for different heights of the first peak. In one condition the slope of the baseline co-varied with the height of the first peak so that the distance between the peak and the baseline remained constant on a logarithmic scale. In another condition the pitch contours of the stimuli were synthesized with flat baselines so that the distance between the first peak and the baseline co-varied with peak height. If the slope of the baseline is irrelevant, the same frequency difference between the first and the second peak should be needed for the perception of equal prominence in the two different baseline conditions. It was found that the presence or absence of baseline declination indeed influenced the frequency difference between the two peaks that was needed to give equal prominence: larger frequency differences between the peaks were needed in the utterances with declination than in the utterances without declination, or – stated otherwise – if the distance between the first peak and the baseline decreased, the distance between the second peak and the baseline should also decrease to maintain equal prominence. However, the details of the results ruled out a simple model that assumes a direct relation between the distance between the peak and the baseline and the perceived accent strength. Instead, the results suggested that if the beginning of the contour was higher in the speaker's overall pitch range, a given excursion size resulted in a greater accent strength than if the beginning of the contour was lower in the speaker's pitch range. The results inspired a model in which the excursion size for the second accent needed to give the same prominence as the first accent is obtained by the following function (cf. Fig. 5):

$$D2 = a.D1 + b.E1$$

where D1 and D2 are the distances from the first and the second peak, respectively, to the baseline, and E1 is the distance between the baseline below the first peak and the speaker's floor (which was assumed to be the end frequency of the utterance), and a and b are constants. A linear regression analysis provided values of 0.9 and .23 for a and b, respectively. This equation also predicted the findings of Terken (1994) rather closely. Thus, these findings complicated the generalization of conclusions drawn from experiments with single-peak utterances to utterances containing several accents, because they further complicated the mapping between the size of F0 changes and perceived prominence.

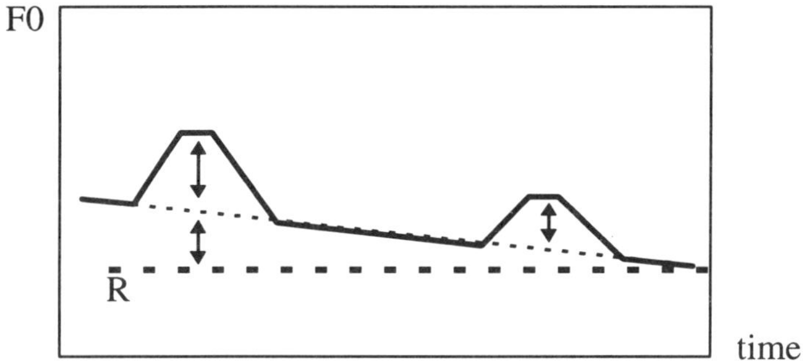

Figure 5. Graphical representation of model for prominence perception. The size of the second accent needed to obtain the same strength as the first accent is a function both of the size of the first accent and the elevation of the first accent above the speaker-specific reference value R.

The precise effect of the slope of the baseline on the frequency difference between two peaks that is needed for equal prominence ($\Delta_{PIP2/EqProm}$) was investigated in follow-up experiments conducted by Repp and Rump (reported in Gussenhoven et al. 1997). Rather than just varying the presence or absence of declination as in Terken (1991) they assessed ($\Delta_{PIP2/EqProm}$ for a given P1 frequency at different slopes of the baseline (which always ended at the same frequency). The frequency difference was assessed for different heights of the first peak. Again, the earlier finding that the second peak should be lower than the first peak to give equal prominence was confirmed. Furthermore, it was found that ($\Delta_{PIP2/EqProm}$ varied in relation to the distance between the first peak and the baseline (cf. Fig. 6). If the onset of the baseline was raised while keeping the frequency of the first peak constant (thus decreasing the distance between the peak and baseline), ($\Delta_{PIP2/EqProm}$ also decreased; i.e. the second peak should be somewhat higher than in a condition with a lower onset of the baseline. Likewise, if the peak was lowered while keeping the slope of the baseline constant (again decreasing the distance between the peak and baseline), ($\Delta_{PIP2/EqProm}$ decreased again.

In a separate series of experiments conducted by Gussenhoven and Rietveld (Gussenhoven et al. 1997), it was found that the compensation for declination that is generally observed in utterances with two accents, also applies in utterances with single accents: a peak with a given frequency is judged as more prominent when it occurs farther away from the onset of the utterance than when it occurs early in the utterance. Since this finding was obtained with utterances with level pitch before the peak, the effect could not have been caused by actually declining pitch, so that it was concluded that the listeners judge the prominence of the peaks relative to an abstract declination line ('abstract' because this line is not actually present in the utterance's contour, but is imposed by the listener as part of a mental frame of reference). Furthermore, it made clear that the expectation that a peak later in the utterance should be lower than a peak early in the utterance does not require the presence of an early peak; i.e., expectations about declination are not projected from early peaks to subsequent peaks, but are derived from a phrasal frame of reference, which is generated by the listener to interpret the actually occurring contour. Furthermore, since the height of the initial part of the contour influenced the prominence of accent peaks whereas the height of the final low did not, it was concluded that the abstract reference line is anchored in the initial part of the utterance; that is, it appears that listeners estimate the position of the abstract reference line on the basis of information situated near the onset of the utterance.

Combining the findings of the two groups of experiments, Gussenhoven et al. (1997) suggested that the perception of prominence relations is influenced by two effects. In the first place, there is the compensation for expected declination, which stems from the listeners' knowledge that peaks later in the utterance are generally lower than peaks early in the utterance. One possibility to account for this lowering is in terms of the decrease of subglottal pressure and/or laryngeal tonus, which causes a lowering of the base level in the course of the utterance (cf. Fujisaki and Hirose 1984; 't Hart et al. 1990). In the second place, the findings so far suggest that greater differences between peaks are required to give the impression of equal accent strength when the excursion size of the initial peak gets larger. That is, it seems as if listeners assume that, in addition to the declination that affects the base level, the movements needed to realize pitch peaks are also subject to some kind of declination which is stronger as the movements are larger. The model that emerges from this interpretation would take the following general form:

$$\Delta_{PIP2/EqProm} = f(D1),$$

where D1 is determined relative to the abstract baseline, the position of which is estimated on the basis of information in the beginning of the utterance or phrase. The slope of this baseline still remains to be determined, and a weak point in this explanation is that it remains unclear from which data the listener obtains an estimate of the slope of this baseline. Furthermore, the model is rudimentary; in order to be made more precise it should at least include a temporal component. Moreover, it should specify the margins for Δ_{PIP2} within which listeners will still perceive two accents to be equally prominent at different temporal intervals: only then can the model also be used to determine when two peaks will give rise to the perception of unequal prominence.

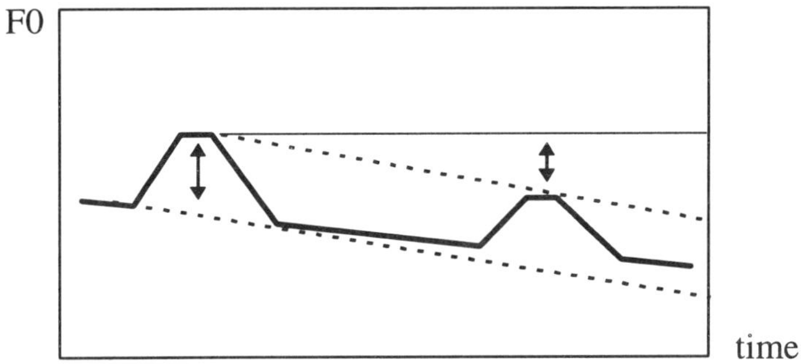

Figure 6. Schematic representation of revised model for perception of relative accent strength. The slope of the upper dashed line, representing the equal-strength curve for the two peaks, is supposed to vary as a function of the magnitude of the first peak, P1, relative to an abstract reference line (represented by the lower dashed line).

4.4 Mutual effects of the prominence of two accents on each other's perceived prominence

The results summarized in section 4.2 focused on the question as to what properties make an accent to be perceived relatively more or less prominent than another accent in the same phrase, i.e., on the question of how the effect of declination on the perceived relative prominence of accents should be modeled. A further complication of the mapping between excursion size and accent strength is provided by the apparent mutual effects of the prominence of accents on each other's perceived prominence. This aspect has not yet been thoroughly explored, but at least the available evidence suggests that, in utterances containing two accent peaks P1 and P2, the relation between F0 and the perceived prominence for the second peak (P2) is influenced by the frequency of the first peak (P1) (Gussenhoven and Rietveld 1988; Ladd 1994; Ladd et al. 1994).

There is some evidence from these experiments that variation in the height of P1 affects the perceived prominence associated with a given P2. Furthermore, it has been suggested that the direction of the effect may depend on the height of P2. In an experiment reported in Ladd (1994) it was found that an increase in the height of P1 resulted in an increase in the prominence associated with P2, but only for low P2; for high P2, an increase in P1 height resulted in a decrease of the prominence associated with P2. However, the reliability of these findings needs to be further established: the sizes of the effects vary considerably across experiments; in addition, the findings for high P2 in Ladd (1994) were inconsistent with those in Gussenhoven and Rietveld (1988), where the effect for high P2 was in the same direction as for low P2.

Evidently, as long as the inconsistency has not been cleared up, it makes no sense to look for an explanation. Both sets of results agree, however, in that they provide a further complication of the relation between F0 variation and perceived prominence.

5. RELATION BETWEEN THEORY AND PRACTICE: TRANSCRIPTION OF PROMINENCE, ACCENT, AND ACCENT STRENGTH IN DATABASES

Natural speech displays variation in prominence, in particular accent strength. If this kind of variation is not modeled in speech synthesis, the resulting synthetic speech will sound boring, the listeners will be distracted and they will no longer be able to process the information in an efficient way. Therefore, developments in the area of text-to-speech conversion have given a new impetus to the study of prominence variation: efforts have been taken to analyze speech corpora with respect to the factors governing variation in accent strength, with the aim of including the obtained rules in speech synthesis systems (among others O'Shaughnessy 1976; Horne 1987).

The analysis of speech corpora requires that accent strength be transcribed in some way. In the ideal case, models for variation in accent strength such as those summarized in the previous sections provide the theoretical foundations for transcription systems, i.e. the models should specify which prominence categories can be distinguished and how accent strength is perceived, and accordingly dictate in which way such prosodic characteristics of speech should be transcribed. In practice, this situation is not realized. Investigators are still in the middle of a debate about the adequate treatment of variation in accent strength.

Two main classes of models for variation of accent strength have been proposed. One class of models assume that accent strength is perceived in terms of different categories, along the lines of Trager and Smith (1951) (for instance Ladd 1994). Another class of models assume that accent strength may vary in a gradient way. The application of methods borrowed from work on categorial perception of phonemes has not been successful so far in solving the issue. In addition, there is a further complication in attacking the problem. As was shown in the previous sections, variation in perceived accent strength primarily relates to variation in excursion size of peak height. However, these properties are also relevant to variation of pitch range, which is a property of phrases or utterances (and is used by speakers to signal speaker-involvement, expressiveness, and discourse structure). Thus, even though accent strength might vary in a categorial way, pitch range may vary in a gradient way, so that the two aspects are very difficult to separate experimentally, and the phrasing of utterances used as stimuli may strongly influence the outcomes. The least one can say is that this situation precludes the possibility of solving the issue by experiments employing utterances containing single accents. With respect to utterances containing multiple accents, there is another complication. For these utterances, it is unclear whether accent strength may vary at all from accent to accent within a single phrase. For instance, there might be a constraint saying that accent strength should remain fixed within a phrase, and that the speaker, if he wants to produce an accent with a different accent strength than the preceding accent, he must insert a phrase boundary in between. This reasoning may seem far-fetched, but it should be kept in mind that, if a phrase contains more than one accent, the pre-final ones are pre-nuclear, and it is unclear what communicative function might be served by variation in accent strength, either categorially or gradiently, of pre-nuclear accents.

The question of whether accents may vary in accent strength within a single phrase was addressed by Terken (1997). He presented phrases containing multiple accents taken from spontaneous descriptive monologues to fifteen listeners who judged the accent strength of the accented words on a ten point scale. The patterns of relative prominence obtained by taking the mean prominence judgments per word across listeners showed that indeed accents within a phrase may vary in accent strength, and that, in addition, the difference in relative prominence between two accents was variable as well. Furthermore, it was observed that there was neither a simple pattern of decreasing accent strength in the course of a phrase (which might be expected on the basis of a simple phonological constraint), nor a straightforward relation between patterns of accent strength and surface syntactic properties such as lexical category or grammatical role. Thus, so far there is no evidence that accent variation is constrained in any way, although it must be admitted that the set of utterances to be judged was rather small; analyses of larger corpora might at least suggest certain trends in the relation between patterns of accent strength and surface structure characteristics such as the ones mentioned above.

Further light on the question as to whether accent strength can be best treated categorially or gradiently and the function the variation of accent strength might play comes from a study by Rump (1996), also reported in Rump and Collier (1996). In this study, the sentence *Amanda gaat naar Malta* 'Amanda goes to Malta' was synthesized with different contours, differing with respect to the height of the accent peaks on *Amanda* and *Malta* (both in their absolute and relative heights). In principle, each utterance might provide the answer to each of the following questions: 'What is going on' (broad focus), 'Does Amanda go to Cyprus' (contrastive, narrow focus on 'Malta'), 'Does John go to Malta' (contrastive, narrow focus on 'Amanda'), and 'Does John go to Cyprus' (multiple contrastive, narrow focus, both on 'Amanda' and 'Malta'). Listeners were asked to select for each utterance presenting a particular pattern of absolute and relative accent strengths, the question that would provide the best-matching question-answer pair. The outcomes are summarized in Fig. 7. It was found that large excursion sizes were interpreted as signaling narrow focus. For instance, if both *Amanda* and *Malta* were marked by large excursion sizes, listeners selected the question providing the multiple contrastive question-answer combination. If either *Amanda* or *Malta* were marked by large excursion sizes, the listeners selected the question providing narrow focus on *Amanda* and *Malta*, respectively. Interestingly, in the case with narrow focus on *Malta*, listeners preferred contours in which the accent on *Amanda* had almost the same excursion size as in the broad focus over those in which it had no excursion size at all. This is in line with the findings of Horne (1987), and suggests that content words in pre-nuclear position should have at least some prominence, even if they are not considered to be in focus; furthermore, it suggests that the absolute height of a pitch peak does not determine whether a word is considered to be focused, but rather its height relative to the height of other peaks in the utterance. Finally, the fact that the presence of extra-high peaks gave rise to narrow-focus interpretations, weakens the view that narrow-focus interpretations do not depend on absolute peak height but are primarily conveyed by the lowering of peaks on non-focal items. Thus, they appear to support Ladd's plea for the revaluation of Trager and Smith's (1951) level approach (Ladd 1994).

In sum, there is still no agreement about the adequate treatment of variation in accent strength. Much experimental work seems to assume implicitly that accent strength may vary in a gradient way and the results may seem to support this assumption. On the other hand, the results of Rump and Collier, which more directly address the communicative function of variation in accent strength, appear to support a more categorial treatment, but it is still too early to draw definite conclusions.

Of course, those who are interested in modeling variation of accent strength in natural speech in order to improve the acceptability of synthetic speech, do not wait until the question of the adequate treatment has been solved by experimentation. Therefore, it is interesting to see how different transcription systems deal with variation of accent strength.

One system which has drawn much attention in recent years is the ToBI system (Silverman et al. 1992; Beckman and Ayers 1994), which has been designed primarily for the annotation of the prosodic-phonological characteristics of American English. In the ToBI system, no conventions have been included for the transcription of accent strength. The system allows the possibility to mark the point of highest pitch within a phrase, which can be used as a measure of global (phrase-level) pitch range variation; perceptually relevant variation in the scaling of accent peaks within the phrase is not transcribed independently. The same applies to the transcription system that has been proposed by Hirst et al. (This volume).

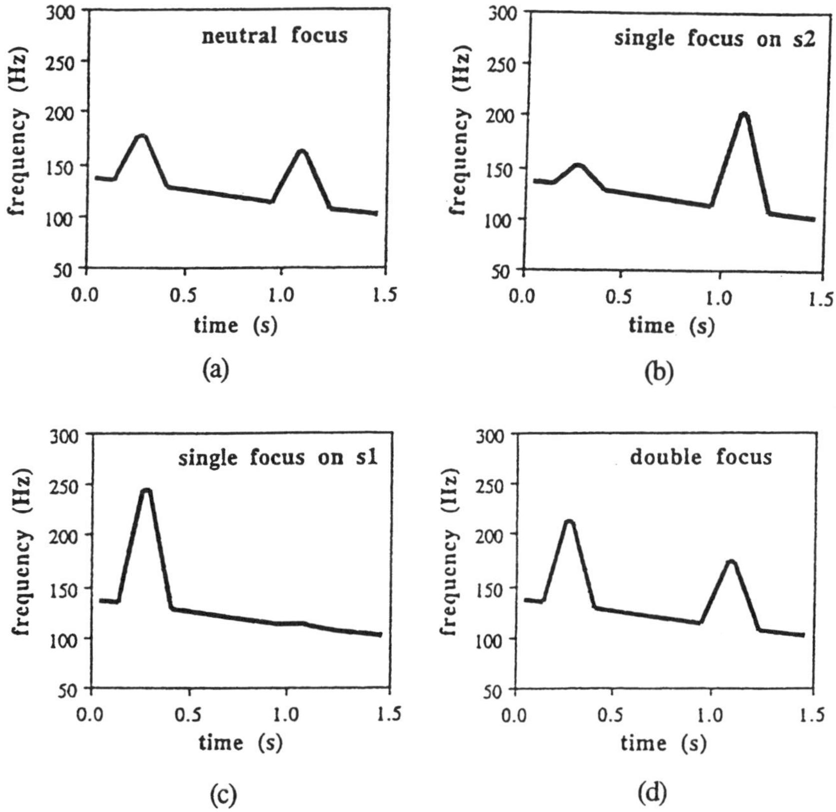

Figure 7. Average pitch contours for Amanda gaat naar Malta *('Amanda goes to Malta') (accent peaks on* Amanda *and* Malta*) as answer to the following questions: (a) 'What is going on' (broad or neutral focus), (b) 'Does Amanda go to Cyprus' (contrastive, narrow focus on Malta (s2)), (c) 'Does John go to Malta' (contrastive, narrow focus on* Amanda *(s1)), and (d) 'Does John go to Cyprus' (multiple contrastive, narrow focus, both on* Amanda *and* Malta*) (with courtesy from H.H. Rump).*

Another well-known system, with a claim to respectability, is the transcription system of the International Phonetic Association (IPA). Under the auspices of the IPA, a working group was organized to provide transcription conventions for suprasegmental categories (Bruce 1989). With respect to prominence, in particular accent strength, it was agreed to have a four-level distinction between no, weak, strong, and extra-strong prominence (called stress in the proposals of the working group), to be annotated as [X], [,X], ['X], and [''X], respectively. However, the proposals do not stem from a thorough theoretical analysis, but more from an inventory of the needs of the Working Group members. This pragmatic attitude is defensible given on the one hand the need to model accent variation in natural speech in speech synthesis, while on the other hand agreement about the proper theoretical treatment is still lacking. It is also evident from the recommendation to indicate, when the need is felt to do so, additional degrees of extra-strong stress by means of additional marks [’’’X], [’’’’X], and so forth. Clearly, this once more underlines the impressionistic and subjective character of the transcription of accent strength that still prevails. Until now, the only way to satisfy certain minimal requirements of scientific methodology and to guarantee the reliability of the observations, is to aim for intersubjectivity and elicit judgments from a panel of listeners rather than from a single transcriber, although this may be prohibited by practical considerations (the only alternative so far is to describe directly the link between F0 and grammatical characteristics, but this is undesirable for well-known reasons: it will certainly result in lack of abstraction, and lead to the impression that there is a direct mapping from grammatical properties onto phonetic ones). Clearly, we are in need of further insights into the way variation in accent strength is perceived, so as to come up with adequate treatments that provide the interface between communicative function on the one hand and physical properties on the other.

6. CONCLUSION

We have summarized available data concerning the perception of prominence, and discussed the way prominence should be treated in our view, given the evidence available so far. We started to define prominence in a syntagmatic way, i.e., in relation to the properties that make syllables stand out from their environment. In section 2, we concluded that the available evidence supports a four-level distinction of prominence categories: reduced, full, stressed, accented. It is evident, however, that the categorial distinctions also involve paradigmatic differences (i.e. the properties of items in relation to those of the other members of the same class), such as the distinction between reduced and full vowels. Likewise, current approaches to the automatic detection of prominence (Campbell 1995; Leenhardt 1996) employ normalized phoneme durations to locate stressed syllables, i.e. the duration of phonemes relative to the durations of all instances of the same phonemes in a large corpus. The syntagmatic and paradigmatic perspectives may be in conflict, however: given three consecutive syllables of which the first and the third are accented, the middle one may be classified as stressed from the paradigmatic perspective, but it will be relatively non-prominent in its context, so that listeners may not be able to decide whether it is stressed or not. A relevant example would be *John sees Pete* with *John* and *Pete* accented. Available data suggest that listeners would not classify *sees* as prominent (cf. Grover and Terken 1995). In our view, the best way out would be to determine in what way the different prominence categories contribute to the processing of spoken language (cf. for instance Cutler and Norris 1988, with respect to the way listeners may use stress).

In sections 3-4, we concentrated on the treàtment of accent strength, i.e. perceptually relevant variation in the prominence of accented syllables, and in section 5 we discussed how the results of experimental work on accent strength might guide the transcription of accent strength in speech corpora. It was concluded that more experimental work is needed to settle crucial issues. In particular, the question as to whether different degrees of accent strength are to be treated in a categorial or gradient way has not yet been answered satisfactorily. Thus, for the time being, the conventions for transcribing accent strength cannot yet be justified by adequate theories. Again it appears that real progress can be made only if we relate potential theoretical distinctions to their function in the processing of spoken language, i.e., if we look at the way listeners might exploit variation in accent strength to process and interpret spoken messages.

REFERENCES

Beckman, M.E. 1986. *Stress and Non-Stress Accent*. Dordrecht: Foris.

Beckman, M.E. and Ayers, G. J. 1994. Guidelines for ToBI transcription. Columbus, OH: Ohio State University, Linguistics Department (available by ftp from kiwi.nmt.edu)

Beckman, M.E. and Edwards, J. 1994. Articulatory evidence for differentiating stress categories. In P.A. Keating (ed), *Phonological Structure and Phonetic Form. Papers in Laboratory Phonology* III. Cambridge, UK: Cambridge U.P., 1-33.

Beckman, M.E. and Pierrehumbert, J.B. 1986. Intonational structure in Japanese and English. *Phonology Yearbook* 3, 255-309.

Bolinger, D. 1958 A Theory of Pitch Accent in English. *Word* 14, 109-149.

Bolinger, D. 1961. *Generality, Gradience, and the All-or-None*. The Hague: Mouton.

Bruce, G. 1977. *Swedish Word Accents in Sentence Perspective*. Lund: Gleerup.

Bruce, G. 1987. On the phonology and phonetics of rhythm: evidence from Swedish. In W.U. Dressler et al. (eds), *Phonologica* 1984. Cambridge: Cambridge University Press, 21-31.

Bruce, G. 1989. Report from the IPA Working Group on suprasegmental categories *Working Papers* 35 (Dept.of Linguistics, Lund University), 25-40.

Campbell, W.N. 1995. Prosodic influence on segmental quality. *Proc. Eurospeech 95*, 1011-1014.

Chomsky, N. and Halle, M. 1968. *The Sound Pattern of English*. New York: Harper & Row.

Cruttenden, A. 1986. *Intonation*. Cambridge: Cambridge University Press.

Currie, K.L. 1980. An initial search for tonics. *Language and Speech* 23, 329-350.

Currie, K.L. 1981. Further experiments in the search for tonics. *Language and Speech* 24, 1-28.

Cutler, A. and Darwin, C.J. 1981. Phoneme-monitoring reaction time and preceding prosody: Effects of stop closure duration and of fundamental frequency. *Perception and Psychophysics* 29, 217-224.

Cutler, A. and Norris, D.G. 1988. The role of strong syllables in segmenting for lexical access. Journal of Experimental Psychology: *Human Perception and Performance* 14, 113-121.

Edwards, J. and Beckman, M.E. 1988. Articulatory timing and the prosodic interpretation of syllable duration. *Phonetica* 45, 156-174.

De Jong, K. 1995. The supraglottal articulation of prominence in English: Linguistic stress as localized hyperarticulation. *Journal of the Acoustical Society of America* 97, 491-504.

Fant, G. and Kruckenberg, A. 1988. Some durational correlates of Swedish prosody. *Proc. Speech '88, 7th FASE Symposium*, Edinburgh.

Fear, B.D., Cutler, A. and Butterfield, S. 1995. The strong/weak syllable distinction in English. *Journal of the Acoustical Society of America* 97, 1893-1904.

Fletcher, H. 1940. Auditory patterns. *Reviews of Modern Physics* 12, 47-61.

Fry, D.B. 1955. Duration and intensity as physical correlates of linguistic stress. *Journal of the Acoustical Society of America* 27, 765-768.

Fry, D.B. 1958. Experiments in the perception of stress. *Language and Speech* 1, 126-152.

Fujisaki, H. and Hirose, K. 1984. Analysis of voice fundamental frequency contours for declarative sentences of Japanese. *Journal of the Acoustical Society of Japan* (E) 5, 233-242.

Glasberg B.R. and Moore B.C.J. 1990. Derivation of auditory filter shapes from notched-noise data. *Hearing Research* 47, 103-138.

Graddol D. 1986. Discourse specific pitch behavior. In C. Johns-Lewis (ed), *Intonation in Discourse*. London: Croom Helm, 221-237.

Grover, C. and Terken, J. 1995. The role of stress and accent in the perception of rhythm. *Proc. XIIIth Int. Congress of Phonetic Sciences*, Vol 4, 356-359.

Gussenhoven, C. and Rietveld, A.C.M. 1988. Fundamental frequency declination in Dutch: testing three hypotheses. *Journal of Phonetics* 16, 355-369.

Gussenhoven, C. and Rietveld, A.C.M. 1996. Estimating the location of the 'Graph paper': On the speaker-dependence of the perceived prominence of F0 peaks. Ms., Nijmegen University.

Gussenhoven, C., Repp, B.H., Rietveld, A.C.M., Rump, H.H. and Terken, J. 1997. The perceptual prominence of fundamental frequency peaks. *Journal of the Acoustical Society of America*, 102, 3009-3022.

Hasegawa, Y. and Hata, K. 1992. Fundamental frequency as an acoustic cue to accent perception. *Language and Speech* 35, 87-98.

't Hart, J. 1981. Differential sensitivity to pitch distance, particularly in speech. *Journal of the Acoustical Society of America* 69, 811-821.

't Hart, J. and Cohen, A. 1973. Intonation by rule: A perceptual quest. *Journal of Phonetics* 1, 309-327.

't Hart, J. and Collier, R. 1975. Integrating different levels of intonation analysis. *Journal of Phonetics* 3, 235-255.

't Hart, J., Collier, R. and Cohen, A. 1990. *A Perceptual Study of Intonation: An Experimental-Phonetic Approach to Speech Melody*. Cambridge: Cambridge University Press.

Hermes, D.J. 1995. Timing of pitch movements and accentuation of syllables. *IPO Annual Progress Report* 30, 38-44.

Hermes, D.J. and Van Gestel, J. 1991. The frequency scale of speech intonation. *Journal of the Acoustical Society of America* 90, 97-102.

Hermes, D.J. and Rump, H.H. 1994. Perception of prominence in speech intonation induced by rising and falling pitch movements. *Journal of the Acoustical Society of America* 96, 83-92.

Hirst, D., Di Cristo, A., and Espessser, R. This volume. Levels of representation and levels of analysis for the description of intonation systems.

Horne, M. 1987. *Towards a Discourse-Based Model of English Sentence Intonation. Working Papers* 32. (Dept. of Linguistics, Lund University).

Horne, M. 1990. Empirical evidence for a deletion formulation of the rhythm rule in English. *Linguistics* 28, 959-981.

Huss, V. 1978. English word stress in the Post-nuclear position. *Phonetica* 35, 86-105.

Ladd, D.R. 1993. On the theoretical status of the baseline in modelling intonation. *Language and Speech* 36, 435-451.

Ladd, D.R. 1994. Constraints on the gradient variability of pitch range, or, Pitch level 4 lives! In P. Keating (ed), *Phonological Structure and Phonetic Form. Papers in Laboratory Phonology* III. Cambridge, UK: Cambridge University Press, 43-63.

Ladd, D.R., Verhoeven, J., and Jacobs, K. 1994. Influence of adjacent pitch accents on each other's perceived prominence: two contradictory effects. *Journal of Phonetics* 22, 87-99.

Leenhardt, G. 1996. Automatic Labelling of Prosodic Events. *Technical Report,* Institute for Perception Research, Eindhoven NL.

Liberman, M., and Pierrehumbert, J. 1984. Intonational invariance under changes in pitch range and length. In M. Aronoff and R. Oehrle (eds), *Language, Sound and Structure*. Cambridge, MA: MIT Press, 157-233.

Liberman, M. and Prince, A. 1977. On stress and linguistic rhythm. *Linguistic Inquiry* 8, 249-336.

Lieberman, P. 1965. On the acoustic basis of the perception of intonation by linguists. *Word* 21, 40-54.

Moulines E. and Laroche J. 1995. Non-parametric techniques for pitch-scale and time-scale modification of speech. *Speech Communication* 16, 175-205.

Needham, W.P. 1990. Semantic structure, information structure, and intonation in discourse production. *Journal of Memory and Language* 29, 455-468.

O'Shaughnessy, D.1976. *Modelling Fundamental Frequency, and its Relationship to Syntax, Semantics, and Phonetics*. Ph.D. thesis, Cambridge: MIT.

Pierrehumbert, J.B. 1979. The perception of fundamental frequency declination. *Journal of the Acoustical Society of America* 66, 363-369.

Patterson, R.D. 1976. Auditory filler shapes derived with noise stimuli. *Journal of the Acoustical Society of America* 59, 640-654.

Pierrehumbert, J.B. 1980. *The Phonology and Phonetics of English Intonation*. Ph.D Thesis, MIT, Distributed by the Indiana University Linguistics Club.

Pierrehumbert, J.B. 1981. Synthesizing intonation. *Journal of the Acoustical Society of America* 70, 985-995.

Pierrehumbert, J.B. and Hirschberg, J. 1990. The meaning f international contours in the interpretation of discourse. In P. Cohen, J. Morgan, and M. Pollock (eds), *Intentions in Communication.* Cambridge, MA: MIT Press, 271-311.

Rietveld, A.C.M. and Gussenhoven, C. 1985. On the relation between pitch excursion size and prominence. *Journal of Phonetics* 13, 299-308.

Rump, H.H. 1996. *Prominence of Pitch-Accented Syllables.* Ph.D. Thesis, Eindhoven University of Technology. Eindhoven (NL).

Rump, H.H. and Collier, R. 1996. Focus conditions and the prominence of pitch-accented syllables. *Language and Speech* 39, 1-17.

Rump, H.H. and Hermes, D.J. 1996. Prominence of pitch-accented syllables and the ERB-rate scale, *IPO Manuscript no.* 1170 (Institute for Perception Research, Eindhoven).

Rump, H.H. and Hermes, D.J. 1996. Prominence lent by rising and falling pitch movements: testing two models. *Journal of the Acoustical Society of America* 100, 1122-1131.

Selkirk, E. 1984. *Phonology and Syntax: the Relation between Sound and Structure.* Cambridge: MIT Press.

Selkirk, E. 1995. Sentence prosody: Intonation, stress, and phrasing, In J. Goldsmith (ed), *Handbook of Phonological Theory.* Oxford: Blackwell, 550-569.

Shattuck-Hufnagel, S. 1994. 'Stress shift' as early placement of pitch accents: a comment on Edwards and Beckman. In P. Keating (ed), *Phonological Structure and Phonetic Form. Papers in Laboratory Phonology* III. Cambridge, UK: Cambridge University Press, 34-42.

Shattuck-Hufnagel, S. 1995. The importance of phonological transcription in empirical approaches to stress shift versus early accent: comments on Grabe and Warren, and Vogel, Bunnell and Hoskins. In B Connell and A. Arvaniti (eds), *Phonology and Phonetic Evidence. Papers in Laboratory Phonology* IV. Cambridge, UK: Cambridge University Press, 128-140.

Silverman, K., Beckman, M., Pitrelli, J., Ostendorf, M., Wightman, C., Price, P., Pierrehumbert, J., and Hirschberg, J. 1992. ToBI: A Standard for Labelling English Prosody. *Proc. International Conference on Spoken Language Processing* (Banff), 867-870.

Sluijter, A.M.C. 1995. *Phonetic Correlates of Stress and Accent.* Ph.D. Thesis, Leiden University. The Hague: Holland Academic Graphics (HIL Dissertations 15).

Sluijter, A.M.C. and Van Heuven, V.J. 1996. Spectral balance as an acoustic correlate of linguistic stress. *Journal of the Acoustical Society of America* 100, 2471-2485.

Stevens S.S., Volkmann J. and Newman E.B. 1937. A scale for the measurement of the psychological magnitude pitch. *Journal of the Acoustical Society of America* 8, 185-190.

Stevens, S.S. and Volkmann, J. 1940. The relation of pitch to frequency: a revised scale. *American Journal of Psychology* 53, 329-353.

Takeda, S. and Ichikawa, A. 1994. Analysis of prominence in spoken Japanese sentences and application to text-to-speech synthesis. *Speech Communication* 14, 171-196.

Terken, J. 1989. Reaction to C. Gussenhoven and A. Rietveld: Fundamental frequency declination in Dutch: testing three hypotheses. Letter to the Editor. *Journal of Phonetics* 17, 357-364.

Terken, J. 1991. Fundamental frequency and perceived prominence of accented syllables. *Journal of the Acoustical Society of America* 89, 1768-1776.

Terken, J. 1993. Baselines revisited: Reply to Ladd. *Language and Speech* 36, 453-459.

Terken, J. 1994. Fundamental frequency and perceived prominence of accented syllables II: Non-final syllables. *Journal of the Acoustical Society of America* 95, 3662-3665.

Terken, J. 1997. Variation of accent prominence within the phrase: Models and spontaneous speech data. In Y. Sagisaka, W. Campbell and N. Higuchi (eds), *Computing Prosody*. New York: Springer, 95-116.

Trager, G.L. and Smith, H.L. 1951. *An Outline of English Structure Studies in Linguistics. Occasional Papers* 3. Washington: American Council of Learned Societies.

Traunmüller, H., Branderud, P. and Bigestans, A. 1989. Paralinguistic speech signal transformations. *Phonetic Experimental Research, Institute of Linguistics, University of Stockholm (PERILUS)* 10, 47-64.

Traunmüller, H. and Eriksson, A. 1993. F0-excursions in speech and their perceptual evaluation as evidenced in liveliness estimations. *Phonetic Experimental Research, Institute of Linguistics, University of Stockholm (PERILUS)* 17, 1-34.

Traunmüller, H. and Eriksson, A. 1995. The perceptual evaluation of F0-excursions in speech as evidenced in liveliness estimations. *Journal of the Acoustical Society of America* 97, 1905-1915.

Van Heuven,V. 1987. Stress patterns in Dutch (compound) adjectives: Acoustic measurements and perception data. *Phonetica* 44, 1-12.

Van Katwijk, A. 1974. *Accentuation in Dutch*. Assen: Van Gorkum.

Zwicker, E. 1961. Subdivision of the audible frequency range into critical bands. *J. of the Acoustical Society of America* 33, 248.

Zwicker, E., Flottorp, G. and Stevens, S.S. 1957. Critical bandwidth in loudness summation. *Journal of the Acoustical Society of America* 29, 548-557.

Center for Research on User-System Interaction, Eindhoven, The Netherlands

CARLOS GUSSENHOVEN

THE LEXICAL TONE CONTRAST OF ROERMOND DUTCH IN OPTIMALITY THEORY

1. INTRODUCTION

The intonational systems of the dialects of Dutch and German in an area covering a large part of the former German Rhineland (the northern half of Rhineland-Palatinate and the southern half of North Rhine-Westphalia), Luxemburg, the northeast of Belgium, and the southeast of the Netherlands resemble those of Norwegian and Swedish: in addition to the tones contributed by the intonation, there is an opposition between two tonal word accents.* The purpose of this chapter is to give an account of the way in which the lexical tone contrast in one of these dialects, that of the city of Roermond in the Dutch province of Limburg, is realised under different intonational conditions.[1] The intonation tones with which the lexical tones combine should be divided into tones that mark the focus of the sentence, which appear in the syllable with primary stress in focused words, and tones that signal discoursal meanings, which appear at the boundaries of intonation phrases. Bruce (1977) showed that by making the appropriate comparisons, the lexical tones can be separated from both types of postlexical (intonational) tones:

> "By comparing F0-contours of words in final and non-final position, in and out of focus and with contrasting word accent out of focus, the individual F0-contributions of terminal juncture [*sc.* boundary tones, C.G.], sentence accent [*sc.* focus-marking tone, C.G.] and word accent (accent I and accent II) could be isolated." (Bruce 1977: 37).

In (1), a Swedish minimal pair is shown in a focused context (Bruce 1977, Pierrehumbert and Beckman 1988:243, Gussenhoven and Bruce 1999; for a general introduction to the phonological model presupposed here, see Ladd 1996.) Both Accent I and Accent II consist of a H+L melody, of which the L-tone is aligned with the stressed syllable (here: *an*) in the case of Accent I, while the H-tone is so aligned in the case of Accent II. Focus is marked by a H-tone, which is pronounced immediately after the lexical tones of the focused word. The contours are completed by the boundary tones of the intonation phrase, indicated by the subscript 'i'. An initial and a final boundary L_i mark the edges of the intonation phrase; the initial L_i is only realised if there is a

M. Horne (ed.), Prosody: Theory and Experiment, 129–167.

syllable available for it. As will be clear, the big dots in the contours represent the phonetic F0-targets of the phonological tones.

(1) a.

[med a n den]

| | |

L_i H+L H L_i

'with the duck'

b.

[med a n den]

| | |

L_i H+L H L_i

'with the ghost'

The focal H can be removed from the nouns by switching the focus to a preceding predicate, as shown in (2). Example (2a), for instance, is well-formed as a reply to *What are you doing with the duck?*.

(2) a.

[jag s k i l d r ar anden]

| | | |

L_i H+L H H+L L_i

'I'm DEPICTING the duck'

b.

[jag s k i l d r ar anden]

| | | |

L_i H+L H H+L L_i

'I'm DEPICTING the ghost'

Roermond Dutch can likewise be described in terms of a (movable) focal tone, a number of intonational boundary tones signalling discoursal meanings, and a lexical tone. An important general difference with respect to Swedish is that the tonal targets in the intonation contours cannot always be directly related to the tone strings arising from the concatenation of the underlying tonal morphemes. In the dialect of Roermond, phonological adjustments will need to be made to express four phonological generalisations. First, unlike Swedish, which retains the lexical tones on words pronounced without focal tone, Roermond Dutch loses its lexical tone in nonfocused, nonfinal positions, causing a neutralisation of the lexical contrast in those contexts. Second, the final boundary tones of the intonational phrase spread leftward into the phrase, creating both phrase-internal and phrase-final targets. Third, a lexical H-tone undergoes assimilation to L after a focal L-tone in the same syllable. Fourth, and most spectacularly, there is a pattern of tonal infixation which causes the phrasal boundary tones to be realised before the lexical tone, if this tone occupies the last mora of the intonational phrase.

The purpose of this chapter is to account for these generalisations in Optimality Theory (henceforth OT, Prince and Smolensky 1993, McCarthy and Prince 1993, 1995). The main motivation for choosing this theory as the descriptive framework for our data is that it would seem to be the only theory which can account for the infixation referred to above as generalisation 4. It will be shown that it is very difficult to incorporate the generalisation in a derivational description using the model of Chomsky and Halle (1968). Before presenting the analysis, the assumptions of OT are briefly laid out, together with two modifications of the theory by Pierrehumbert (1993) and Beckman (1996).

A crucial aspect in the analysis concerns the distinction that Pierrehumbert and Beckman (1988) make between the association of a boundary tone to the prosodic constituent node and the additional association of such a tone to a tone bearing unit ('secondary association').[2] It will be argued that this distinction is reproduced in OT by exploiting the two interpretations that have been given to alignment constraints, one applying in the case of morphemes and the other applying in the case of segmental features. If, in the case of boundary tones, alignment is interpreted as for morphemes, the edge-alignment of a tone makes demands on the serial order of the tone, and need not imply the association of a tone to a tone bearing unit. The adoption of this interpretation of alignment will allow us to account for the pattern of tonal infixation mentioned above.

The organisation of this chapter is as follows. First, some general phonological characteristics of the dialect and an overview of the data are presented in §2. In that same section, the underlying representations of the lexical tone, the focal tones and the discourse-marking boundary tones are given and briefly motivated. In §3, the tonally transcribed data are presented, together with the four exceptionless generalisations. §4 introduces OT as well as the two proposals by Pierrehumbert (1993) and Beckman (1996). In §5, an OT account will be offered which explains the four generalisations in four consecutive sections. The last of these, §5.4, lays out our assumptions about the role of alignment constraints. A conclusion is offered in §6.

2. GENERAL CHARACTERISTICS OF THE ROERMOND TONAL SYSTEM

Within the word, two types of restriction conspire to confine the tonal contrast to certain locations. Firstly, the contrast is restricted to the syllable with main stress. The dialect of Roermond has the same stress system as standard Dutch. Main stress falls on either of the rightmost two syllables of the word if the penult is closed, and on one of the last three rightmost syllables if the penult is open. A final syllable without main word stress can be strong, i.e. form a foot. Secondly, the contrast is restricted to syllables with two sonorant moras. Thus, the syllable rime must contain a long vowel, a diphthong, or a lax vowel followed by a sonorant consonant ([m, n, ŋ, l, R]) in order for the contrast to exist. As a result, words whose main-stressed syllable contains only a single sonorant mora have neither Accent I nor Accent II.

The distribution of the word accents is to some extent predictable from prosodic and segmental properties of the word, a survey of which is presented in Hermans (1994).[3]

In (3) some tonal minimal pairs are given. As in Norwegian and Swedish, minimal pairs are not numerous. An interesting feature of Roermond Dutch, as well as of other dialects in central and southern parts of the province of Limburg, is that some nouns distinguish plural from singular forms by having Accent I for the former and Accent II for the latter (cf. (3d,e,f)). Only the (identical) segmental transcriptions are given: their tonal characteristics depend on the intonation used (cf. below).

3)	Accent I		Accent II	
a.	[ʃo:n]	'beautiful'	[ʃo:n]	'shoe (SG or PL)'
b.	[ha:s]	'hare'	[ha:s]	'glove'
c.	[ˈɛRɣəR]	'worse'	[ˈɛRɣəR]	'annoy'
d.	[kni:n]	'rabbit-PL'	[kni:n]	'rabbit-SG'
e.	[bɛin]	'leg-PL'	[bɛin]	'leg-SG'
f.	[ɛRm]	'arm-PL'	[ɛRm]	'arm-SG'

We present an overview of the tonal data in Table I. Each cell of the matrix shows the contrast between Accent I (solid line) and Accent II (interrupted line). The two rows represent the two discoursal meanings that are expressed in the intonation of the dialect, referred to as 'Declarative' and 'Interrogative'.[4] The three columns represent different positions of the word-accented syllable in the intonation phrase. These are nonfinal focused position, illustrated in (4a), final focused position, illustrated in (4b), and final nonfocused position (4c). The vertical lines mark off the boundaries of the syllable. Thus, when occurring in focused nonfinal position (cf. (4a)), the word [ʃo:n] (Accent II) 'shoe(s)' is pronounced with a rise from mid to high, and is followed by a fall (interrupted line, first row, first column), while the same word in focused final position (cf. (4b)) is pronounced with a falling-rising intonation (interrupted line, first row, second column). Obviously, words that do not have the main stress on the final syllable, like those in (3c), lack the patterns given in the second and third columns.

	FOCUS NONFINAL	FOCUS FINAL	NONFOCUS FINAL
Declarative	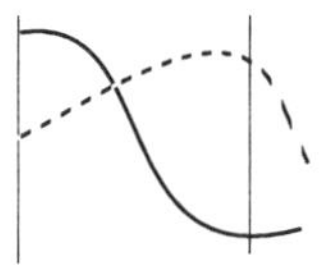	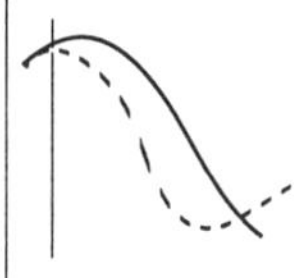	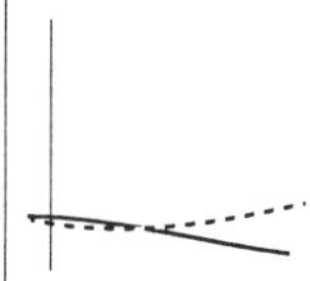
Interrogative	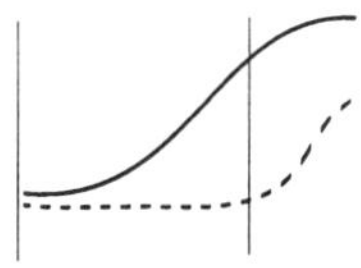	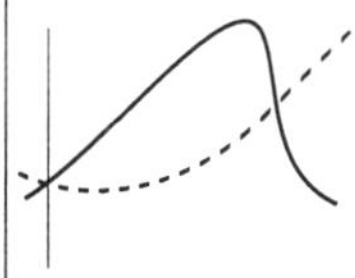	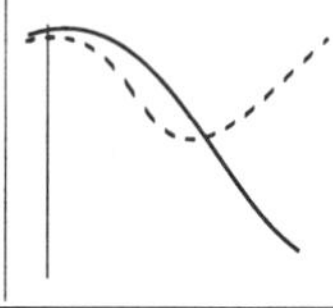

Table I. *Schematic representations of the contrast between Accent I (solid line) and Accent II (interrupted line) with two intonation patterns (rows) and three positions in the sentence (columns)*

(4) a. Ich höb 'ne SJOON gevònje *(focus nonfinal)*
I have a shoe found
'I have found a SHOE'
b. Ich höb twee SJOON *(focus final)*
'I have two SHOES'
c. Ich HÖB gein sjoon *(nonfocus final)*
'I HAVE no shoes'

The fact that a second sonorant mora needs to be present in the syllable for the word accent contrast to be possible leads one to expect that this mora is somehow involved in the representation of the contrast. If we consider the further fact that stressed syllables with one sonorant mora pattern like bimoraic syllables with Accent I, a reasonable assumption is that syllables with Accent II have a lexical tone on the second mora, while syllables with Accent I, just as syllables that have no second sonorant mora, are lexically toneless. Inspection of the 'Declarative' pitch patterns in Table I suggests that the lexical tone of Accent II is H. This representation is shown in (5) which is identical to those proposed by Hermans (1985, 1992) for the closely related dialect of Maasbracht and by Gussenhoven and van der Vliet (1999) for the more distantly related dialect of Venlo.

(5) Accent I: ${}_{F}({}_{\sigma}(\mathrm{m}\quad \mathrm{m})_{\sigma}$

Accent II: ${}_{F}({}_{\sigma}(\mathrm{m}\quad \mathrm{m})_{\sigma}$
|
H

(where m represents a sonorant mora)

The two 'discoursal' morphemes are expressed by the boundary tones L_i for 'Declarative' and H_iL_i for 'Interrogative'. Accent I, which is lexically toneless, consistently has a low ending in the 'Declarative' intonations, and a final high fall in the 'Interrogative' intonations, which patterns form the obvious motivation for assuming these boundary tones.[5] To mark focus, the dialect of Roermond uses a high tone in 'Declarative' sentences, but a low tone in 'Interrogative' sentences. These focal tones, symbolised H* and L*, respectively, associate with the first (or only) sonorant mora of the main stressed syllable of the focused word.[6] In (6), these combinations of focal tones and boundary tones are given. The 'Declarative' intonation is typically used for statements and WH-questions, and 'Interrogative' for yes/no questions.

(6)	Declarative:	H*	L_i
	Interrogative	L*	H_iL_i

Finally, an initial boundary L_i is used at the beginning of the intonation phrase. The next section illustrates the data summarised in Table I and the tonal representations introduced in this section. In the course of presenting those data, we will identify four generalisations, which we will accommodate in §5.

3. THE DATA

3.1. The neutralisation of the contrast in nonfocused, nonfinal position

The first observation to be made here is that the lexical tone contrast is neutralised unless the syllables concerned have a focal tone or if those syllables are final in the intonation phrase (Gussenhoven and van der Vliet 1999). Words with Accent II lose the lexical H in nonfocused, nonfinal positions, which makes them indistiguishable from words with Accent I. Examples (7ab) illustrate this point. While [bɛin] *bein* 'leg' forms a minimal pair with [bɛin] *bein* 'legs' when spoken in isolation, they are homophonous in the sentences given. The focused words [vø:t] *veut* 'feet' (7a) and [vo:t] *voot* 'foot'(7b) both have Accent I, while the final adjective [vas] *vas* 'attached' only has one sonorant mora, so that the example has no lexical tone.

(7) a.

[Mien VEUT^I zitte aan mien bein vas]

L_i H* L_i

my feet sit on my legs attached

'my feet are attached to my legs'

b.

[Miene VOOT^I zit aan miene bein vas]

L_i H* L_i

my foot sits on my leg attached

'my foot is attached to my leg'

We give the generalisation in (8).

(8) Generalisation 1: The lexical H is deleted unless T* or T_i is present in the same syllable.

Sentences minimally have one focus-marking pitch accent; prefinal focus-marking pitch accent will be ignored in this chapter. The data, therefore, conform to the formula in (9). The elements in brackets are optionally present: (a) the syllable with the focus-marking T* may or may not have a lexical H on a second mora, (b) a final syllable may or may not have a lexical H on a second mora, and (c) the sentence either has 'Declarative' or 'Interrogative' intonation.

(9) [… m m … m]

L_i T* (H) (H) (H_i)L_i

3.2 The nonfinal 'Interrogative' contours

The third generalisation concerns the assimilation of the lexical H-tone of Accent II in focused 'Interrogative' contexts. When L* precedes the lexical tone in the same syllable, the pitch contour is not a rise from low to high, as we might expect from the representation of the focal tone as L* and the lexical tone as H, but is fully low throughout the duration of the syllable. One possible analysis is that the lexical H changes to L when preceded by L* in the same syllable. An alternative assumption is that the L* spreads to the second mora, and that the lexical H is delinked, and hence deleted. In §5.2 it is argued that the former solution is the correct one. Example (10a) gives the contour with Accent I, which has just the L* in focal position. The representation of Accent II with lexical L instead of H is shown in (10b).

(10) a.

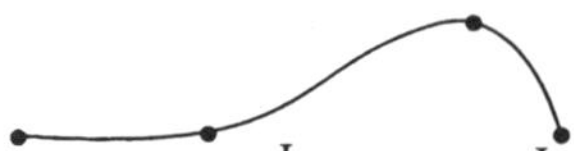

[Zitte dien BEIN^{I} aan dien veut^{I}?]

L_i L* H_i L_i

sit your legs on your feet

'are your legs attached to your feet?'

b.

[Zit diene BEIN^{II}aan diene voot^{I}?]

L_i L*L H_i L_i

sits your leg on your foot

'is your leg attached to your foot?'

The third generalisation is given in (11). According to (11), any syllable with Accent II has its lexical H on the second mora transformed into a L-tone whenever it receives the focal L* of the 'Interrogative' on its first mora.

(11) Generalization 2: Underlying 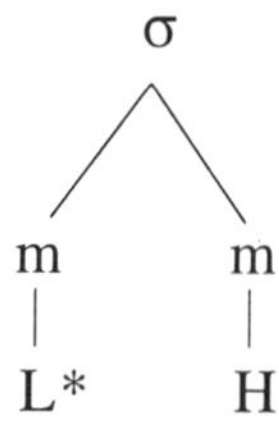

changes to 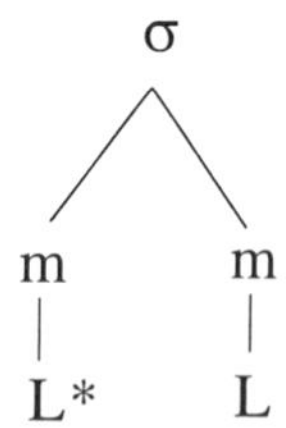

3.3 The nonfinal 'Declarative' contours

In nonfinal syllables with 'Declarative' H*, Accent I is realised as an early fall during the focused syllable (12a), while Accent II has a fall immediately after the focused syllable (12b). (The contours are well-formed replies to *What is (are) attached to your hand(s)?*) A crucial aspect in the contour of Accent I is the early, fairly steep fall. It is this feature that, in this context, signals the distinction with Accent II. However, on the assumption that tonally unspecified stretches of speech are provided with interpolations between one tone and the next, the H* on *erm* and the L_i at the end of the sentence would lead one to expect a more gradual fall. In fact, the Accent I fall is steeper than that for Accent II. This, of couse, is not predicted by the representation which we assume for Accent I in (7a,b), which merely suggests that the fall for Accent I is earlier, but not steeper, than that for Accent II. To account for the steep fall of Accent I, the final L_i is assumed to spread to the free second mora of the focal syllable. This has been shown in (12a), where the Accent I syllable [εRm] 'arms' has a free sonorant mora, and contrasts with the Accent II syllable [εRm] in (12b), which does not.

(12) a.

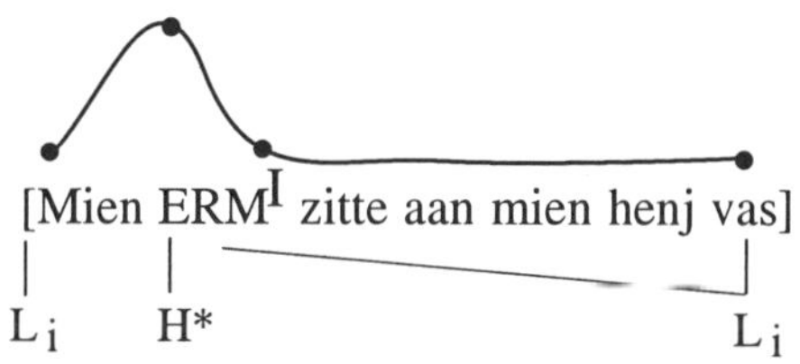

b.

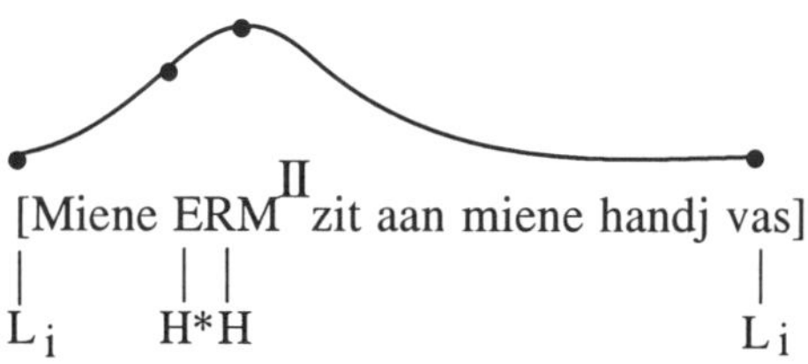

The third generalisation that we need to account for, therefore, is (13).

(13) Generalisation 3: LEFTWARD TONE SPREADING

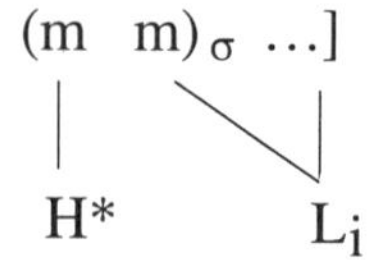

Two observations about the shape of the contour in (12b) are made. First, the focal syllable has a rise from low or mid to high, rather than a level high pitch, as would be expected on the basis of the representation with the two H-tones. It is assumed that the phonetic implementation of H* before H in the same syllable is a mid target. Second, the fall from the accented syllable does not describe a linear interpolation between H and L_i, but tends to descend more rapidly in the initial portion. The slope of this fall is less steep, however, than that in focal syllables with Accent I shown in (12a). The issue of the slope is further discussed in §5.3.

3.4 The final focused 'Declarative' contours

In final focused position, the 'Declarative' contour for Accent I, shown in (14a), has the same shape as the nonfinal contour shown in (12a): a steep fall inside the accented syllable. The contour for Accent II, however, differs from the corresponding nonfinal contour (12b): final Accent II is realised with a falling-rising contour, of which the rise only rises to mid (cf. (14b)). Clearly, a line-up of the focal H*, the lexical H and the boundary L_i does not explain why a falling-rising pattern is observed. A number of possible analyses would seem to be available. Gussenhoven and van der Vliet (1999) analyse a similar case in Venlo Dutch as the result of a tonal assimilation rule, LEXICAL H-LOWERING, which lowers a lexical H on the last mora to L before L_i. This sequence of two L-tones is held to be responsible for the lowish plateau which characterises final 'Declarative' Accent II in the dialect of Venlo. That analysis is independently supported by the behaviour of a final lexical H before the 'Interrogative' L_iH_i of that dialect, which also triggers LEXICAL H-LOWERING. However, for the Roermond dialect, that solution is not available. First, the final part of the contour is typically a rise to mid, and it is therefore different from the more level pitch of the Venlo dialect. Second, there is no independent evidence for the lowering of the lexical H, since the dialect of Roermond does not have a L_iH_i boundary sequence before which the same rule could be operative. The correct generalisation for Roermond appears to be that the boundary tone(s) are realised **before** a lexical H on the last mora. In (14b), the tones have been arranged in that order. In spite of the seeming unorthodoxy of this order, the generalisation, given in (15), is exceptionless. In addition to the final focused 'Declarative' in (14b), the final nonfocused 'Declarative' as well as the final 'Interrogative' forms with H_iL_i show the same pattern (cf. the solid contours in the third and fourth columns of Table I, as well as the examples in sections 3.5-3.7 below.)

Two comments are in order. First, example (14b) will strike speakers of most intonation languages as an interrogative intonation. It is stressed that (14b) has the **same** 'Declarative' intonation pattern as (14a); in addition to WH-questions, statements also typically have this pattern, like *D'r is get gebeurd mit diene KNIEN*^{II}*!* 'Something has happened to your RABBIT!'. Second, Roermond Dutch has a downstepping pattern which is frequently used in declarative utterances with two accented positions, in which case the second has a reduced range. Since we are dealing with one-accent contours only, downstep has been left out of our analysis.

(14) a.

[Waat is d'r gebeurd mit dien KNIEN^{I}]

L_i H* L_i

what has-there happened with your rabbits

'what has happened to your rabbits?'

b.

[Waat is d'r gebeurd mit diene KNIEN^{II}]

L_i H* L_i H

what has-there happened with your rabbit

'what has happened to your rabbit?'

(15) Generalisation 4: If the final mora of the intonation phrase has a lexical H, the boundary tones (L_i and H_iL_i) are realised before it.

Generalisation 4 cannot be accommodated in the model of tonal alignment of Pierrehumbert and Beckman, who stipulate that boundary tones associated to some node N are pronounced on the outside of tones associated to nodes dominated by N (1988:158). However, the stipulation does not follow directly from their conception of boundary tone association, but is rather based on what is known about the behaviour of boundary tones. It should be kept in mind that the number of languages that have been described in the general target-and-interpolation model of Pierrehumbert (1980) is still limited, and descriptions have not concentrated on the interaction between tone and intonation at the ends of phrases. It is conceivable, therefore, that more cases of tonal 'reversals' will be discovered. In §5, it will be shown that Generalisation 4 can very naturally be accounted for in Optimality Theory.

We have now introduced the four relevant generalisations about Roermond tonal structure. The remaining data, to be introduced in §3.5, §3.6, and §3.7, provide further evidence for Generalisations 3 and 4.

3.5 The final focused 'Interrogative' contours

In intonation-phrase final position, the focused realisation of Accent I is a rise-fall (16a). The low beginning of the rise is located inside the syllable, which distinguishes this contour from the 'Declarative' pronunciation of final Accent I illustrated in (14a), in which the rising part is located before the final syllable. The contour is readily explained by the expected tonal sequence L* H_iL_i. By contrast, the contour for Accent II is a plain rise, a contour which obviously does not result from an implementation of L* H H_iL_i which tone sequence would at first sight seem to result from the concatenation of the underlying tones. However, we have seen in the previous section that the boundary tones are located to the left of a lexical tone on the last mora. According to Generalisation 4 (15), therefore, the correct line-up is L* H_iL_i H. Secondly, we saw in §3.2 that Roermond does not tolerate a nonfinal LH contour on the two moras of a syllable. Generalisation 2 (11), therefore, stipulates that instead of L*H_iL_i H, the representation should be L*L_iL_i H, where the H_i of the boundary sequence has been replaced with L_i. This representation is shown in (16b).

General well-formedness considerations, like Clements and Keyser's (1983) TWIN SISTER CONVENTION, according to which more than one association of some element to the same site are automatically reduced to one, will simplify the tonal representation of 'leg' in (16b) to L*L_iH. Since the boundary tones evidently associate in the final syllable, we will from now on by default represent them with moraic associations instead of an association with the boundary. In section 5.4, we present a motivated case of boundary association.

3.6 The final nonfocused 'Declarative' contours

In intonation-phrase final position, the lexical H of Accent II is retained even if the syllable is not focused, and the word accent contrast can therefore be observed in nonfocused, final position. The contours in this context follow the generalisations identified so far. Example (17a) contrasts with (17b) in its final syllable only. (In both examples, the focus is on a word with Accent I, [vøːt] *veut* 'feet' and [voːt] *voot* 'foot'.) The contour with the plural form [bɛin]^{II}

(16) a.

[Is d'r get gebeurd mit dien BEIN ^{I}]

L_i L^* $H_i L_i$

has-there something happened with your legs

'has something happened to your legs?'

b.

[Is d'r get gebeurd mit diene BEIN ^{II}]

L_i $L^* L_i L_i H$

has-there something happened with your leg

'has something happened to your leg?'

shows a rise to mid pitch. This is explained by the location of the final L_i before the lexical H, in accordance with Generalisation 4.

3.7 The final nonfocused 'Interrogative' contours

The contrast in the final unfocused 'Interrogative' condition can likewise be predicted on the basis of the data discussed above. Example (18a) is like (16a). The focused [ʒw*a*R tə] *zjwarte* 'black' has a stressed syllable with a single sonorant mora, since [R] is devoiced before voiceless consonants (Kats 1939:85). The final word has Accent I. In (18b), we observe a high-mid-high contour on the final syllable, which is explained by the infixation of the boundary tones before the final lexical tone.

(17) a.

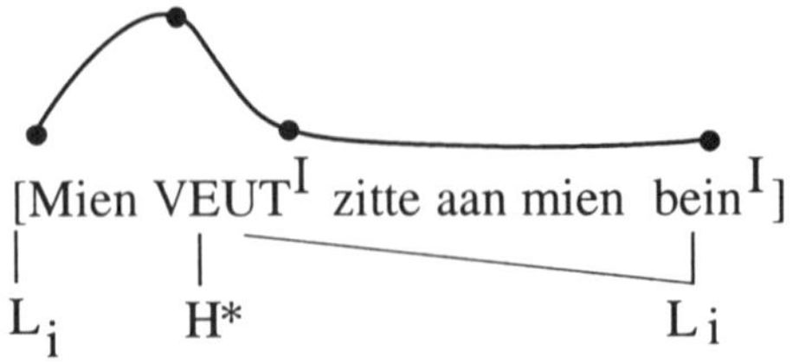

my feet sit on my legs attached

'my feet are attached to my legs'

b.

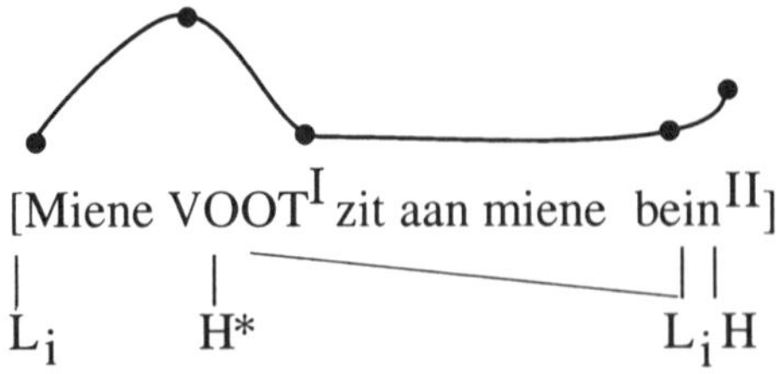

my foot sits on my leg attached

'my foot is attached to my leg'

This concludes our survey of the tonal data of Roermond Dutch. In the next paragraph, we will briefly explain why these facts do not lend themselves to a derivational account.

3.8 SPE's failure to deal with Generalisation 4

As observed in the Introduction, Generalisation 4 provides the chief motivation for the adoption of the constraint-based conception of OT. In order to show why a rule-based description is not available, it is sufficient to consider the interaction between the only two feature changing rules that would be needed to account for the data. First, Generalisation 2 (11), which requires the lexical tone to be L if L* occurs in the same syllable, could be formulated as (19).

(18) a.

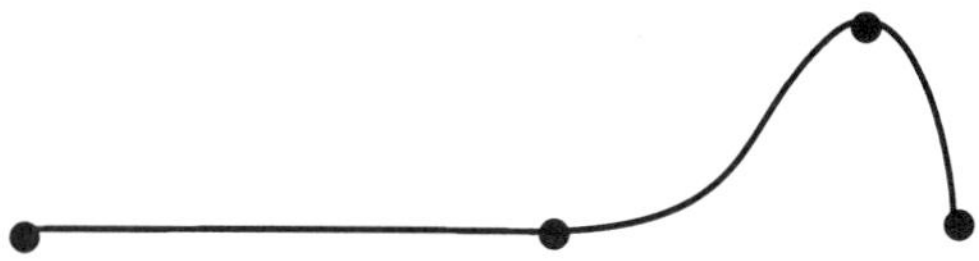

[Höbse dan ZJWARTE knien^{I}?]

L_i L* $H_i L_i$

have-you then black rabbits

'so do you have black rabbits?'

b.

[Höbse dan ene ZJWARTE knien^{I} ?]

L_i L* $H_i L_i H$

have-you then a black a rabbit

'so do you have a black rabbit?'

(19) H-ASSIMILATION $H \rightarrow L / (L^{*}\_\_\_)_{\sigma}$

Giving an SPE formulation of Generalisation 4 (15) is more difficult. We would need a metathesis rule, which changes a final tone string HHL into HLH, as in (20).

(20) METATHESIS

H	H	L	]						
1	2	3	4	$\Rightarrow$	1	3	2	4	

METATHESIS would correctly apply in three of the four contexts in which Accent II is in final position. A final focused 'Declarative' with Accent II, underlyingly H^*HL_i, is duly transformed into the correct $H^*L_i\ H$ (cf. (14b)). If the leftmost H-tone in the structural description in (20) occurs in a different syllable, the rule also applies in nonfocused syllables. These have HL_i underlyingly, and have either a lexical H to their left (if the preceding focused word has Accent II) or a focal H* (if the preceding focused word has Accent I). The output would be L_iH. Final nonfocused Accent II syllables with 'Interrogative' intonation, underlyingly HH_iL_i, are also correctly dealt with: they are transformed to HL_iH_i, accounting for the falling-rising pattern observed (cf. (18b)).

The problem, however, is posed by the final focused 'Interrogative', underlyingly $L^*HL_iH_i$. This form requires the application of both METATHESIS and H-ASSIMILATION. Derivation (21) shows that if METATHESIS applies before H-ASSIMILATION, the output is correctly derived, as shown in (21a), where parentheses stand for syllable boundaries.

(21)	Input	a.$(L^*H\ H_iL_i)$	b. $(L^*H)(H_iL_i)$
	METATHESIS	L_iH_i	L_iH_i
	H-ASSIMILATION	L	L
	Output	$(L^*L\ L_iH_i)$	$^*(L^*L)\ (L_iH_i)$

However, in (21b), where the focal syllable is not phrase-final, and underlying L*H is in a different syllable than H_iL_i, METATHESIS would apply incorrectly so as switch round the two boundary tones, this time in a syllable with Accent I. The order METATHESIS before H-ASSIMILATION, therefore, is not usable. Since the reverse order produces the wrong result in the final focused syllable with Accent II, as shown in (22a), we are stuck with an ordering paradox: while the counterbleeding order of (21) is required for a tautosyllabic input of $L^*H\ H_iL_i$, a bleeding order would be required for a correct derivation of the heterosyllabic $(L^*H)\ (H_iL_i)$, as shown in (22). It is not possible to restrict the rule to HHL-strings in final syllables, as it would then fail to apply to the nonfocused final 'Declarative' contour $(H^*H)(HL_i)$ or $(H^*)(HL_i)$, where the leftmost H is in a different syllable.

(22) Input	a.$(L^*H\ H_iL_i)$	b. $(L^*H)(H_iL_i)$
H-ASSIMILATION	L	L
METATHESIS	*not appl.*	*not appl.*
Output	$^*(L^*L\ H_iL_i)$	$(L^*L)\ (H_iL_i)$

To salvage the derivational account, METATHESIS would have to be provided with a condition 'If either the first or the second H is associated to the final mora', a description which patently fails to put the finger on the spot.

Before moving on to the constraint-based account, the principles of Optimality Theory are briefly presented in §4.

4. SOME BACKGROUND: OPTIMALITY THEORY

4.1 General

Optimality Theory (henceforth OT; Prince and Smolensky 1993, McCarthy and Prince 1993, 1995) holds that phonological surface forms (the output forms) optimally satisfy a ranked series of constraints. Rather than serving as a structure to be modified by successive applications of phonological rules, the underlying form (the input form) serves as the structure that is to be maximally preserved, subject to the prescriptions and proscriptions of a set of universal constraints. The interactions among three components in the theory ensure this result. First, a set of universal operations (GEN) generates a very large set of candidate output forms, by freely improvising on the input form, subject to general wellformedness constraints, like Goldsmith's (1976) NO CROSSING CONSTRAINT on associations. Second, a number of faithfulness constraints stipulate that the output form should be identical to the input form. Third, a set of phonological constraints specify universal phonological unmarkedness, according to which output forms must be maximally unmarked. The constraints are universal, but their ranking is language-specific. The set of output forms generated by GEN is vetted by the ranked constraints: any form that violates a constraint where other form(s) satisfy it is discarded from the set of candidates. If more than one form violates the same constraint, these all proceed to the following constraint, for as long as is needed for a single, winning candidate to emerge. Because the faithfulness constraints are typically interleaved with the phonological unmarkedness constraints, surface forms will deviate from underlying forms wherever a phonological unmarkedness constraint is in conflict with a lower-ranking faithfulness constraint. In addition, OT employs alignment constraints, which are used for instance for locating morphemes relative to their base.

GEN, a relevant formulation of which is given in Myers and Carleton (1996), is here taken for granted. The remaining paragraphs contain very brief discussions of faithfulness, markedness, positional faithfulness, and alignment.

4.2 Faithfulness

Faithfulness is expressed in terms of correspondences between the elements in the input and elements in the output (McCarthy and Prince 1995). There are three constraint 'families'. The first two are MAX-IO (Maximality: 'Every element in the input has a correspondent in the output') and DEP-IO (Dependence:

'Every element in the output has a correspondent in the input'), which militate against the deletion and insertion of phonological elements, respectively (cf. Myers and Carleton 1996). Third, in addition to the requirements that every output tone should have a correspondent in the input and that every input tone should have a correspondent in the output, we need to stipulate that a tone's input value should correspond to its output value. IDENT(T) (25) expresses this.

(23) DEP-IO: Insertion of tones is prohibited.

(24) MAX-IO: Deletion of tones is prohibited.

(25) IDENT(T): $$\begin{array}{l} *\alpha T_{\textit{input}} \\ \quad | \\ -\alpha T_{\textit{output}} \end{array}$$

These faithfulness constraints potentially come in as many forms as there are features.

4.3 *Markedness*

A general descriptive schema is that phonological markedness constraints forbid the parsing of a (valued or privative) feature F (expressed as *F, known as an anti-association constraint (Prince and Smolensky 1993)). These anti-association constraints are pitted against MAX (Feature), which requires the parsing of the input feature specified. Tableau (26) gives the input form in the top lefthand corner, lists possible output forms in the rows (in any order), and gives the constraints in the columns, ordered left to right according to dominance. In the cells, * indicates a constraint violation, ! indicates that a constraint violation is fatal, while ☞ points at the winning candidate. The effect of ranking MAX (+nasal) above *[+nasal] is that [+nasal] is parsed, i.e. that the language has nasal segments. If the reverse ranking were to obtain, the language would not have nasal segments.

(26)

[+nasal]	MAX (+nasal)	*[+nasal]
a. ø	*!	
☞ b. [+nasal]		*

4.4 Positional faithfulness

Phonological features are frequently restricted to specific locations in prosodic structure. Dutch, for instance, which contrasts voiced and voiceless obstruents in syllable onsets, does not permit voiced obstruents in the coda (e.g. Booij 1995). Likewise, the vowel system of English is largely restricted to stressed syllables, with only three vowels occurring in unstressed syllables (Bolinger 1986:347), while Shona has a five-vowel system, which is fully employed only in word-initial syllables: in non-initial syllables, mid vowels do not occur, unless preceded by a mid vowel (Beckman 1995). In like manner, prosodic features tend to shun weak and nonperipheral locations. Some Chinese dialects have tone on both weak and strong syllables in the foot, some only on strong syllables (Yip 1996). The significance of prosodic heads has been emphasized at least since Liberman and Prince (1977), while the special status of peripheries is evident, for instance, from Selkirk's (1986) treatment of vowel lengthening in Chi Mwiini or Prince's (1983) End Rules.

To account for such positional neutralisations, Beckman (1996) proposes a family of positional faithfulness constraints. She observes that the onset is a privileged position in the syllable, and that the set of privileged syllables consists of stressed syllables and peripheral syllables. The general idea is that faithfulness constraints can be restricted to these privileged locations. Thus, in Guaraní, where nasal vowels may only occur in stressed syllables, a constraint MAX-'σ[+nasal] would ensure that [+nasal] is banned from all positions except stressed syllables, provided it is ranked above *[+nasal], which in turn would be ranked above the generic MAX (+nasal). This is illustrated in Tableau (27).

(27)

'Cĩ Cĩ	MAX - ́σ [+nasal]	*[+nasal]	MAX(+nasal)
☞ a. 'Cĩ Ci		*	*
b. 'Cĩ Cĩ		*!*	
c. 'Ci Cĩ	*!	*	*
d. 'Ci Ci	*!		**

Evidently, these positional faithfulness constraints can be used to restrict the association of tones to focal and final syllables, so as to account for Generalisation 1 (8).

4.5 Alignment

An important role in our account of Roermond Dutch tone is played by Alignment constraints (McCarthy and Prince 1993). Alignment constraints determine the location of a phonological element or a (tonal) morpheme relative to the phonological or morphological structure of the expression. They do this by stipulating that the right/left edge of some element (foot, affix, tone, etc.) should coincide with the right/left edge of some other morphological or phonological constituent. For instance, the English plural suffix [z] is subject to the constraint in (28), according to which the suffix's right edge coincides with the right edge of the derived word (cf. McCarthy and Prince 1993:102).

(28) ALIGN-Z Align([z]$_{PLUR}$,R,Word,R).

Since the location of tones is determined by constraints, it is conceivable that alignment constraints for different tones compete for the same position in the output form, and that their ranking determines which tone is given precedence.

This brief outline enables us to begin our account of the data.

5. THE ANALYSIS

5.1 The first generalisation: Licensing the lexical tone

The first generalisation about our data is that the lexical tone contrast is suspended in nonfinal positions outside the focus. In such positions of neutralisation, both Accent I-words and Accent II-words act as if they are toneless, and the F0-contour is just interpolated through them. Apparently, the lexical tone must be licensed by another tone in the same syllable. There are different interpretations of the notion 'licensing' in the literature. In one view, licensing is done by 'strong' structural positions, as in Lombardi's (1991) proposal that final devoicing in Dutch, according to which no voiced obstruents can occur in the coda of the syllable, can be interpreted to mean that the feature combination [+voice, -son] is licensed by the syllable onset. Another view, proposed by Steriade (ms), is that licensing amounts to the requirement that for a phonological element to be legitimate in some position, some phonetic property should be uniquely available in that position which enhances the perception of the phonological element. The 'positional' view suggests we should account for the neutralisation of the lexical tone contrast with the help of positional faithfulness constraints (Beckman 1995), while Steriade's conception would suggest a constraint that makes the tone on a weak mora dependent on the presence of a tone on the strong mora. We opt for the conceptually simpler solution of making Max-constraints sensitive to position, as proposed by Beckman.

The account thus follows the schema outlined in §4. Constraint (29) stipulates that tone should be parsed in focused syllables, or the Designated Terminal Element (DTE) of the focussed constituent (cf. Liberman and Prince 1977), while (30) does the same for intonation phrase final positions. If these positional faithfulness constraints are ranked above a constraint banning tone (*T), which in turn is ranked above the more general faithfulness constraint MAX(T), the effect is that tones survive only in the desired syllables. Tableau (31) shows the deletion of the lexical H, where the input is a nonfinal, nonfocal syllable with Accent II.

(29) MAX-DTE(T): Associate tones in focused syllables.

(30) MAX -σ](T): Associate tones in phrase-final syllables.

(31)

(H) ...]	MAX-DTE(T)	*T	MAX (T)

a. (m m) ...]
        |
        H
		*!	

☞ b. (m m) ...]			*

The retention of the lexical tone is achieved by the same constraints. For the sake of completeness, we first introduce the alignment constraints that are responsible for locating the lexical tone and the focal tone. The presence of the lexical tone on the second mora is interpreted as resulting from ALIGNLEXRT, according to which the lexical H is aligned with the right edge of the syllable.

(32) ALIGNLEXRT - (H,R,Syl,R): The right edge of Lexical H coincides with the right edge of the syllable.

The alignment of the focal tone is governed by ALIGN-T*. Pierrehumbert (1993) proposes that in addition to edges, alignment constraints may refer to prosodic heads. Pitch accent assignment in English is one of the phenomena that is cited in support of this proposal: the starred tone of the pitch accent is aligned with the head of the foot. A similar characterisation is possible of the location of the focal tone of Roermond Dutch. Instead of saying that the tone is left-aligned in the foot, we can say it is aligned with the foot head. In fact, if we make the reasonable assumption that no language aligns its tones with the nonprominent edge of the foot, head alignment is to be preferred, since a statement about the edge alignment of the focal tone would duplicate the independently needed statement of the foot prominence edge of the language (iamb or trochee). Therefore, the Align-T* is given by (33), where 'pitch accent' is a variable ranging over H* and L* in Roermond Dutch.[7]

(33) ALIGN-T*: Align (<pitch accent>,DTE,Foot,DTE)

In Tableau (34), the two alignment constraints are included; their ranking with respect to MAX-DTE(T) cannot be established.

(34)

(H* H) ...]	ALIGN-T*	ALIGNLEXRT	MAX-DTE(T)	*T	MAX(T)
☞ a. (m m) ...] | | H* H				**	
b. (m m) ...] | | H H*	*!	*		**	
c. (m m) ...] | H*			*!	*	*
d. (m m) ...]			*!*		**

5.2 *Generalisation 2: Banning rises in nonfinal position*

Ohala (1978) cites evidence that F0 rises take longer to produce than F0 falls, and tentatively suggests that falls are perceptually more salient and make better tonal contrasts than rises. Indeed, it is not uncommon for languages to favour falls over rises. In Hausa, no syllable has rising tone (Leben 1978:206), while falls are common on bimoraic syllables. Hermans (1996) gives an account of the remarkable case of Toura (Bearth 1971), which has ten distinctive tones on bimoraic syllables. Of these, four are level tones and six are falling tones: the language does not have a single rising contour within the syllable.

If rises are less efficient than falls in both production and perception, we may expect a constraint *RISE to emerge in the constraint hierarchies of languages. It is reasonable to assume that rises, and indeed contours in general, are more readily excluded from unfavourable positions than from positions in which more time is available for their execution. Thus, rises should be more probable in phrase-final position, where the availability of final lengthening favours the execution of the F0-rise, than in phrase non-final position. Following a suggestion independently made by Jaye Padgett and Lisa Selkirk, such positionally determined degrees of markedness should not be included in the phonological markedness constraints themselves (e.g. by assuming a constraint *NONFINALRISE), but be expressed in positional faithfulness constraints, in line with Beckman's (1996) proposal. Accordingly, we split off IDENTFINAL(T), which demands the faithful parsing of the last tone of the IP and is given in (35), from the general IDENT(T) given in (25), and position *RISE between them. This has the effect of disallowing rises in all positions except when the rise is (wholly or in part) defined on the last TBU of the IP. (In (35), the square bracket stands for the IP-boundary, and in (36), the parentheses stand for syllable boundaries, as usual.)

(35) IDENTFIN(T):

$$
\begin{array}{l}
* \ \alpha T]_{input} \\
\quad | \\
-\alpha T]_{output}
\end{array}
$$

(36) *RISE:

$$
\begin{array}{cc}
(m & m)_{\sigma} \\
| & | \\
L & H
\end{array}
$$

The fully low contour in nonfinal focused 'Interrogative' syllables with Accent II, illustrated in (10b) and repeated here as (37), thus results from the ranking IDENTFIN(T) >> *RISE >> IDENT(T). Tableau (38) illustrates this interaction. In (38a), the LH contour is rejected, because it violates *RISE. MAX(T) has been included to show how deletion of the lexical tone, shown in (38b), is prevented by ranking this constraint above IDENT(T). Thus, the lexical tone is preserved, even though it can only be parsed as a L-tone. Or, put differently, instead of a deletion of the feature [-H(ightone)], shown in (38a), the correct form, shown in (38c), is characterised by its replacement with the feature [+H(ightone)]. Observe that this account implies that there are indeed two L-tones, rather than a single L* which has spread to the second mora.

(37)

[Zit diene BEIN^{II}aan diene voot?]

L_i L* L $H_i L_i$

sits your leg on your foot

'is your leg attached to your foot?'

(38)

(L* H) ...]	IDENT FIN(T)	*RISE	MAX(T)	IDENT(T)
a. (m m) ...] L* H		*!		
b. (m m) ...] L*			*!	*
☞ c. (m m) ...] L*L				*
d. (m m) ...] H			*!	

There is a fourth possible output form, not listed in Tableau (38) which is not explicitly excluded in our analysis: H*H. It is assumed that L* is more highly valued by IDENT(T) than the lexical tone: changing the value of the T* constitutes a worse violation of this constraint than changing the value of the lexical H. This indicates the need to separate MAX(T) into constraints that are sensitive to the morphological identity of a tone. This is of course no different from the morpheme specificity of alignment constraints for segmentally encoded affixes.

We now move on to an account of the third generalisation, the leftward spreading of the boundary tone.

5.3 Generalisation 3: Spreading the boundary tones

In §3.3, it was observed that the boundary L_i will spread left to a nonfinal syllable with Accent I, so as to create a low target in the focused syllable. To create the effect of leftward spreading, the boundary tones are assumed to obey two seemingly contradictory alignment constraints. ALIGNT$_i$LEFT, given in (39), requires the postfocal boundary tone to be aligned leftmost, without violating the universally undominated constraint against crossed associations. The other, ALIGNT$_i$RT, given in (40), ensures that boundary tones will always appear at the end of the intonational phrase.

(39) ALIGNT$_i$LEFT - (Ti,L): The left edge of the Phrasal boundary tones is aligned leftmost.

(40) ALIGNT$_i$RT - (T_i,R,Phrase,R): The right edge of the Phrasal boundary tones coincides with the right edge of the Phrase.

In the case of 'Declarative' L_i, the leftward alignment can be honoured with an association to the free second mora of a focal syllable with Accent I, which results in a low target inside the focal syllable. However, in focused syllables with Accent II no such association can take place, because there is no free TBU available. Strictly speaking, ALIGNT$_i$ LEFT is violated for every sonorant mora to the left of the final syllable to which the boundary tone fails to spread. Of course, the undominated positional faithfulness constraints effectively restrict the tonal phonology to the focal and final syllables, and in order not to clutter up our Tableaus, we will ignore violations of ALIGNT$_i$ LEFT that are due to a failure of T_i to spread to nonfocal syllables. As observed in §3.3, the difference between these two representations is observable in the slope of the fall, which is quite steep in (12a) (repeated here as (41a)), but less steep in (12b) (repeated here as (41b)). A further prediction is that if the focal syllable has one sonorant mora, as in words like [kɑt] 'cat' and [taːˈbɑk] 'tobacco', the slope should resemble that for Accent II, i.e. it should not be steep. This is because both Accent II and monomoraic syllables lack a free mora for the boundary tone to associate with. Phonetic evidence for these predictions is presented in Gussenhoven (1996).

In Tableau (42), the spreading of L_i to focal syllables with Accent I is shown to be the preferred output.

(41) a.

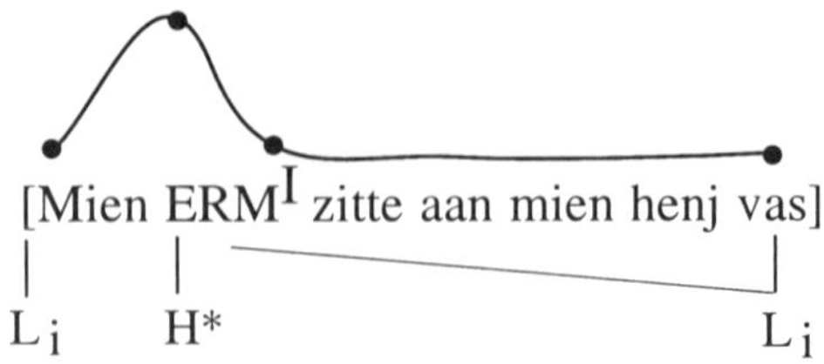

b.

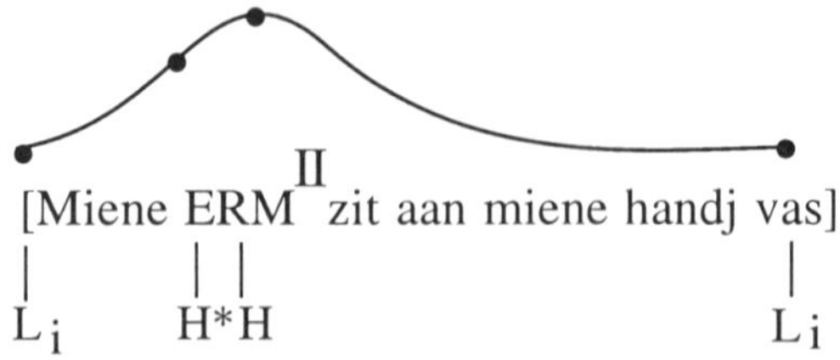

(42)

(H*) ...]L_i	ALIGNT_iRT	ALIGNT_iLEFT
☞ a. (m m) ...] H* L_i		
b. (m m) ...] H* L_i		*!
c. (m m) ...] H* L_i	*!	

It is not entirely self-evident that in a contour with Accent II or in a contour with a monomoraic focal syllable, left-alignment of L_i is impossible. After all, we do not observe a linear interpolation between the target for the H-tone and L_i, but a decidedly 'drooping' contour. Arguably, this form could be attributed to the working of ALIGNT_iLEFT, with the left-hand target creeping up to the focal syllable in a gradient fashion. Hayes and Lahiri (1991), however, make it clear in their treatment of Bengali that such early realisation of boundary tones should be left to the phonetic implementation rules. This would appear to be the correct position. That is, we should not see phonetic implementation as somehow mimicking phonological constraints. Rather, the relationship is the other way around: phonological constraints may arise when phonetic implementation gets codified in the grammar. The motivation for phonetic behaviour is efficiency in production and perception, and it is not difficult to see why an early, steeper fall is a more efficient perceptual cue that a gradual, drawn out fall. For the speaker, too, controlling steep falls may be easier than controlling gradual falls, because the time span over which the control is to be exercised is shorter. The 'droop', therefore, is the expected situation: this is what the phonetic implementation typically does. Only if such behaviour leads to categorical effects do we need the services of the ontologically derivative phonological constraint.

The contrast between Accent I and Accent II in 'Declaratives' is interestingly different from that in 'Interrogatives'. A focused syllable with Accent I followed by the 'Interrogative' boundary sequence H_iL_i does not find its free mora occupied by a left-spreading H_i. The rising movement observed in this position typically extends across the postfocal syllable(s), and is clearly not finished within the focused syllable. If left-aligned L_i associates as in (41a), then why does H_i not do so in the corresponding 'Interrogative' contour (10a), repeated here as (43)?

(43)

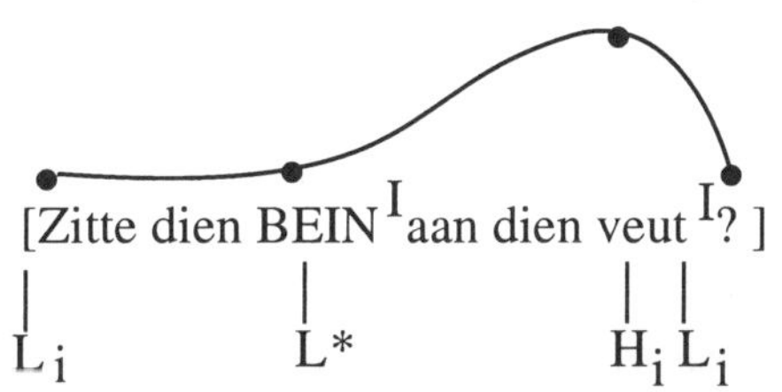

Spreading of the H_i of the 'Interrogative' boundary tones H_iL_i is prevented by *RISE. This is shown in Tableau (44), where candidate (44a), in which H_i does not associate, is preferred over the b-candidate, where it does. We cannot be happy with the c-candidate either, in which the H_i has been changed to L_i: while it escapes the censure of *RISE, it incurs a gratuitous violation of IDENT(T). Recall that in Tableau (38), the form that violates IDENT(T) was the optimal form, because it is the only one that succeeds in satisfying MAX(T). However, in Tableau (44), the a-candidate does not incur a violation of MAX(T), because H_i is preserved by virtue of its right alignment. We thus see that *RISE has two rather different effects in nonfinal syllables: in focused Accent II syllables, it forces the parsing of H as L, while in focused Accent I syllables, it prevents the H_i from spreading to the free second mora.

(44)

(L*) ...] H_iL_i	*RISE	MAX(T)	ID(T)	ALT$_i$RT	ALT$_i$LT
☞ a. (m m) …] | /| L* H_iL_i					*
b. (m m) …] | \\/| L* H_iL_i	*!				
c. (m m) …] | \\/| L* L_iL_i			*!		
d. (m m) …] | | L* L_i		*!			*

In this section, we have seen that the observance of ALIGNT$_i$LEFT causes the L_i to spread to the free mora of a nonfinal focused syllable with Accent I. The corresponding 'Interrogative' contour, in which H_i is not associated to the free mora, shows that, in addition to the earlier established ranking *RISE >> MAX(T) >> IDENT(T), it must be the case that IDENT(T) outranks ALIGN-T$_i$LEFT, which prevents H_i from changing to L_i in order to satisfy ALIGNT$_i$ LEFT. No ranking between ALIGNT$_i$RT and ALIGNT$_i$LEFT could be established.

5.4 Generalisation 4: Competing Alignment constraints

As observed earlier, the most striking feature of the tonal system of Roermond is the location of the intonational boundary tones: when the final syllable has Accent II, and therefore the last mora of the intonational phrase is occupied by a lexical tone, the boundary tones are positioned before the lexical tone, which appears as the last tone of the utterance. We intend to account for this pattern by ranking ALIGNLEXRT above ALIGNT$_i$RT. To see how this works, we need to be more explicit about what we mean by 'alignment' in these constraints. There are two possible interpretations, one modelled on the alignment of affixes or feet and one modelled on the alignment of segmental features. As an example of the former, consider the alignment constraint governing the position of the English plural suffix again, given in (28). The requirement here is that the right edge of the derived word should coincide with the right edge of the affix. In this interpretation, ALIGNT$_i$RT requires that the end of the IP and the right edge of H_iL_i and L_i should coincide. Featural alignment, however, implies not just coincidence of edges - assuming we can speak of the 'edge' of a feature – but also association to the appropriate node. If rightward spreading of [+nasal] is described by means of an alignment constraint, then that constraint is satisfied if the [+nasal] is associated with the supralaryngeal node in the feature tree of the last segment of the word. In the case of tone, both interpretations are available. If only edge coincidence is required, the tones need not be associated with the final mora(s) for the alignment constraint to be obeyed, and in this interpretation, a separate constraint is therefore needed to demand association. This ambiguity in fact accurately reproduces the distinction drawn by Pierrehumbert and Beckman (1988), who point out that tones may have more than one type of association. Boundary tones are associated with the relevant prosodic constituent node, but they also recognise a configuration whereby a boundary tone, in addition to its association with the constituent node, acquires an association with a TBU, a configuration known as 'secondary association'. Following Hayes and Lahiri (1991), I have shown association to the constituent node as association to the constituent bracket. It is proposed that it is this type of association which is demanded by an alignment constraint. Association to a TBU, required by the constraint TBU, is shown as such, but without the additional association to the constituent node. Thus, the task of alignment is to determine the location of each tone, while association to a TBU will follow if a legitimate TBU is available in the location concerned. In this view, all tones are aligned, but only some are associated. Constraint TBU (45) stipulates that tones should be associated with a sonorant mora in a stressed syllable in the output.[8]

(45) TONEBEARINGUNIT (TBU): [+son]
|
$(\ldots\mu\ldots)_{'\sigma}$

(45) can be interpreted as an abbreviation of two constraints, one requiring that tones should associate to TBUs, and one that requires that TBUs should be associated with tone. For the purposes of our description, it seems unnecessary to separate these. This means that free TBUs in focal or final stressed syllables will demand the association of tones that are aligned so as to be near them. Our assumptions vindicate the tonal representations presented in §3, because wherever boundary tones are aligned with stressed syllables, they will associate with the free sonorant moras in these syllables. For instance, in (18a), repeated in (46), the representation not only satisfies ALIGNT$_i$RT, but also TBU.

(46)

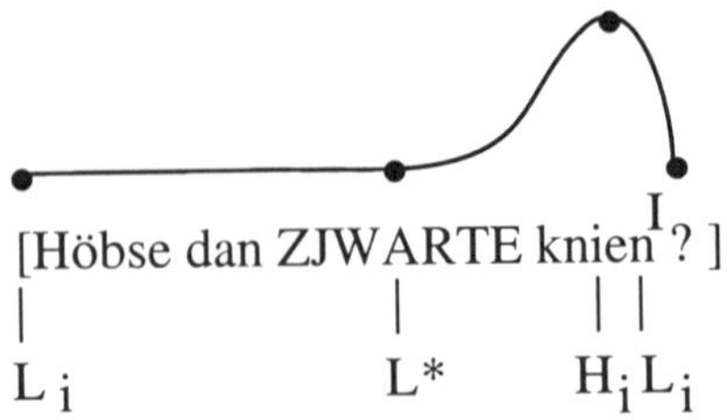

Following Antilla and Bodomo (1996), we assume that the unmarked relation between TBUs and their tone bearers is one-to-one. That is, both spreading of tones and contouring of tones need to be enforced by constraints that rank above the constraints forbidding these many-to-one configurations. These are given, again somewhat crudely, as *SPREAD (47) and *CROWD (48).

(47) *SPREAD: Do not spread tone.
(48) *CROWD: A TBU is associated with at most one tone.

We now return to the issue of the serial position of the tones in final syllables with Accent II. When the lexical H appears on a nonfinal syllable, ALIGNT$_i$RT and TBU can always be satisfied, since there are no other tones that seek to be in the same position. However, when the lexical H is on the last mora of the intonational phrase, its final position is compromised by the presence of the phrasal boundary tones. If ALIGNLEXRT is to be satisfied in phrase-final position, the lexical tone must appear in rightmost position in the tonal string. This is because the edges of phonological constituents are themselves edge-aligned with the corresponding edges of the dominating constituents in which they have a peripheral position. If a tone is to be rightmost in a phrase-final syllable, it will have to be rightmost in the phrase. Therefore, if we rank ALIGNLEXRT above ALIGNT$_i$RT, the effect is that in intonation phrase-final position, the boundary tones are inserted before the lexical tone, which is exactly the configuration we have observed in the data. Consider the final syllable in (17b), repeated here as (49), which needs to accommodate the 'Declarative' L_i in a situation in which the lexical tone is demanding to be the last tone of the syllable.

(49)

[Miene VOOT^{I} zit aan miene bein^{II}]

L_i H* L_iH

Tableau (50) shows that (50a) emerges as the winning candidate: it is the only one with the lexical tone in final position (making it better than the c and d-candidates) in which both moras are provided with tone (making it better than the b and d-candidates). The violation of ALIGNT$_i$RT is unavoidable.

(50)

(H)] L_i	ALIGN LEXRT	ALIGN T_iRT	TBU
☞ a. (m m)] L_iH		*	
b. (m m)] L_i H		*	*!
c. (m m)] H L_i	*!		
d. (m m)] H L_i	*!		*

A somewhat more complicated situation arises when three tones need to be accommodated in the final syllable, which occurs in the 'Interrogative' counterpart of (49). This nonfocus 'Interrogative' with Accent II, given in (18b) and repeated in (51), is treated in Tableau (52).

(51)

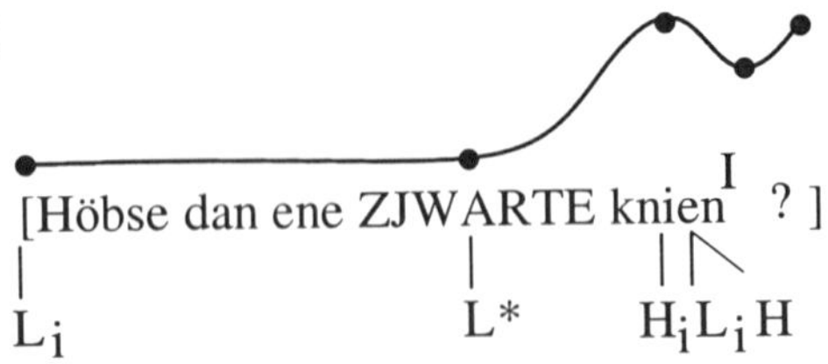

Even though there are only two moras in the final syllable, and there is a free mora available in the focal syllable, the three tones are parsed in the final syllable. Although the winning a-candidate violates *RISE in the final syllable, it manages to satisfy *RISE in the focal syllable, which constraint removes (52b) from the candidate set. The fact that this form is better than the winner by a number of lower ranking constraints, like *CROWD (not shown) and TBU, cannot save it. The c-candidate, which avoids a violation of *RISE in the focal syllable by parsing the H_i as L_i, cannot offer any relief, as it violates IDENT(T) for a tone which can safely be parsed outside the focal syllable. Candidate (52d) attempts to win by avoiding a *RISE violation altogether, but it is frustrated by IDENTFIN(T), an undominated constraint in Roermond Dutch. And avoiding any violation of *RISE by ordering the lexical H before the boundary tones of the intonational phrase, as attempted by the e-candidate, is futile, because ALIGNLEXRT is undominated.[9]

(52)

(m m) ... (m m) (L*) ... (H)] H_i L_i	LEXRT	ID FIN(T)	*RISE	ID(T)	T_iRT	T_iLT	TBU
☞ a. (m m) ... (m m)] L* H_i L_i H			*		*	*	*
b. (m m) ... (m m)] L* H_i L_i H			**!		*		
c. (m m) ... (m m)] L* L_i L_i H			*	*!	*		
d. (m m) ... (m m)] L* H_i L_i L		*!		*	*	*	*
e. (m m) ... (m m)] L* H H_i L_i	*!					*	*

There remain the final focused syllables to be accounted for. The final focused 'Declarative' with Accent II, underlyingly H*HL$_i$, will come out as H*L$_i$H, because ALIGNLEXRT >> ALIGNT$_i$RT. The treatment of final focused 'Interrogative' with Accent II represents an interesting confirmatory case for our analysis. Underlyingly, this syllable has no less than four tones to accommodate, as shown in (16b), repeated here as (53). The form shows the effects of complying with *RISE in nonfinal position, as well as ALIGNLEXRT.

(53)

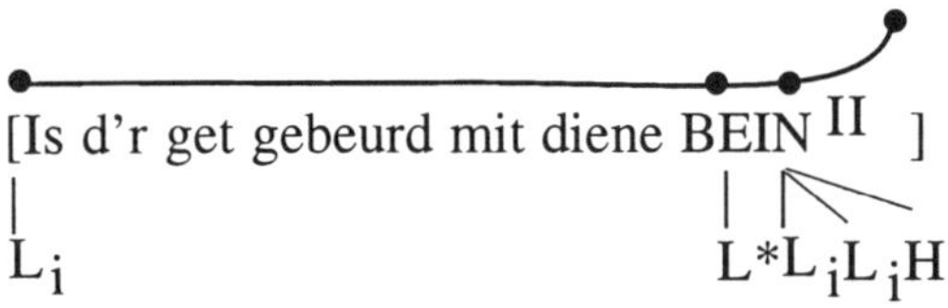

Our analysis straightforwardly accounts for this form, as shown in Tableau (54). The winning a-candidate satisfies ALIGNLEXRT, and is therefore better than (54b,c). The d-candidate has the lexical tone in the right location, but violates *RISE twice, thereby losing to (54a).

(54)

(m m) (L* H)] H$_i$ L$_i$	LEXRT	ID FIN(T)	*RISE	ID(T)	T$_i$RT	T$_i$LT	TBU
☞ a. (m m)] L* L$_i$L$_i$H			*	*	*		
b. (m m)] L* H H$_i$ L$_i$	*!*		*				
c. (m m)] L* L H$_i$ L$_i$	*!		*	*			
d. (m m)] L* H$_i$ L$_i$ H			**!		*		

As observed in §3.5, the tonal specification of winning candidate (54d) is equivalent to L^*L_iH, it being meaningless to repeat information (TWIN SISTER CONVENTION, Clements and Keyser 1983). The maximum number of tones associated to a final TBU is thus two, while the maximum for a nonfinal TBU is one. Constraint *CROWD is therefore violated in final moras, in deference to MAX(T), which outranks it. An output in which the first mora rather than the second mora of the final syllable has two tones associated to it is ruled out by ALIGNT$_i$RT, which penalises gratuitous non-rightmostness. There are of course further conceivable outputs for this form, for instance those in which the boundary tones do not associate to a mora, but are only aligned with the phrase end. All of these are ruled out by ALIGNLEXRT.

Finally, there is the final focused 'Declarative' contour to account for, given in (16a), repeated here as (55).

(55)

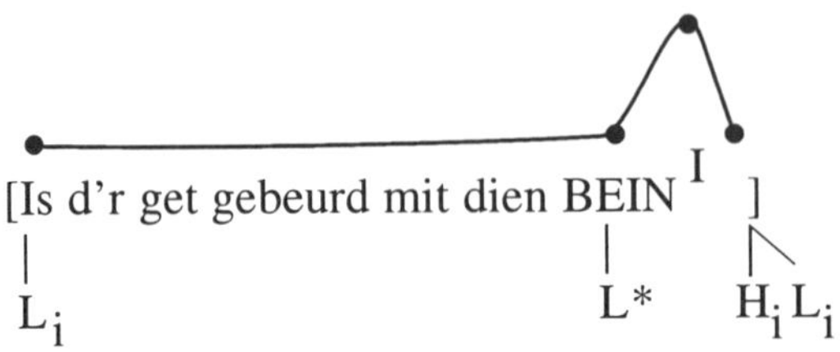

Our account forces us to assume that the boundary sequence H_iL_i here fails to associate to the final TBU. Constraint *RISE was taken to be violated if the rising movement was defined on the last mora of a syllable. In the correct a-candidate in Tableau (56), *RISE is satisfied, because H_i does not have an association with a TBU. That is, the b-candidate is to be rejected on theory-internal evidence, there being no empirical phonetic predictions to be derived from the different representations. The final focused 'Interrogative' syllable with Accent I, then, provides additional theory-internal evidence for the need to separate the alignment of tones from the association of tones to TBUs.

6. CONCLUSION

A characterisation of the surface intonational patterns of Roermond Dutch requires an account of a number of phonological adjustments of the underlying representations (our four generalisations). One result of our treatment is that a derivational description is not available. The reason for this appeared to be that the rules that express the two most important generalisations, the assimilation of H to L after L* (Generalisation 3) and the infixation of the boundary tones of the intonation phrase (Generalisation 4) or (H-ASSIMILATION and METATHESIS), would have to apply in different orders in different forms to obtain the

(56)

(m m) (L*)] H_i L_i	*RISE	MAX(T)	ID(T)	ALT_iRT	ALT_iLT	TBU
☞ a. (m m)] L* H_i L_i					*	*
b. (m m)] L* H_i L_i	*!					
c. (m m)] L* L_i L_i			*!			

correct output, an ordering paradox which is inconsistent with the premise that rules apply in the same order to all forms.

A description in the framework of OT was shown to give the desired outputs. We adopted the position of Beckman (1996), who argues that faithfulness constraints can be sensitive to locational information, such that faithfulness is valued more highly in more privileged phonological positions than in less privileged ones, where order of 'privilege' is universally given. Our analysis has yielded at least two interesting results. First, we characterised spreading as the simultaneous satisfaction of two alignment constraints with opposite directions. The distinction between boundary tones that undergo leftward spreading to the focal syllable and boundary tones that do not was put at the door of the interaction with other constraints, notably *RISE, when L* appeared in a syllable with an empty second mora, and *CROWD, when no free mora was available in the focal syllable.

Moreover, we reproduced the representational difference drawn by Pierrehumbert and Beckman (1988) between association to the prosodic constituent node and association to a TBU by assuming that an alignment constraint may make a purely string positional demand on a tone and to have the association with a TBU enforced by a separate association constraint like TBU. This enabled us to explain the pattern of boundary tone infixation given as Generalisation 4. The facts yielded to a natural analysis in which two alignment constraints seek to place their tones in the same location, with one of them winning. This conception of competing constraints makes the infixation appear as a very ordinary consequence of the organisation of the grammar. That is, OT not only makes a description available, it also in a sense explains how this situation can arise.

NOTES

*I would like to thank Lianne Mestrom for supporting my efforts to understand the phonology of the Roermond tonal system throughout the duration of this investigation. She acted as an informant, found representative speakers, and collected most of the data presented here for her MA thesis. I thank Nancy Gijsen for recording the corpus on which the investigation was based, and Stephanie van Elven for recording the examples included in this article. I am grateful to Esther Grabe, Haike Jacobs, René Kager, Will Leben, Tomas Riad and particularly to Jaye Padgett and Lisa Selkirk for their comments on earlier versions of this text. All the numbered examples of the contours in this article can be heard on the Internet (http://lands.let.kun.nl/staff/gussenhoven.en.html).

1 A map with the approximate geographical distribution of the tonal opposition is given in Gussenhoven and Bruce (1999).

2 The OT use of the term 'alignment' should be kept distinct from the term referring to the phonetic timing of tonal targets relative to anchor points in the syllable, like the voice onset or the sonorant rhyme offset.

3 There is probably also a good deal of variation, both across speakers and within speakers. Some speakers of the closely related dialect of Montfort, for instance, pronounce certain monosyllabic words (e.g. [tu:] *toe* 'closed') with Accent II in nonfinal position and with Accent I in final position (Bakkes 1996:52).

4 The 'Continuative' intonation, typically used for nonfinal intonation phrases, is the same as the 'Interrogative' intonation with the exception of the pitch on focused phrase-final syllables with Accent II, which have a fall to mid rather than a fall-rise. This difference is most probably to be attributed to a phonetic implementation rule which is sensitive to the utterance boundary, where the utterance is the prosodic constituent dominating the intonation phrase.

5 A bitonal HL boundary sequence is also used for interrogative intonation in Bengali (Hayes and Lahiri 1991) as well as in Greek, Hungarian and Romanian (Ladd 1996).

6 The 'star' is here used to indicate the morphological category 'focal' tone. Its more conventional task is that of indicating the aligning tone in a bitonal pitch accent (Pierrehumbert 1980).

7 Tones that do not have an association with either the TBU or higher-level constituents may still survive. Bruce (1987) and Gussenhoven and Bruce (1999) show that the focal H of Stockholm Swedish does not have either type

of association, but nevertheless surfaces. This is illustrated in examples (1) and (2), where the focus-marking H is shown as 'floating'. The main motivation for this representation is that the tone does not align consistently with a specific location in the prosodic or segmental structure. Rather, its location is determined by the distance to the preceding lexical H (and is therefore later after Accent I than after Accent II), and, in the case of a phrase-final word, by the distance to the end of the Intonation Phrase. The constraint MAX(T) is satisfied if the tone is realised, regardless of its association.

[8] Again, this constraint is one of a number of increasingly restrictive versions of the TBU, running from any moraic segment to moraic segments in focused syllables only. Our data do not allow us to answer the question if sonorant moras in final unstressed syllables are TBUs in Roermond. The assumption is that they are not.

[9] A conceivable output form in which the H_i and the L_i are split up so as to allow the lexical H to come between them can be ruled out by the assumption that the tones of multi-tonal morphemes are edge-aligned with each other, or by the constraint CONCAT(enate), proposed by Riad (1996) in his treatment of Scandinavian tone.

REFERENCES

Antilla, A. and Bodomo, A. 1996. Tonal correspondences in Dagaare. Ms, Stanford University and University of Trondheim.

Bakkes, P. 1996. *Variatie en Verandering in het Montforts*. Amsterdam: P.J. Meertens-Instituut voor Dialectologie.

Bearth, T. 1971. *l'Enouncé Toura*. PhD dissertation. University of Geneva.

Beckman, J. 1995. Shona height harmony: markedness and position identity. In J. Beckman, L. Walsh Dickey and S. Urbanczyk (eds), *Papers in Optimality Theory*. Univ. of Mass. *Occasional Papers in Linguistics* 18, 53-75.

Beckman, J. 1996. Positional faithfulness and the distribution of features. Handout: Univ. of Mass. (Amherst).

Bolinger, D. 1986. *Intonation and its Parts. Melody in Spoken Language*. Stanford: Stanford University Press.

Booij, G. 1995. *The Phonology of Dutch*. Oxford: Oxford University Press.

Bruce, G. 1977. *Swedish Word Accents in Sentence Perspective*. Lund: Gleerup.

Bruce, G. 1987. How floating is focal accent? In K. Gregersen and H. Basbøll (eds), *Nordic Prosody IV*. Odense: Odense Univ. Press, 41-49.

Chomsky, N. and Halle, M. 1968. *The Sound Pattern of English*. New York: Harper and Row.

Clements, G.N. and Keyser, S.J. 1983. *CV Phonology*. Cambridge (Mass): MIT Press.

Fromkin, V.A. (ed).1978. *Tone. A Linguistic Survey*. New York: Academic Press.

Goldsmith, J. 1976. *Autosegmental Phonology*. Doctoral dissertation MIT, Cambridge, Massachusetts. Published by Garland, New York, 1979.

Gussenhoven, C. 1996. The boundary tones are coming: On the non-peripheral realisation of boundary tones. Paper presented at Labphon5. Northwestern University, Evanston, 6-8 July 1996.

Gussenhoven, C. and Bruce, G. 1999. Word prosody and intonation. In H. van der Hulst (ed), *Word Prosodic Systems in the Languages of Europe*. Berlin: Mouton-De Gruyter, 233-271.

Gussenhoven, C. and van der Vliet, P. 1999. The phonology of tone and intonation in the Dutch dialect of Venlo. *Journal of Linguistics* 35, 1-47.

Hayes, B. and Lahiri, A. 1991. Bengali intonational phonology. *Natural Language and Linguistic Theory* 9, 47-96.

Hermans, B. 1985. Het Limburgs en het Litouws als metrisch gebonden toontalen. *Spektator* 14, 48-70.

Hermans, B. 1992. On the representation of quasi-long vowels in Dutch and Limburgian. *Linguistics in the Netherlands 1992*. Amsterdam: John Benjamins.

Hermans, B. 1994. *The Composite Nature of Accent: With Case Studies of the Limburgian and Serbo-Croatian Pitch Accent*. Tilburg: Katholieke Universiteit Brabant.

Hermans, B. 1996. Tone in Toura. Paper presented at the Linguistics in the Netherlands Meeting. Utrecht, 20 January 1996.

Kats, J. 1939. *Het phonologisch en morphologisch systeem van het Roermondsch dialect.* Roermond: Romen.

Ladd, D.R. 1996. *Intonational Phonology*. Cambridge: Cambridge University Press.

Leben, W.R. 1978. The representation of tone. In V. Fromkin (ed), *Tone a Linguistic Survey.* London: Academic Press, 177-219.

Liberman, M. and Prince, A. 1977. On stress and linguistic rhythm. *Linguistic Inquiry* 8, 249-336.

Lombardi, L. 1991. *Laryngeal Features and Laryngeal Neutralization.* PhD dissertation. Univ. of Mass., Amherst.

McCarthy, J.J. and Prince, A. 1993. Generalized alignment. In G. Booij and J. van Marle (eds), *Yearbook of Morphology*. Dordrecht: Kluwer, 79-153.

McCarthy, J.J. and Prince, A. 1995. Faithfulness and reduplicative identity. In J. Beckman, L. Walsh Dickey and S. Urbanczyk (eds), *Papers in Optimality Theory.* Univ. of Mass., *Occasional Papers in Linguistics* 18.

Myers, S. and Carleton, T. 1996. Tonal transfer in Chichewa. *Phonology* 13, 39-72.

Pierrehumbert, J. 1980. *The Phonetics and Phonology of English Intonation.* PhD dissertation MIT. Published 1990, New York: Garland.

Pierrehumbert, J. 1993. Alignment and prosodic heads. *Proc. Eastern States Conference on Formal Linguistics Graduate Student Association.* Cornell, New York, 268-286.

Pierrehumbert, J. and Beckman, M. 1988. *Japanese Tone Structure*. Cambridge, Mass.: MIT Press.

Prince, A. 1983. Relating to the grid. *Linguistic Inquiry* 14, 19-100.

Prince, A. and Smolensky, P. 1993. *Optimality Theory. Constraint Interaction in Generative Grammar*. Rutgers University and University of Colorado.

Riad, T. 1996. Remarks on the Scandinavian tone accent typology. *Nordlyd* 24, 129-156.

Selkirk, E. 1986. On derived domains in sentence phonology. *Phonology Yearbook* 3, 371-405.

Steriade, D. (Ms.) Positional Neutralisation. UCLA. Incomplete version October 1995.

Yip, M. 1996. Feet, tonal reduction and speech rate at the word and phrase level in Chinese. Paper presented at the Phrasal and prosodic phonology Workshop, Utrecht, July 1-2, 1996.

Dept. of English, University of Nijmegen, Nijmegen, The Netherlands

MARY E. BECKMAN AND K. BRETONNEL COHEN

MODELING THE ARTICULATORY DYNAMICS OF TWO LEVELS OF STRESS CONTRAST

1. SPEECH TIMING CONTROL FOR PROSODIC PROMINENCE CONTRASTS

One of the most intractable problems in the description of utterance prosody is that of understanding speech timing control. Gösta Bruce's seminal monograph on Swedish word accents in sentence context (Bruce 1977) highlighted an aspect of this problem that is crucially important for the perception of intonational contrasts in many languages — that of modeling what speakers do when they utter a given tune for a sentence and coordinate it with the string of words that constitute its text. By decomposing the tune into independent contributions from word accent, focal phrase accent, and final boundary events, Bruce was able to show that a robust correlate of the contrast between Accent I and Accent II involves a difference in timing, relative to rhythmically critical events in the text, of tonal events that are common to both word accent types. While much work remained to be done (cf. Bruce 1990 and the literature reviewed there), Bruce's careful attention to prosodic structure, both at the level of lexical contrasts and at the level of focal prominence contrasts for the phrase or utterance as a whole, allowed him to build a phonological model of Swedish intonation contours that could generate precise, testable quantitative predictions. This work was a foundation stone for Pierrehumbert's later phonological model for English intonation (Pierrehumbert 1980; Liberman and Pierrehumbert 1984; Beckman and Pierrehumbert 1986), a model which produced a synthesis system capable of generating by rule the full range of grammatical intonations for English (Anderson et al. 1984). Summarizing such developments in intonation synthesis over the last fifteen years, we can say that Bruce's work on the Swedish model has inspired, directly or indirectly, implementable phonological models of intonation systems for many languages (see reviews in Ladd 1992, 1996; Pierrehumbert, This volume). The English model, in particular, underscores the importance of accurate tonal timing, since there are pairs of accent types that are like the Swedish word accents in contrasting just in tune-text alignment (Pierrehumbert and Steele 1989), and at the same time, there are differences in timing for the same accent type when it is coordinated with rhythmic events at different structural positions in the text (Silverman and Pierrehumbert 1990).

M. Horne (ed.), Prosody: Theory and Experiment, 169–200.

In this paper, we will discuss another important aspect of this problem of speech timing control — that of modeling what speakers do to produce the rhythmically critical events of the text to which they align the tonal elements. We will examine the timing of consonants and vowels in three different types of English syllable. We will compare the stronger first syllable of a trochaic word with its weaker second syllable, and we will compare the strong first syllable when the word is in focus to the same syllable when the focal accent is elsewhere in the sentence. This second, higher level of prominence contrast is one that is marked by the placement of tonal elements analogous to the Swedish phrase accent. Thus, the hierarchy of rhythmic prominences at these two levels is roughly analogous to the hierarchy that Bruce (1977) proposed to accommodate the different prominences of word accent versus sentence accent in Swedish. Restating the problem in terms that would have been more familiar to speech scientists in 1977, we will be concerned here with the control of segment duration as a correlate of English stress contrasts at the two levels of "word stress" and of "sentence stress", or the two levels of [±heavy] and [±intonation] in Vanderslice and Ladefoged's (1972) typology.

It has long been known that there are marked differences in segmental timing control associated with differences in prosodic prominence for at least one of these levels. Already by the time of Bruce (1977), there had been many studies showing that "stress" can affect duration. That is, when acoustic intervals are measured from spectrograms or waveforms, the interval associated with a vowel is typically longer when the vowel is in the stressed syllable of its word. For example, Fry (1955) showed that in natural citation form productions of noun-verb pairs such as *súbject* versus *subjéct*, the ratio of first syllable duration to second syllable duration was considerably larger in the noun.

Several studies also demonstrated the perceptual effectiveness of these durational effects. For example, Fry (1958) synthesized tokens of words such as *subject*, in which he lengthened or shortened the steady state portions of the vocalic intervals to match the mean vowel duration ratios observed earlier in the natural productions, and presented these to speakers to identify as "noun" or "verb". In one set of presentations, he kept the fundamental frequency at a constant 120 Hz, and crossed the manipulation of vowel durations with an analogous manipulation of dB levels for stylized square-wave vowel RMS amplitude contours. In another, he combined the duration manipulation with sixteen different "intonations" created by concatenating low or high F0 levels and rising or falling glissandi over the two syllables. In the first experiment, the duration manipulation had a greater effect than the amplitude manipulation on the percentage of responses of "noun" (i.e. first syllable more stressed). In the other experiment, patterns that could be interpreted as stylized versions of actual common intonation patterns strongly biased the responses to the duration continuum. For example, pattern B had a low level F0 on the first syllable followed by a high falling glissando over the second. This shape might have been interpreted as a stylized version of a rise-fall nuclear configuration on the second syllable (i.e., L+H* L- L% in the ToBI transcription system — see Beckman and Hirschberg 1994), and it elicited no more than 38% "noun" judgments even for the duration ratio that matched the mean for the noun tokens in Fry (1955). By contrast, Pattern E had a rising glissando over the first syllable and a high level F0 on the second, which might have been interpreted as a list-like rising configuration on the first syllable (i.e., L* H- L% in the ToBI system), and it elicited more than 50% "noun" judgments, even for the most verb-like duration ratio.

Fry interpreted his results as evidence that the duration ratio "can swing listeners' perception from the first to the second syllable" in these one-word utterances, but that the "sentence intonation is an over-riding factor in determining the perception of stress and that in this sense the fundamental frequency cue may outweigh the duration cue" (Fry 1958: 151). Bolinger concluded similarly from the results of his own less controlled experiments "that in the duration-pitch complex it is pitch that primarily signals accent," although he went further to interpret the durational difference as only an indirect cue, saying: "I therefore assume that duration is ancillary. Figuratively speaking, it is there IN ORDER TO make room for the accent" (Bolinger 1958: 45). Others, however, were not so careful as Fry to differentiate the two levels of word "stress" and sentence "intonation" (or "stress" and "accent" in Bolinger's terms). Since Fry's more controlled studies used words produced in isolation, his experiments were easily misinterpreted as showing that a higher fundamental frequency or a more pronounced frequency rise is a direct correlate of a word's stress pattern. And there has not been an extended direct test of Bolinger's more radical suggestion that duration itself is only an indirect correlate to word stress in English, with longer durations originally being produced in citation forms because a pitch accent needs "room" to be realized [although see Lyberg (1979) and Öhman et al. (1979) versus Bannert (1982) and Bruce (1983) for the comparable claim for Swedish].

Many later studies of the durational correlates of stress also examined words produced in isolation or in fixed citation frames and thus could not differentiate between the two levels of stress contrast any better than did Fry's results. That is, it is impossible to tell from these experiments whether the first vowel in *súbject* is longer relative to the second because the first syllable bears the nuclear accent in the intonation contour, or because the prosodic word consists of a single bisyllabic stress foot headed by an initial heavy syllable. And it was not until decades later that Huss (1978) showed that the durational effects are measurable (albeit not reliably perceptible) in the intonational tail of a utterance when narrow focus of attention places the sentence stress on an earlier word. Thus, from Huss's study, we could conclude that there is a durational effect associated with the lower level of stress contrast that is defined by syllable weight, although this did not rule out the possibility of an additional lengthening effect of prominence at the higher level of stress contrast that is defined by association to the nuclear pitch accent. There were suggestive observations of the latter from small anecdotal studies of "emphatic stress" for speech synthesis (e.g. Coker et al. 1973) and from some later, more careful, reiterant speech studies by Nakatani and his colleagues (see Nakatani and Aston 1978, described in Beckman 1986: 60-62, and Nakatani et al. 1981). However, since none of these studies controlled for the intonation patterns actually produced by the talkers, it is unclear how (or even whether) English speakers manipulate segmental durations to make the rhythmic prominences which anchor the focal pitch accents of the intonation contour to the text.

2. THE PROBLEM OF DURATION AS A PHONETIC REPRESENTATION OF TIMING CONTROL

The situation is further complicated by the fact that acoustic interval durations measured for consonants and vowels differ for many other reasons as well, and not just as a correlate of the rhythmic prominences that anchor focal accents in the intonation contour. For example, in many languages, other things being equal, low vowels tend to be inherently longer than high vowels (see Lindblom 1967 for an important early account of the dynamics of this effect). Also, in English and several other languages, vowels are substantially longer before voiced obstruents, particularly if the triggering consonant is in the same syllable or stress foot (see, *inter alia*, House and Fairbanks 1953; Port 1981; Laeufer 1992). Segment intervals are also longer for the vowel and any following consonants of a phrase- or utterance-final syllable relative to those of a matched non-final syllable (see, e.g. Lehiste 1972), and the effect is larger at discourse segment boundaries (e.g. Lehiste 1975).

Klatt (1976) catalogued these and more than a dozen other robustly documented durational effects, many of which he incorporated directly into a set of synthesis rules for proportionally stretching and shrinking minimum acoustic interval durations inherently specified for each consonant and vowel phoneme (Klatt 1979). Klatt decided which durational effects to incorporate into his rule system by looking at perception tests which had manipulated durations in the same way as his synthesis model (e.g. Fry 1958; Denes 1955; Nooteboom 1973), and also by comparing the effects to the just noticeable differences that psychoacousticians have documented for simple stimuli such as noise bursts and steady state tones (e.g. Creelman 1962; Abel 1972). Thus, implicit in Klatt's rules is the notion that human speech producers control the durational effects of stress and phrasing as a direct manipulation of segment interval duration. Inasmuch as Klatt's rules have been the basis for nearly all subsequent rule systems for generating durational effects by machine (including the nonce rules that phoneticians use to create experimental stimuli for testing hypotheses about prosodic structure — see, e.g. Gussenhoven and Rietveld 1992), we might call this the dominant paradigm for thinking about how speakers control consonant and vowel timing and how hearers perceive the speech rhythms that result.

We can point to many significant outcomes of work in this paradigm. Paramount among them is a model of the spectral correlates of stress as indirect artifacts of the durational differences between more and less prominent syllables at different levels of the stress hierarchy. The seminal study is Lindblom (1963). Using a regression model that related vowel formant measures to log vowel duration in symmetrical CVC syllables, Lindblom tested the hypothesis that all vowel reduction effects — including the centralization seen in CVC contexts (e.g. Stevens and House 1963) and the morphophonological alternations between unstressed [ə] and stressed full vowel in pairs such as *súbject* versus *subjéct* in English — could be explained as an unprogrammed "undershoot" of a constant underlying vowel target that is isomorphic with the linguistic category. Based on his modeling results, he made the following proposals concerning the spectral and durational targets that the speaker controls:

> As a vowel becomes shorter, there is less and less time for the articulators to complete their 'on-' and 'off-glide' movements within the CVC syllable... [T]he talker does not adjust the control of his vocal tract at fast rates to compensate for its response delay... His strategy of encoding ... presupposes that the listener is able to correct for [such] coarticulation effects... [T]iming is the primary variable in determining the reduction of sounds and... the articulatory imprecision or laxness that may hypothetically be associated with reduced stress can be neglected in this connection.... (Lindblom 1963: 1780-1781)

In other words, the speaker produces an underlyingly invariant systemically-determined vowel formant target within a prosodically variable extrinsically programmed time period. In a CVC context, the transition times between consonant and vowel formant targets are an invariant byproduct of physiologically determined articulator dynamics. Observed variation in the vowel spectrum is then generated as an unprogrammed artifact of shrinking the vowel's interval to a duration less than the CV and VC transition times.

Lindblom's demonstration that vowel reduction can be modeled to a good first approximation as an artifact of speech timing control has had far-reaching consequences for our models of phonological categories and processes in general, including our understanding of how vowel reduction relates to other sorts of fast-speech phenomena (see, e.g. Browman and Goldstein 1990; Beckman 1996). More immediately, Lindblom's assumption of an invariant underlying dynamics with raw duration as the primary variable of timing control was a necessary precondition for the development of the technique of concatenative speech synthesis (e.g. Olive 1977). The specific model assumptions also have allowed us to tap sophisticated stochastic training tools to build speech recognition systems capable of differentiating stress-related durational effects from final lengthening, if the segment identities are known (e.g. Wightman and Ostendorf 1994). On the other hand, results of other experiments suggested that speakers do not behave in the same way as the standard duration synthesis rules, and that hearers do not process the durations that speakers produce in the same way that our recognition models do. For example, Carlson et al. (1979) found that manipulating duration alone to match the observed durational relationships does not always work to create the most natural percept. Thus, duration per se may not be a cue to prosodic structure in human speech recognition. Klatt himself pointed out a "paradox" in thinking of durational effects as cues in human speech perception when he said:

> In order, e.g., to perceive the lengthening at a phrase boundary as a cue to the end of a constituent, it seems necessary to know the identity of the lengthened segments. Durations are lengthened relative to the inherent durations for the segments in question. But a listener can't know the inherent duration until he has identified the vowel, so we have a version of "the chicken or egg" paradox. (Klatt 1976: 1220)

The lesson that Klatt drew from this paradox, however, was not that we should reevaluate duration as a direct phonetic representation of speech timing control. Rather, he was optimistic that the number of alternative interpretations of any ambiguous duration value would be few enough that human speech decoders might develop heuristic strategies for processing all alternatives simultaneously.

Other speech scientists drew a different conclusion. Harris (1978), for example, found that durational differences produced by orthogonal manipulations of "stress" and tempo were not associated with identical differences in vowel formant frequency (or in identical differences in average electromyographic activity level measured from a relevant extrinsic tongue muscle). Harris also cited cineradiographic studies by Kent and Netsell (1971) and Kuehn and Moll (1976) for evidence that stress and speaking rate are not associated with constant articulator velocities. She suggested that a vowel's length is not an ambiguous cue to its stress or to any other linguistic feature of the vowel and its context, because duration is not a primary phonetic parameter to be directly controlled by either the speaker or the listener. Rather, each linguistically relevant durational effect might be produced by a different underlying physiological mechanism, which gives the effect an unambiguous relationship between vowel duration, on the one hand, and formant frequency extremum value and transition velocity, on the other. In other words, what humans generate and recognize is not the static segmental property of acoustic interval duration, but the dynamic gestural property of spectral change over time.

Summers (1987) further tested this notion of a unique "acoustic signature" which a listener can identify directly with the linguistic structures which the speaker aims to produce. He examined jaw height and first formant frequency traces in a study of two durational effects that Klatt had incorporated into his rules — namely, the longer durations of vowels before voiced obstruents and the longer durations of vowels in "stressed" syllables. He found differences between the jaw movement kinematics associated with longer versus shorter vowels. However, the pattern of kinematic differences when the vowel length effect was associated with the stress contrast was not the same as the pattern of differences when the vowel length was associated with the voicing contrast on the following consonant. Moreover, the different patterns for jaw kinematics were related to different patterns for first formant frequency at the middle of the vowel and for first formant slope and cut-off frequency during the CV and VC transitions. The relationship between articulatory and acoustic patterns was complicated, presumably because of the many sources of nonlinearity. For example, jaw height does not predict degree of constriction in the oral cavity exactly because tongue body height is only loosely coupled to jaw height, and oral constriction degree does not predict first formant frequency exactly because there is acoustic coupling between the vocal tract and the trachea as soon as the glottis begins to open in a voiceless stop. Nonetheless, the formant differences associated with the kinematic differences were perceptible as cues to the stress and following stop voicing, as demonstrated in a companion perception test (Summers 1988). Summers's experiments thus suggest that looking at associated measures of articulatory kinematics may be a good way to begin to understand the durational effects associated with stress contrasts at different levels of the prosodic hierarchy.

3. THE KINEMATICS OF LENGTHENING FOR ACCENTED SYLLABLES

Summers (1987) does not report the intonation patterns produced by his talkers. However, the target words were nonsense monosyllables such as *bap* [bæp] versus *bab* [bæb] imbedded in the frame sentence *I'll be at the bus _____ up the street*, and the instructions were to put "sentence stress" on the target or on the preceding *bus*. Therefore, we can be reasonably confident in interpreting the effect of stress in his study as indicative of the contrast in rhythmic prominence between nuclear-accented and postnuclear deaccented positions in the intonation contour. Also, the kinematic patterns that Summers found are very similar to the patterns for accented versus unaccented disyllables in our own later studies using the real words *pop* and *papa* imbedded in mini-dialogues designed to elicit a controlled variation in prosodic context (Beckman et al. 1992; Beckman and Edwards 1994). Our first study compared lengthening for accent to phrase-final lengthening. The target articulations were the jaw opening and closing movements in the [pɑp] sequences of *pop* and *papa* in the frame sentences listed in the top part of Table 1. The first and second sentences allowed us to compare the target sequence in accented syllables that are either final to the intonation phrase or separated from the phrase boundary by the light following syllable. The second and third sentences then let us compare non-final sequence tokens which are either in accented or in unaccented syllables. Our second study compared lengthening for accent to the lengthening associated with a lower level of the prominence hierarchy. The target articulations were the jaw opening and closing movements in the two [pVp] sequences of *papa posed* in the two frame dialogues listed in the bottom part of Table 1. Between the two dialogues we could compare tokens of the word-initial [pɑp] target sequence in nuclear accented versus post-nuclear unaccented positions in the intonation contour. Within the second dialogue, we could compare tokens of the unaccented but heavy [pɑp] target sequence versus the immediately following reduced [pəp] target sequence. Across these two studies, therefore, we had a continuum of durations, from the very long jaw opening and closing movements into and out of the phrase-final accented [pɑp] of *Pop, opposing* ... to the very short movements into and out of the reduced [pəp] of *HER papa posed* ....

As predicted by Lindblom's undershoot model, this continuum of differences in duration was associated monotonically with differences in movement amplitude. That is, looking at the size of the displacement between lowest jaw position in the vowel and highest jaw positions in the adjacent [p] closures, we found that the longer movements of the phrase-final syllable in sentence 1a were not much different from the shorter movements of the non-final syllable in 1b, but much larger than the shortest movements of the [pəp] in dialogue 2b, as if there were undershoot of the jaw minimum when the closing movement into the following consonant came too soon after the initiation of the opening movement into the vowel. However, the concomitant differences in movement velocity were not compatible with this interpretation of the differences in movement amplitude. The shorter jaw opening and closing movements of the non-final target syllable of sentence 1b were faster than those of the final syllable of sentence 1a, but the shorter movements of the unaccented target of sentence 1c were, if anything, very slightly slower than the longer movements of the accented target of sentence 1b, and at the lower level of the stress hierarchy, the shorter movements of the less prominent [pəp] sequence were considerably slower than those of the much longer full-vowel sequence. Thus, shorter vowel lengths were associated with a constant or a smaller jaw movement amplitude, but contra the predictions of Lindblom (1963), the relationship to velocity was not monotonic, so that the velocity differences could not be interpreted in terms of physiologically determined limits on speeding up the articulators for a more closely spaced sequence of segment targets for the vowel and consonant.

L+H* L- H%
Sentence 1a: Pop, opposing the question strongly, refused to answer it.
(1)

L+H* L- H%
Sentence 1b: Papa, posing the question loudly, refused to answer it.
(2)

L+H* H+!H* L-
Sentence 1c: Papa posed the question loudly, and then refused to answer it.
(3)

Dialogue 2a: Q: Did her mom pose a problem as far as their getting married?
L+H* L- L%
A: Her **PAPA** posed a problem.
(4)

Dialogue 2b: Q: Did his dad pose a problem as far as their getting married?
L+H* L- L%
A: **HER** papa posed a problem.
(5) (6)

Table 1. *The three sentences used in Beckman et al. (1992) and the two dialogues used in Beckman and Edwards (1994). The intonation pattern is transcribed above the line using the ToBI system (Beckman and Hirschberg 1994), and the double underlining marks the target [pVp] sequences. The three sentences contrast (1) phrase-final, (2) nonfinal accented, and (3) unaccented sequences, whereas the two dialogues contrast (4) accented, (5) unaccented heavy, and (6) reduced-vowel sequences.*

The task dynamic model of Saltzman and colleagues suggests a different possible explanation for these kinematic patterns (see, e.g. Saltzman 1986, Saltzman and Munhall 1989). Suppose the jaw movements are direct reflexes of phonologically programmed oscillatory "gestures" that settle to a stable target position after a period of movement from the last specified position. That is, suppose that jaw opening and closing movements vary in speed and duration because the speaker controls the targeted endpoints and settling times for two dynamically specified motor routines. These routines might be the stored patterns for alternating between a high jaw position during the labial closure of [p] and a low jaw position during the wide-open oral cavity of [ɑ]. Placing this theory of speech motor control in the context of a psycholinguistic theory of speech production (Levelt 1992), we might think of the broad task specifications of making a labial closure for [p] and a wide-open oral cavity for [ɑ] as invariant phonological features of a word — that is, features which can marshal the relevant motor routines immediately upon lexical retrieval. The rhythmic patterns appropriate for accented versus unaccented and final versus non-final positions, on the other hand, cannot be invoked until the intonation is known. When the invariant lexical specifications for labial closure and so on are composed with the rhythmic structures specified for the intonational frame, the articulatory planning routines provide more specific values for the targeted displacements and durations of the [pɑ] and [ɑp] gestures, values that are appropriate for generating the distinct acoustic signatures that signal such facts as position in the intonational phrase or association to a pitch accent.

Interpreting the observed articulator kinematics in terms of such a contextual specification of gesture-internal dynamics, then, we have not one, but two different mechanisms to explain any given pattern of velocity differences across two prosodic contexts. The first involves the control of settling time, and is manifested in the relationship between movement velocity and movement length. Gestures that are programmed to reach a target constriction in a shorter time should show faster movement velocities. In phrase-final syllables, there is little or no concomitant difference in the movement amplitude. Therefore, the longer durations of phrase-final movements can be interpreted simply as a targeted slowing down of gestures late in the syllable. The other explanation involves the control of the endpoint target, and is manifested in the relationship between movement velocity and amplitude (see Ostry and Munhall 1985). For a given duration, starting at a preceding position that is at a more extensive displacement from target requires a faster movement. We can use this relationship to explain why the jaw movements into and out of accented syllables were not slower than those in unaccented syllables despite being longer. Suppose that the speaker targets a decrease in articulator speed to make the accented syllable longer. That is, suppose that the lengthening of accented vowels is specified in exactly the same way as phrase-final lengthening. But recall that accented syllables show larger movement amplitudes as well as longer movement durations. Since the tongue body rests on the jaw, and the vowel target is lexically specified as an open [ɑ], a larger jaw movement into the vowel could mean that the speaker intends to produce an even more open vowel, closer to the periphery of the vowel space. This is in keeping with de Jong's (1995a) finding that lingual articulations result in constrictions closer to the phonemic target for segments in stressed syllables. Given the relationship between movement amplitude and velocity, we would expect the targeted increase in jaw opening to be accompanied by an increase in movement speed. Thus, one explanation for the kinematic patterns that we observed (i.e. longer, larger and slightly faster movements in accented syllables) is that the increase in velocity to achieve the displacement to the intended vowel constriction target usually is large enough to offset a programmed decrease in velocity to achieve the targeted lengthening of the vowel. This is essentially the explanation that Kelso et al. (1985) give for the patterns in their reiterant speech corpus. The longer "stressed" syllables had generally faster as well as larger opening movements relative to the "unstressed" syllables, but the regression lines relating movement velocity to movement amplitude were different. The data points for the "stressed" syllables lay to the right of those for the "unstressed" syllables, so that using the relationship observed for movements in one type of syllable to model the velocities of movements in the other would have predicted even faster speeds than those actually observed in the stressed syllables or even slower speeds than those actually observed in the unstressed syllables.

However, there are two difficulties with this account of lengthening for accent. First, Kelso et al. (1985) do not report the intonation patterns or the vowel qualities that their talkers produced, but judging from the text that the talkers mimicked in producing the reiterant speech corpus, it seems likely that the set of "unstressed" syllables included both unaccented heavy syllables and reduced syllables. Thus, the observed variation in amplitudes for the "unstressed" syllables must have reflected the even larger difference in jaw heights appropriate for lexically specified [ɑ] versus [ə] constriction targets, and we cannot tell what the velocity-displacement relationship would be if calculated just for the unaccented heavy syllables. Second, if we attribute the kinematic differences between accented and unaccented sequences entirely to the specification of gesture length and size (that is, to the programmed settling time and displacement from programmed endpoint position), then the temporal relationship between the vowel and consonant gestures should be constant. But when Nittrouer, Munhall et al. (1988) tested this prediction explicitly by examining the time of onset for an upper lip lowering movement for a following [p] or [m] relative to the full cycle of jaw opening and closing movements within a [pɑC] sequence, they found that phase angles differed substantially between "stressed" and "unstressed" vowels. While Nittrouer and her colleagues do not report the intonation patterns produced by their talkers, our own first study of accented versus unaccented syllables also showed evidence of a difference in timing relationship between vowel and following consonant gestures. When we plotted the observed jaw cycle durations against the durations predicted by summing the displacement-velocity ratios for the opening and closing movements, we found that the accented sequences had longer durations than predicted by the regression line for the unaccented sequences (see Figure 3.5, Beckman et al. 1992: 76). Since phasing the consonant constriction gesture earlier within the period of the vowel gesture could effectively truncate the end of the movement into the vowel, this alternative mechanism for controlling syllable length also provides an alternative explanation for the observed differences in movement amplitude. In other words, the task dynamic model allows for something like Lindblom's undershoot account as one of several mechanisms for producing the longer and larger jaw movements in accented syllables.

What this plethora of mechanisms means is that the speaker's control of speech timing patterns for different lengthening effects cannot be determined simply by observing differences in movement length and velocity. Rather, to differentiate between the two accounts of lengthening for accent, it is necessary to find a way to compare precise quantitative predictions (generated by models of the length-velocity relationship) against the observed values. Harrington et al. (1995) devised a way to make more precise predictions by manipulating the actual jaw movement traces using very simple model assumptions. Beginning with the traces for accented [ba:b] syllables, they created two sets of model traces for the shorter durations of matching unaccented syllables. In the "truncation" model, they cut the jaw cycle trace into two pieces at the juncture between opening and closing movements, and then overlapped the two pieces without any smoothing. In the "linear rescaling" model, they shrank the jaw cycle trace horizontally and vertically to keep the same aspect ratio. Thus the two models created identically shortened movement durations while maintaining the original movement velocities. Comparing the jaw traces generated by the two models to the observed traces for the actual unaccented syllable productions, Harrington and his colleagues found that the truncation model generated a closer fit to the data, although neither model fit the observed displacements exactly.

Two questions remain outstanding. First, would a hybrid model fit the displacements better? In shortening the accented syllables, both the truncation model and the linear rescaling model left the movement velocities unchanged, while our observations showed unaccented syllables to have slightly slower movements for some speakers. Could this mean that the rhythmic prominence that accompanies accent is produced both by manipulating the between-gesture phasing to increase the vowel duration, and also by exaggerating the more open oral cavity specified for the vowel? Second, does the same explanation work for the kinematic differences that signal the prominence contrast at other levels of the hierarchy? In particular, would we get such a good fit to the data if we used the truncation model to predict the velocity-displacement relationship in reduced syllables by shortening the jaw cycles in the traces for unaccented heavy syllables? We decided to address these questions by applying to a new set of jaw movement data (1) a truncation model which changes only the between-gesture timing relationship; (2) a rescaling model which changes gestural settling time and displacement; and (3) a hybrid model which changes both between- and within-gesture parameters.

4. DATA AND MODELING RESULTS

To get the new data, we replicated the study reported in Beckman and Edwards (1994). We used a similar recording technique to get jaw movement traces for the same two dialogues used earlier (see Table 1 above), and we were able to record again all four of the original talkers for the earlier study. They are all native speakers of regionally unmarked varieties of North American English. KAJ and KDJ are male, and JRE and CDJ are female. Each talker read the dialogues ten times at each of three self-chosen speech tempi. We used the Barlow strain gauge apparatus (see description in Abbs and Gilbert 1973), a device which records displacement from a reference position by measuring the strain exerted on a pair of small metal plates coupled at right angles to each other and mounted on an arm from the side. When one plate of the pair is glued to a surface fleshpoint to rest vertically tangent to the fleshpoint at its reference position, movement away from the reference position bends the plate that is perpendicular to the direction of movement. The apparatus provides three such pairs of plates, which we attached vertically to the vermilion borders of the two lips and to the front of the chin, the last being the point we used to track jaw position. We also recorded the audio signal from an omni-directional microphone 20-30 cm from the mouth. The data from all 7 channels were recorded directly to the hard drive of a PC using a multi-channel sampling board and Cspeech, at a sampling rate of 10 kHz for each channel. We used Cspeech also to record a calibration signal from each metal plate, and to divide up the multiplexed data record for each talker into as many separate files as needed for the different channels and utterance tokens. (Half of the repetitions for KDJ were lost in a computer error in this process of dividing up the large original record file.) We transferred the audio and jaw vertical position files to a SUN Sparcstation, where we generated a smoothed velocity trace for each jaw trace by calculating a moving average over a 35 ms triangular window centered around the target frame. We then used the waves+ program to display the data, and to mark relevant time points in the jaw position and velocity traces for each token.

Figure 1 shows the audio waveform and the jaw position and velocity traces for a typical normal-tempo utterance of the type *HER papa posed a problem.* The bottom window shows the labels for cursor points marked with the waves+ label function. The cursors labeled "p1", "a1", "p2", etc., mark the relevant time points with zero velocity, which we used to segment the jaw trace into the opening and closing movements for the heavy [pɑp] and reduced [pəp] target sequences. The intervening cursors labeled "v" mark the time points where we extracted a peak velocity value for each opening and closing movement. Each movement's duration was calculated by extracting the times for the cursors marking the zero-velocity point at its beginning and end, and then subtracting the earlier time from the later. Movement amplitude was

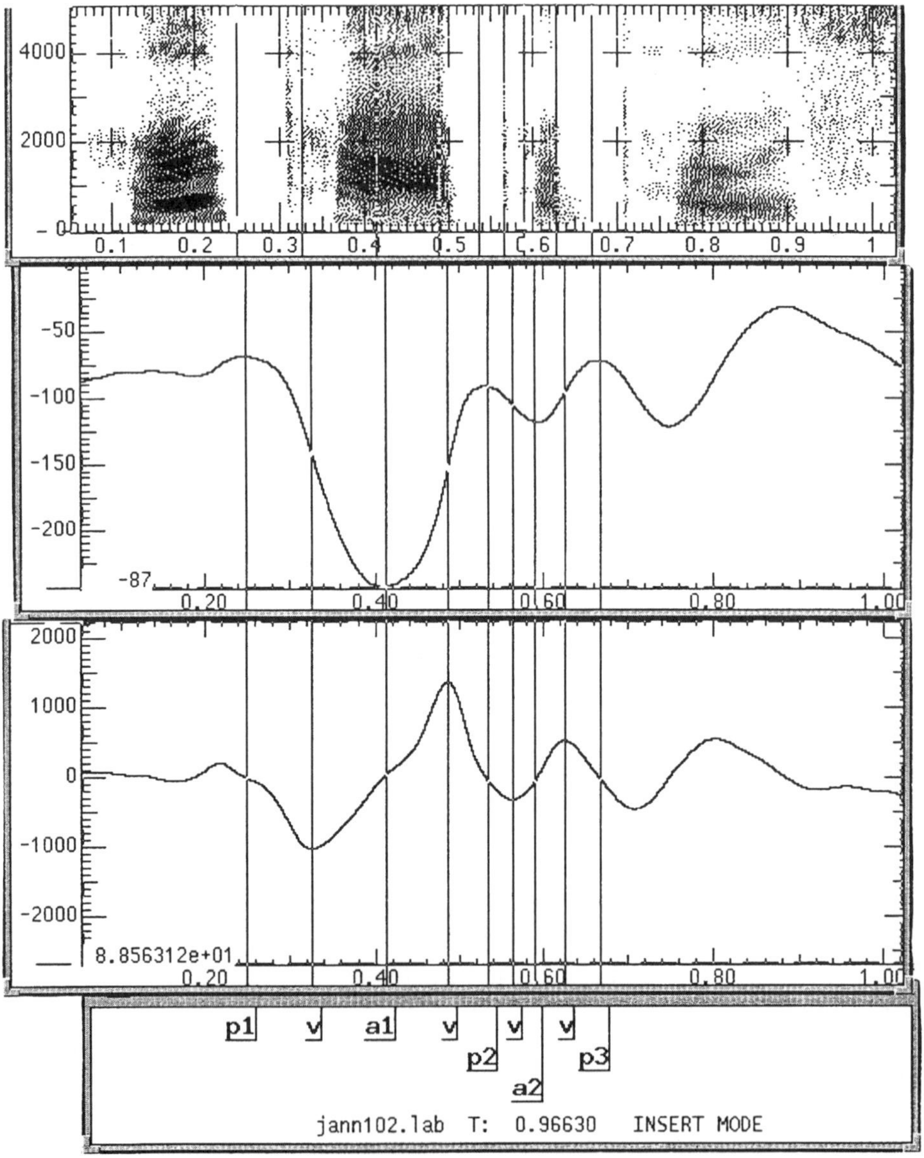

Figure 1. Waves+ display of spectrogram (top window), jaw vertical position trace (second window), velocity trace (third window), and cursor labels for a normal tempo token of HER papa posed a problem *produced by talker JRE.*

similarly calculated by extracting the jaw position values at these points, and then subtracting the value for the vowel from the value for the consonant. The scale values in this figure are in machine representation units, although for the data figures that follow we have used the calibration signal to translate machine units into mm.

Figures 2-5 show mean durations, amplitudes, and peak velocities for the jaw-opening and closing movements for the accented, unaccented heavy, and reduced-vowel sequences produced by the four talkers. Together the figures give an idea of the range of values and of the effects of prominence on the values. Talker CDJ had the largest, fastest movements, and durations that varied considerably in length across the three tempi. Talker JRE also had a large variation in duration across tempi, and she had the longest durations overall. Talker KAJ, on the other hand, had the smallest movement durations and little effect of tempo.

Despite these differences among the speakers in mean values for the movement durations, amplitudes, and velocities, and in the relative size of the effect of the tempo variation, all four figures show the same relational effects of stress at both levels of the prominence hierarchy. Movement durations were longest for the vowels in accented position and shortest for the reduced vowel, with a considerably smaller effect of accent than of syllable weight. The movement amplitudes and peak velocities show the same pattern, with largest and fastest movements for the accented syllable, smallest and slowest movements for the reduced vowel, and the difference between accented and unaccented being less than that between heavy and reduced. Speaker KAJ had somewhat less consistent differences between accented and unaccented syllables for all three measures, but large and consistent differences between unaccented heavy syllables and reduced syllables.

Figure 6 illustrates two different ways of quantifying the relationships among these three kinematic measures, using CDJ's values. The left-hand panel plots the opening movement amplitudes as a function of their velocities, and the right-hand panel plots the sum of the amplitude-velocity ratios for the opening and closing movements as a function of the sum of the observed movement durations. In each plot there are three ellipses showing the results of principal components analyses done in mu+ (Harrington et al. 1993). The center for each ellipse is the mean value for that group of datapoints, and its radii extend to ±2.45 standard deviations along the first two axes of variation, so as to enclose roughly 95% of the datapoints.

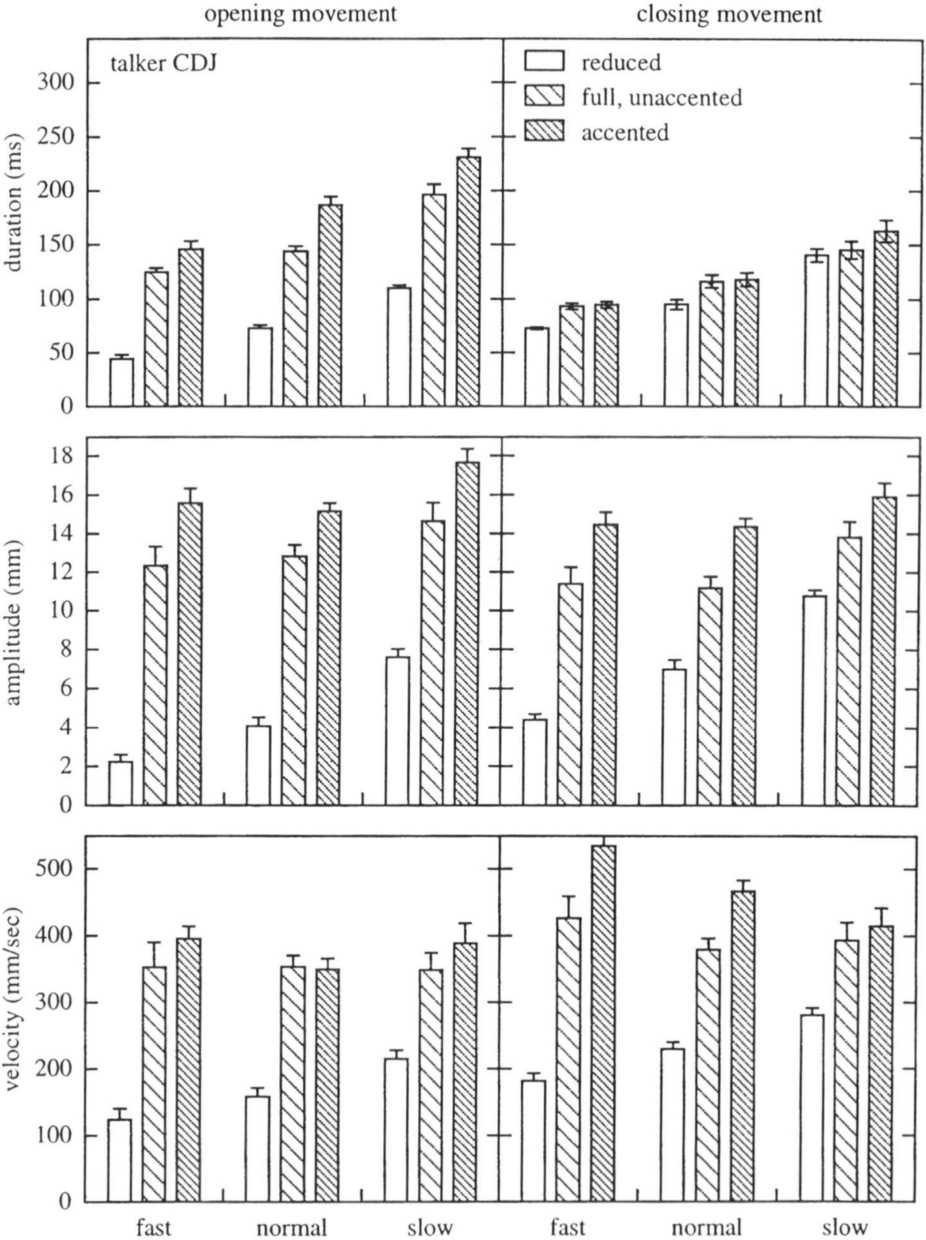

Figure 2. Mean duration, amplitude, and peak velocity for opening and closing movements produced by talker CDJ. Each bar is averaged over all 10 tokens of a given stress and tempo, with error bars for sample standard deviation.

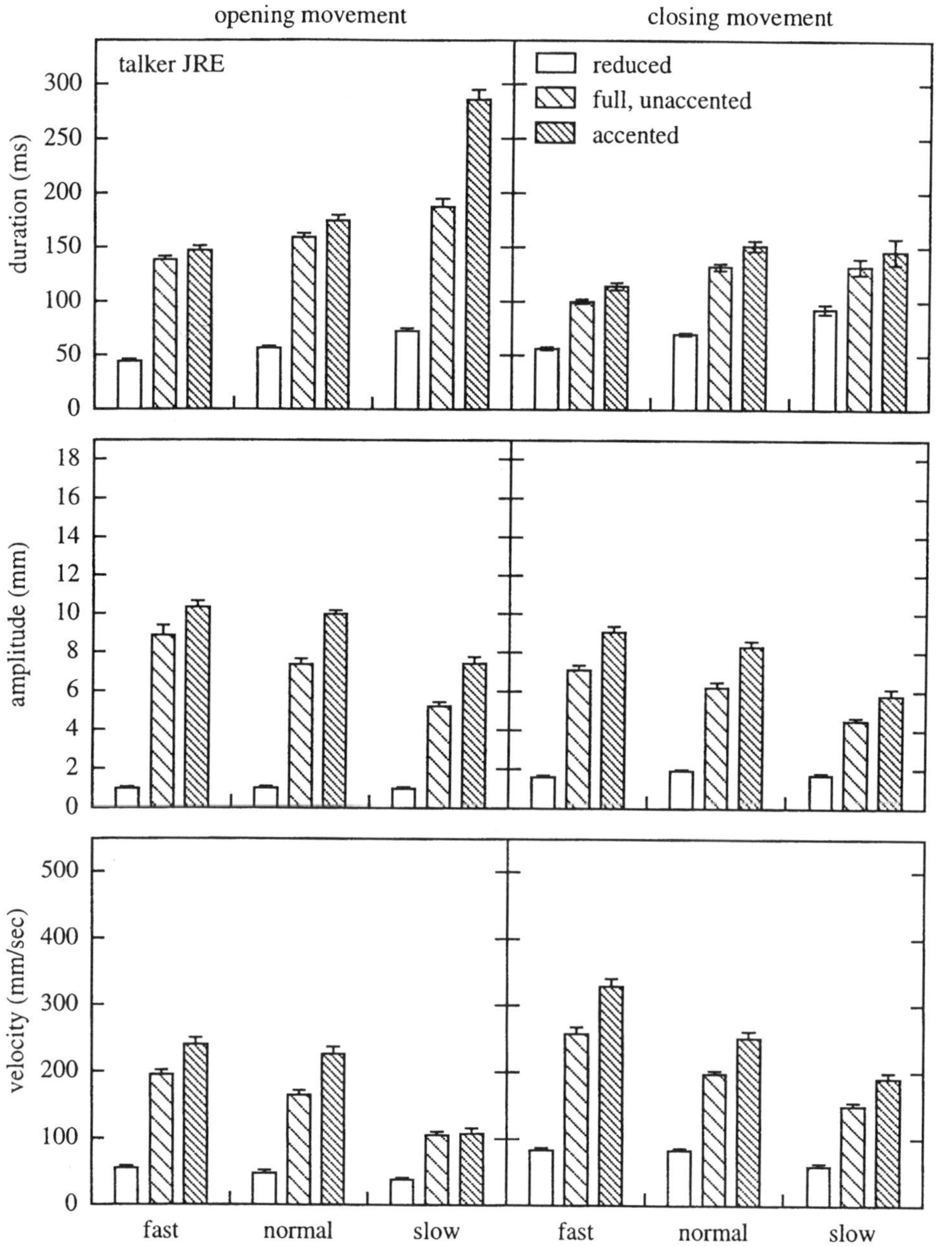

Figure 3. Mean duration, amplitude, and peak velocity for opening and closing movements produced by talker JRE, as in Figure 2.

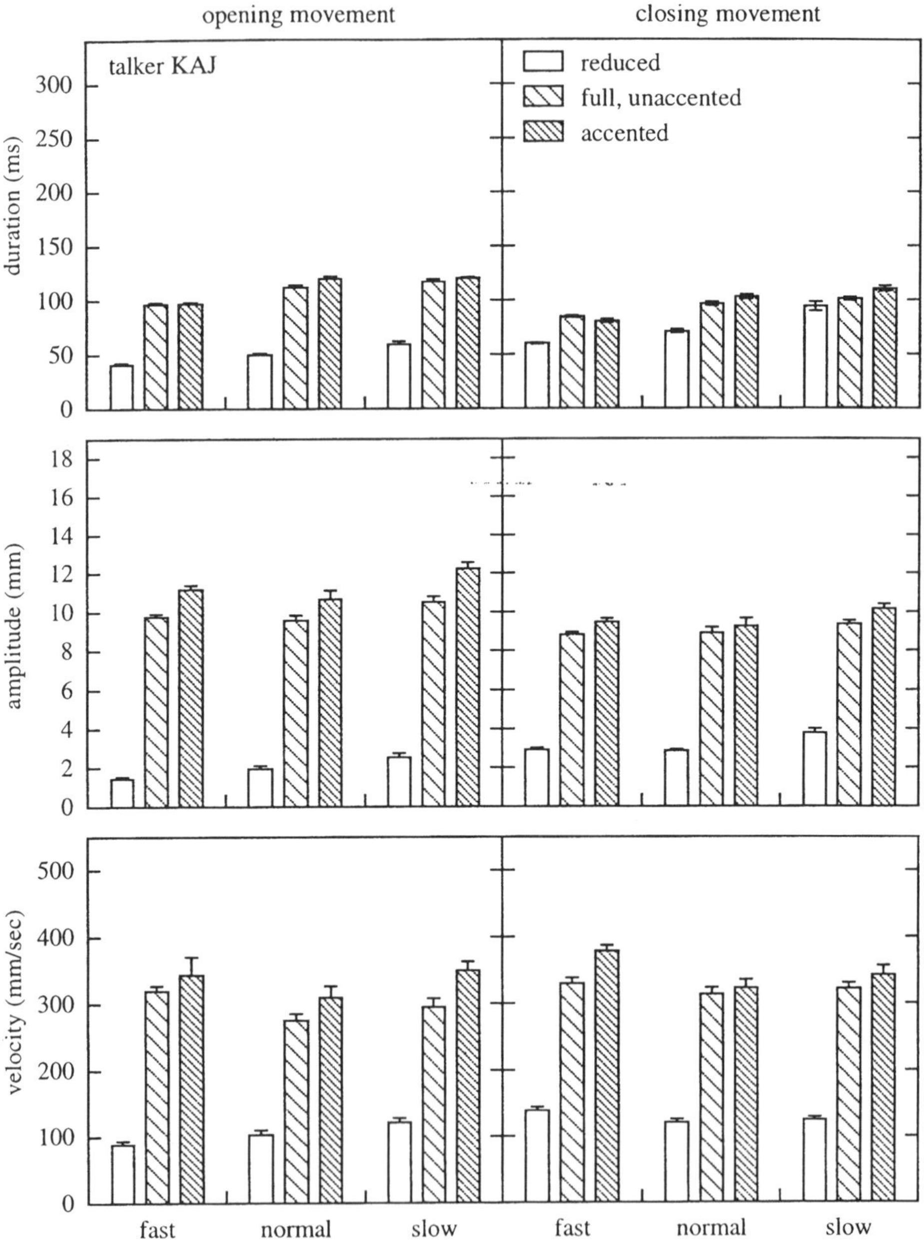

Figure 4. Mean duration, amplitude, and peak velocity for opening and closing movements produced by talker KAJ, as in Figure 2.

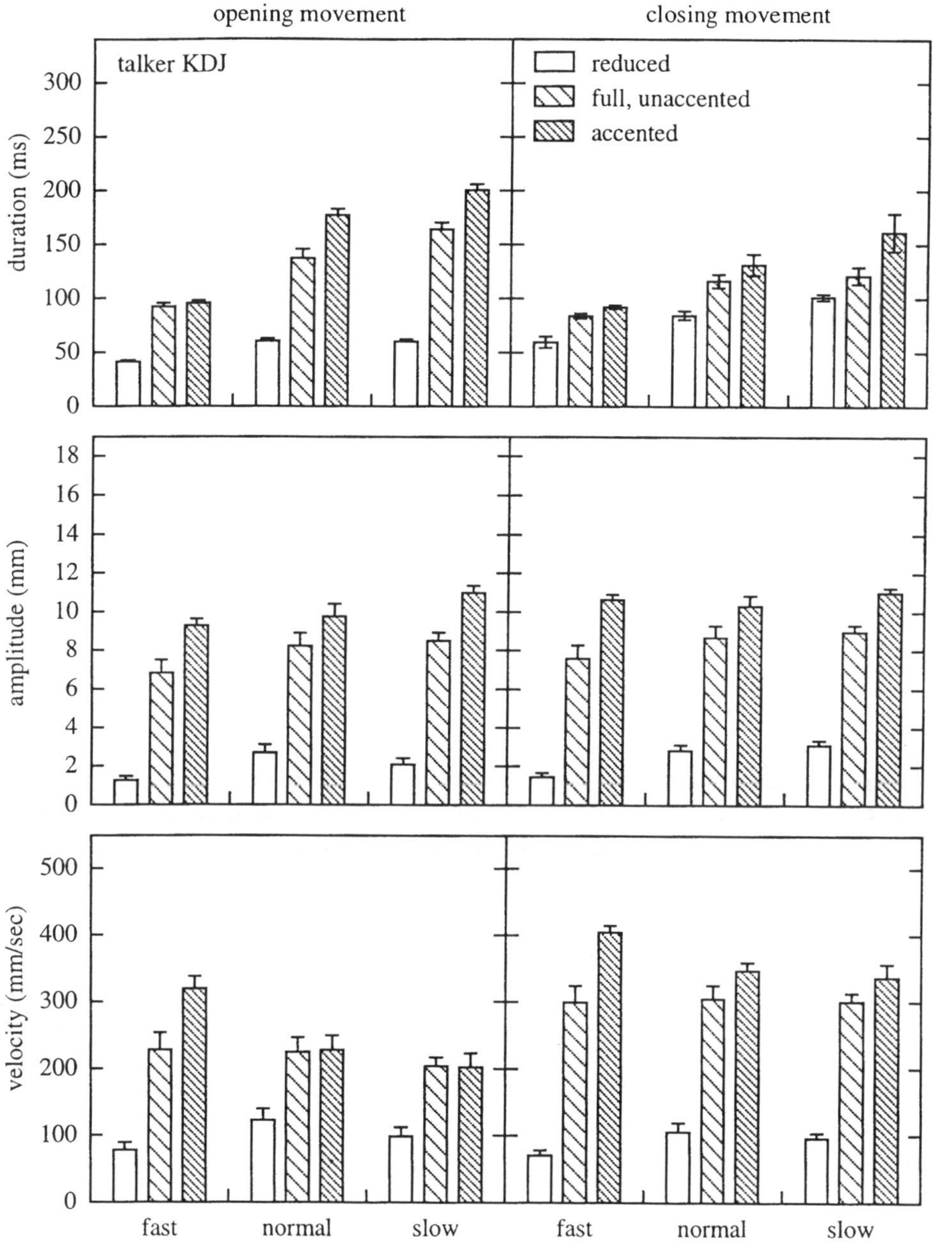

Figure 5. Mean duration, amplitude, and peak velocity for opening and closing movements produced by talker KDJ, as in Figure 2 except that sample size is 5.

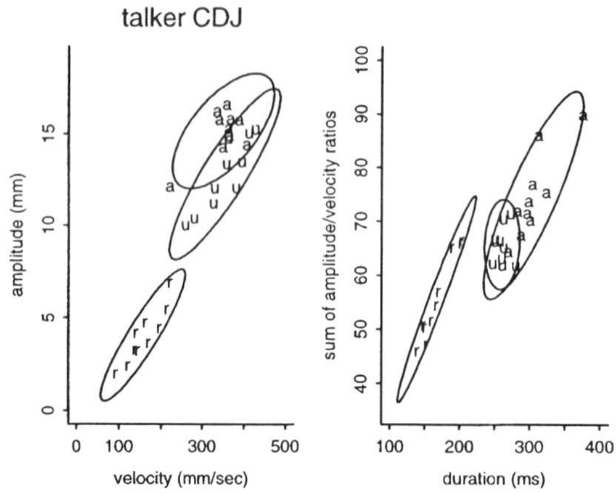

Figure 6. Relationships among kinematic variables depicted in Figure 2 (talker CDJ). Left panel plots opening movement amplitude as a function of peak velocity for normal tempo tokens of each of the three stress types. Ellipses are centered at mean values and extend to ±2.45 standard deviations in the first two principal axes of variation, to enclose 95% of the tokens.

The scatterplot on the left shows that for CDJ, increased prominence had a larger effect on an opening movement's amplitude than on its velocity. The longer radii of the three ellipses run roughly parallel to each other and at a 45% angle in the plot, indicating that faster velocities are associated with larger amplitudes. However, the radius of the ellipse for the accented tokens lies above that of the ellipse for the unaccented tokens, which in turn lies above the line defined by the first principal component for the reduced vowel tokens. In other words, although the more prominent sequences in each stress contrast pair had generally larger velocities than the matching unstressed group, they had smaller velocities than expected for their larger amplitudes. Interpreted in the task dynamic framework outlined in the preceding section, this pattern can be taken as evidence for either of two lengthening mechanisms. If the opening gesture in the more stressed sequence had a longer settling time, then the resulting movement velocities should be smaller than predicted by the velocity-amplitude relationship for the less stressed sequence. Or, if the more stressed sequence was made longer by moving the opening and closing gestures further apart in time, then the observed amplitude might be larger than predicted for the opening movement because the underlying gesture for the vowel is less subject to truncation by the onset of the closing movement into the following consonant.

The scatterplot on the right offers support for the latter interpretation. Since the amplitude-velocity ratio is a measure of relative settling time for a gesture, the sum of the two movement ratios provides a measure of relative predicted duration if between-gesture timing is invariant. At both levels of prominence contrast, the ellipse for the more stressed sequence lies to the right of the line defined by the longer radius of the ellipse for the less stressed sequence. That is, given the relationship between observed and predicted durations for the less prominent sequence, the observed durations in the more prominent sequence were longer than predicted by the amplitude-velocity relationships.

The data for the other three speakers showed similar patterns for both types of plot. Opening movements into less prominent syllables do not show the same relationship between velocity and amplitude as do movements into more prominent syllables, a result that is consistent either with faster settling times or with truncation. The unstressed syllables also show a different relationship between the sum of the amplitude-velocity ratios and the sum of observed durations over the opening and closing movements, a result which is more consistent with a shortening mechanism that truncates the opening movement by beginning the closing gesture earlier.

We made a further test of these two interpretations using the three models depicted in Figure 7. The three models can be used for generating the shorter duration and smaller amplitude typical of a less stressed sequence from the jaw trace recorded for a more stressed sequence. The first is the truncation model of Harrington et al. (1995). It shortens the stressed sequence by overlapping the end of the opening movement into the vowel with the beginning of the closing movement into the following consonant, with the amount of overlap being just the right length to reduce the duration of the full sequence by the difference between the average stressed and unstressed durations. The second model is a gesture rescaling model, which manipulates only within-gesture parameters, but which differs from the linear rescaling model of Harrington et al. in that it does not preserve aspect ratio. The formula for this model first increases velocity just enough to shorten the movements of the stressed syllable by the observed difference in average jaw cycle durations, and then decreases velocity just enough to reduce jaw opening by the observed difference in average movement amplitudes. The third formula in Figure 7 is a hybrid model in which the stressed syllable is shortened by overlapping the two movements, and then the velocity is decreased to reduce the movement amplitude.

All three models were applied at both levels of the stress contrast for all four talkers. For example, in applying the truncation model, the jaw trace for each accented token produced by the talker was used as the basis for truncating by the difference in median durations between accented and unaccented [pɑp] sequences for that talker, and the jaw trace for each of the talker's unaccented tokens was used as a base for truncating by the difference in median durations between the talker's full-voweled [pɑp] and reduced-vowel [pəp] sequences.

Start with **original jaw trace**, such as the unaccented sequence in Figure 1. All three models will reduce the total time by the same amount, to produce something like the observed duration of the following reduced vowel sequence.

Truncation model. Overlap opening and closing movements to produce the observed difference in total sequence duration. The resulting "reduced" token has unchanged velocity and amplitude reduced only by undershoot.

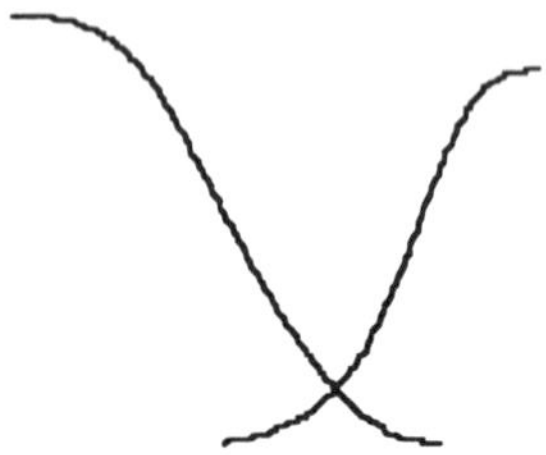

Rescaling model. First, scale velocity upward to produce the observed difference in total sequence duration.

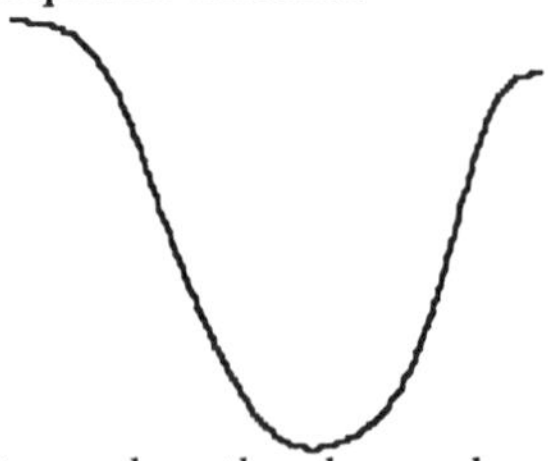

Then, scale velocity downward to produce the observed difference in movement amplitude.

Hybrid model. First, shorten by overlapping opening and closing movements, as in model 1. Then, decrease velocity to produce any further difference between the undershoot of the truncated token and the observed movement amplitude.

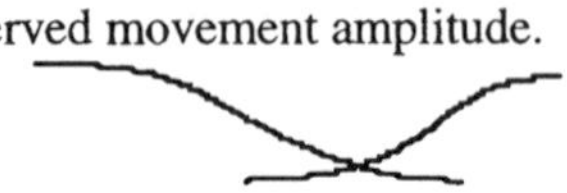

Figure 7. Three models of vowel shortening and reduction.

The model can then be evaluated by comparing amplitude-velocity relationships in the resulting model-generated "unaccented" or "reduced" traces to the relationships observed in the actual unaccented or reduced traces produced by the talker.

Figures 8-11 show these evaluative comparisons. Each figure has two sets of panels, like the scatterplots in Figure 6. The left-hand panels plot principal component analyses for opening movement amplitude against opening movement displacement, and the right-hand panels plot the ellipses for total relative predicted duration against total observed duration. The upper two panels show the ellipses for the three sets of model-generated "unaccented" tokens and (for comparison) the ellipses for the original accented versus unaccented tokens. The lower panels show the same thing for the model-generated "reduced" tokens. In each panel, the long dashed line is for the between-gesture truncation model, the short dashed line is for the within-gesture rescaling model, and the dotted line is for the hybrid model which shortens by truncating across the gestures but also simulates a decrease in the within-gesture displacement.

The left-hand panels for each of the four talkers show that the rescaling model does a better job than the truncation model of accounting for the amplitude-velocity relationship within the opening movement. This is not surprising, since the truncation model leaves gesture-internal parameters unchanged, whereas Figures 2-5 showed all four talkers to have slower opening movements for less prominent sequences at both levels of stress contrast. The right-hand panels, on the other hand, show that the truncation model does a better job than the rescaling model in accounting for total sequence durations. This also is not surprising, at least for the unaccented syllables produced by JRE and KDJ, since these two talkers were the speakers in Beckman et al. (1992) who provided the tokens suggesting that lengthening for accent differs from phrase-final lengthening in affecting the timing relationship between gestures rather than within-gesture settling time. Over both sets of comparisons, however, the hybrid model does the best.

Since the mechanism for shortening is the same as that of the truncation model, it does better than the rescaling model in generating the observed relationship between observed and predicted total sequence duration in the right-hand panels of the figures. Since it also allows for a further reduction in within-gesture displacement to account for the amplitude difference, it does better than the truncation model in accounting for the observed relationship between opening movement amplitude and velocity in the left-hand panels.

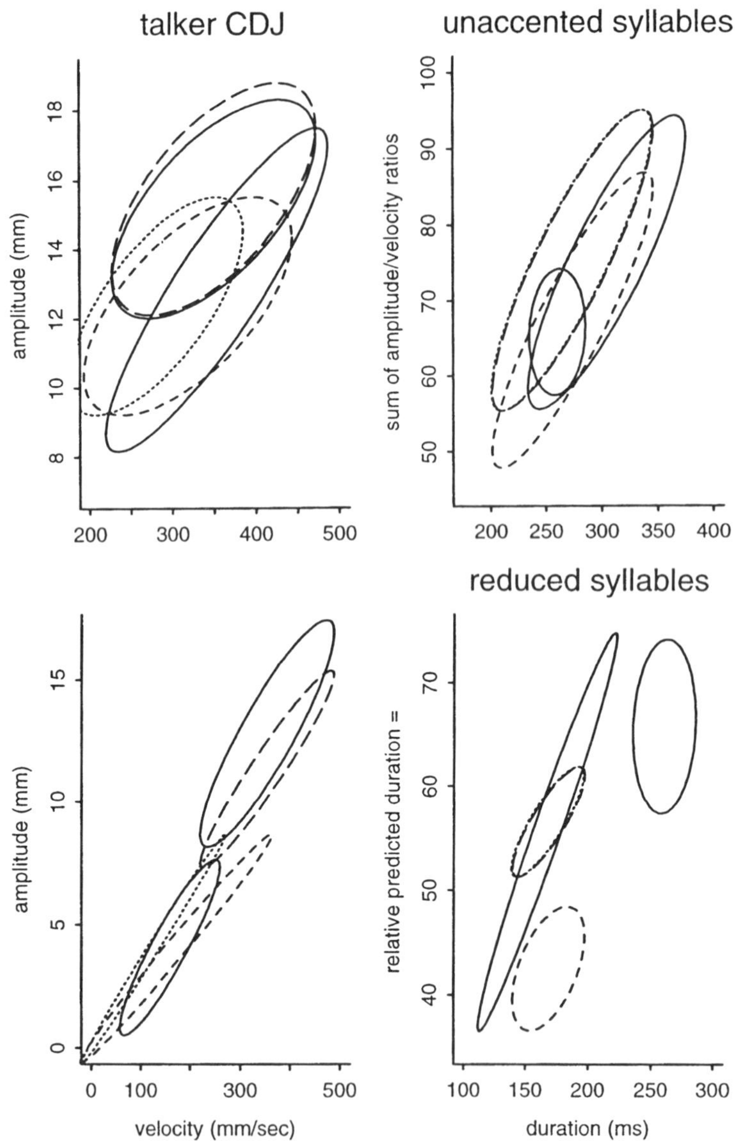

Figure 8. Scatterplots as in Figure 6 for "unaccented" tokens (top) and "reduced" tokens (bottom) generated for CDJ's productions by each of the three models in Figure 7. The solid lines are ellipses for the original tokens — accented versus unaccented (top) or unaccented versus reduced (bottom) — and the broken lines are ellipses for the model-generated tokens, with long dashed lines for the truncation model, short dashed lines for the rescaling model, and dotted lines for the hybrid model.

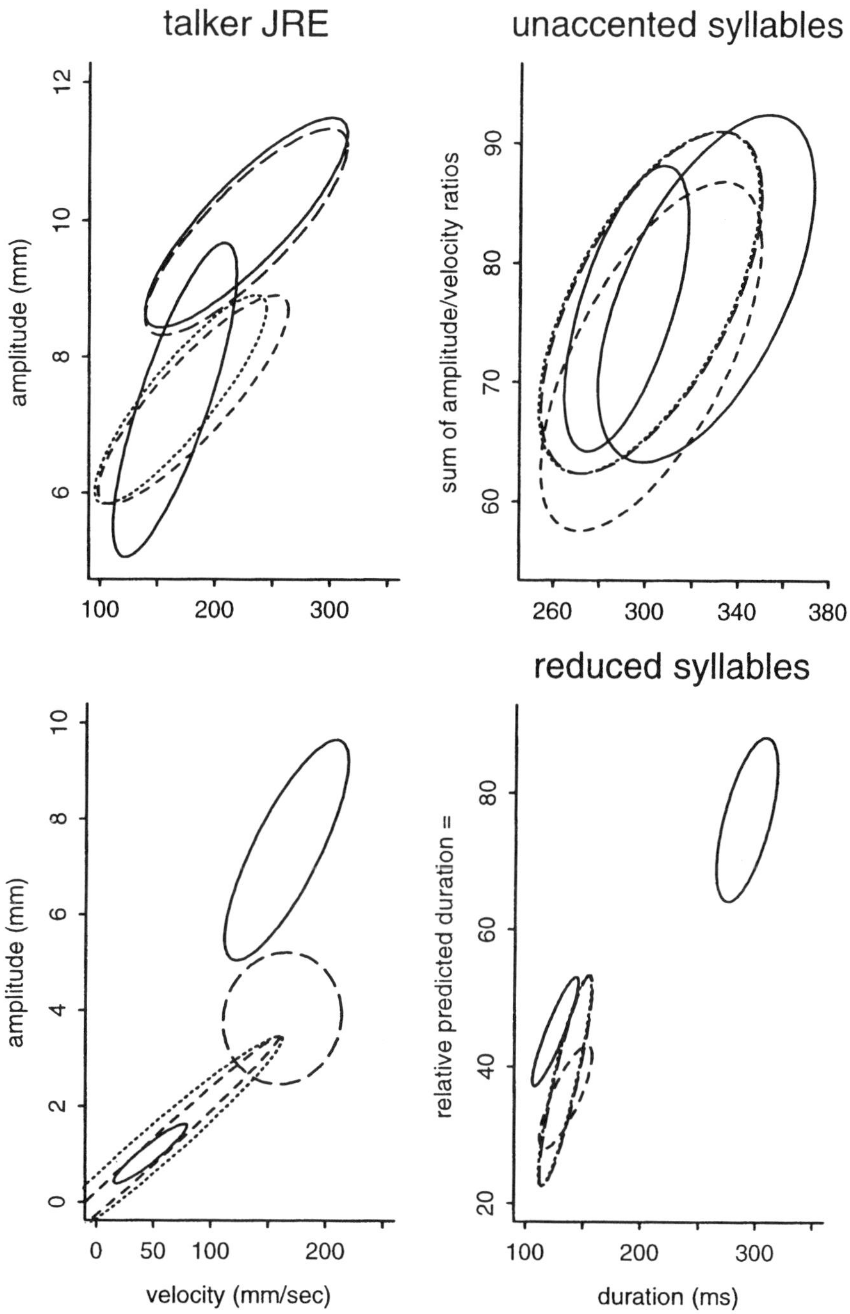

Figure 9. Scatterplots as in Figure 8 for model-generated "unaccented" and "reduced" sequences for talker JRE.

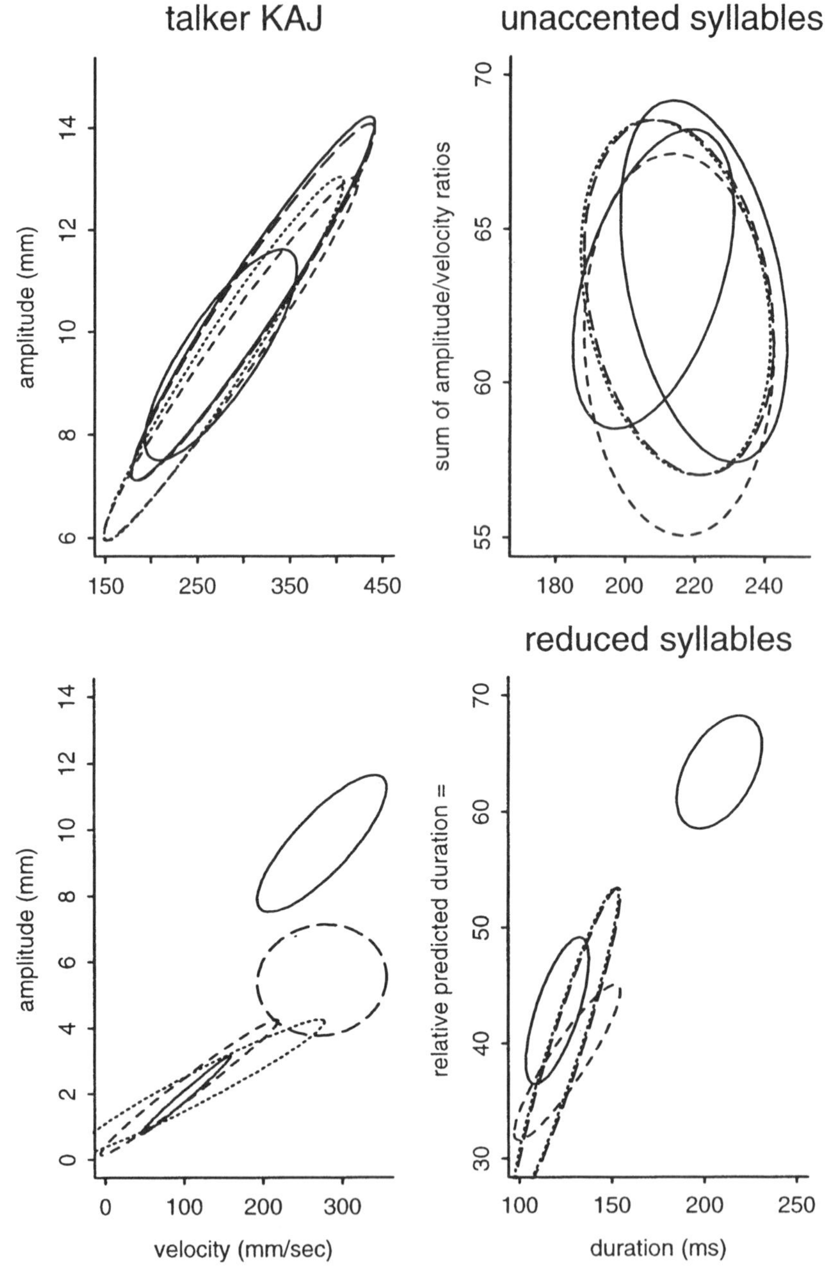

Figure 10. Scatterplots as in Figure 8 for model-generated "unaccented" and "reduced" sequences for talker KAJ.

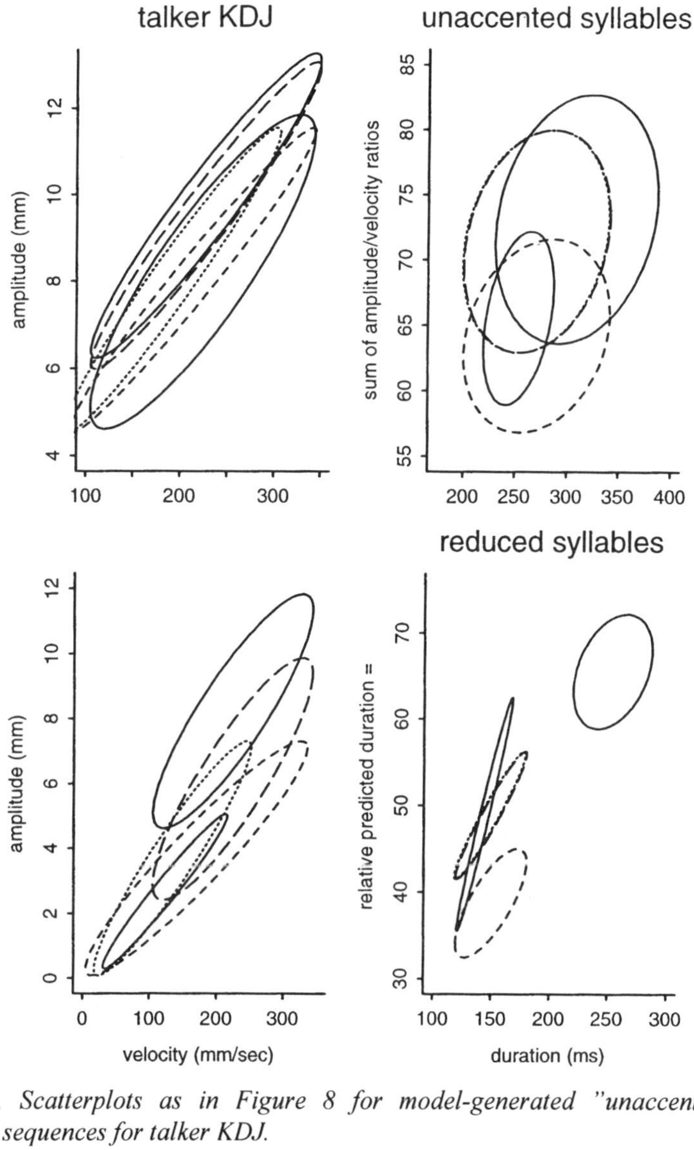

Figure 11. *Scatterplots as in Figure 8 for model-generated "unaccented" and "reduced" sequences for talker KDJ.*

Thus, the current data and model comparisons suggest that lengthening for accent is a considerably more subtle version of the same effect as the lexically specified contrast between a full [ɑ] and a reduced [ə]. Following de Jong et al. (1993), we might call the effects at both levels of prominence contrast a "localized hyperarticulation" that enhances the clarity of the lexical specifications by making a stressed vowel more peripheral and by reducing the amount of coarticulatory overlap between it and neighboring consonants.

While further testing of the hybrid model is necessary before these results can be generalized to other full vowel types, articulatory data in de Jong (1995b) and Harrington et al. (1996) suggest that this account is appropriate at least for high tense vowels.

At the same time, the data further support the idea that not all lengthenings are alike. At both levels of stress, the jaw cycle for the vowel in the more stressed syllable was longer, even though the movements into and out of the vowel were, if anything, faster than for the less stressed syllable. The very simple modeling tests that we conducted showed that the velocity pattern was in accord with the amplitudes — i.e., with the evidence that the more stressed syllable also has a more extreme displacement between the minimum in the vowel and the maximum in the two neighboring stop consonants. However, this makes the durational effect of rhythmic prominence (at both the level of anchoring accents and at the level of full versus reduced syllables) something markedly different from phrase-final lengthening. Beckman et al. (1992) showed that the jaw cycle for a vowel in the last syllable of an intonation phrase syllable is also considerably longer than that for a vowel that is just one syllable away from the phrase end, but the movements into and out of the vowel are not faster. Standard psychoacoustic tests of just noticeable differences in the duration of steady state signals such as noise bursts cannot be applied in any straightforward way to gauge the relative perceptual salience of two so very different lengthening effects.

Another direction for future work, then, is to model more precisely the acoustic consequences of these dynamic differences and to relate them to the role of segmental rhythms in providing the anchor points for intonational events such as pitch accents and boundary tones. Once we have a better understanding of how speakers control speech timing and how their control strategies are realized in the acoustic signature for different durational effects, we will be in a position to apply psychoacoustic measures to sort out which effects are perceptually relevant.

ACKNOWLEDGMENTS

The new production data reported in this paper and part of the modeling results were presented at the 128th and 130th meetings of the Acoustical Society of America, respectively. We owe special thanks to Jan Edwards and Marios Fourakis, who recorded the data with their strain gauge system, and who were our co-authors on those two earlier presentations. We also thank Jan Edwards, Jonathan Harrington and Janet Fletcher for useful discussion of the models, and Jonathan Harrington for providing us with the mu+ system and with helpful advice on how to use it.

REFERENCES

Abbs, J.H. and Gilbert, B.N. 1973. A strain gauge transducer system for lip and jaw motion in two dimensions. *Journal of Speech and Hearing Research* 16, 248-256.

Abel, S.M. 1972. Duration discrimination of noise bursts. *Journal of the Acoustical Society of America* 51, 1219-1223.

Anderson, M.D., Pierrehumbert, J.B. and Liberman, M.Y. 1984. Synthesis by rule of English intonation patterns. *Proc. 1984 IEEE International Conference on Acoustics, Speech and Signal Processing*, vol. 1, 2.8.1-2.8.4.

Bannert, R. 1982. An F0-dependent model for segment duration? *Reports from Uppsala University, Department of Linguistics* 8, 59-80.

Beckman, M.E. 1986. *Stress and Non-Stress Accent*. Dordrecht: Foris.

Beckman, M.E. 1996. When is a syllable not a syllable? In T. Otake and A. Cutler (eds), *Phonological Structure and Language Processing: Cross-Linguistic Studies*. Berlin: Mouton de Gruyter, 95-123.

Beckman, M.E. and Edwards, J. 1994. Articulatory evidence for differentiating stress categories. In P.A. Keating (ed) *Phonological Structure and Phonetic Form: Papers in Laboratory Phonology III*. Cambridge, UK: Cambridge University Press, 7-33.

Beckman, M., Edwards, J. and Fletcher, J. 1992. Prosodic structure and tempo in a sonority model of articulatory dynamics. In Docherty and Ladd (eds), 68-86.

Beckman, M.E. and Hirschberg, J. 1994. The ToBI annotation conventions. Ms. Ohio State University. [Obtain by writing to tobi@ling.ohio-state.edu.]

Beckman, M.E. and Pierrehumbert, J.B. 1986. Intonational structure in Japanese and English. *Phonology Yearbook* 3, 255-309.

Bolinger, D. 1958. A theory of pitch accent in English. *Word* 14, 109-149.

Browman, C.P. and Goldstein, L. 1990. Tiers in articulatory phonology, with some implications for casual speech. In J. Kingston and M.E. Beckman (eds), 341-346.

Bruce, G. 1977. *Swedish Word Accents in Sentence Perspective*. Lund: Gleerup.

Bruce, G. 1983. Accentuation and timing in Swedish. *Folia Linguistica* 17, 221-238.

Bruce, G. 1990. Alignment and composition of tonal accents: comments on Silverman and Pierrehumbert's paper. In Kingston and Beckman (eds), 107-114.

Carlson, R., Granström, B. and Klatt, D.H. 1979. Some notes on the perception of temporal patterns in speech. *Proc. IXth International Congress of Phonetic Sciences*, Vol. II, 260-267.

Coker, C.H., Umeda, N. and Browman, C. 1973. Automatic synthesis from text. *IEEE Transactions on Audio Electroacoustics* AU-21, 293-297.

Creelman, C.D. 1962. Human discrimination of auditory duration. *Journal of the Acoustical Society of America* 34, 582-593.

de Jong, K. 1995a. The supraglottal articulation of prominence in English: linguistic stress as localized hyperarticulation. *Journal of the Acoustical Society of America* 97, 491-504.

de Jong, K. 1995b. On status of redundant features: the case of backing and rounding in American English. In B. Connell and A. Arvaniti (eds). *Phonology and Phonetic Evidence: Papers in Laboratory Phonology IV*. Cambridge, UK: Cambridge University Press, 61-86.

de Jong, K., Beckman, M.E. and Edwards, J. 1993. The interplay between prosodic structure and coarticulation. *Language and Speech* 36, 197-212.

Denes, P. 1955. Effect of duration on the perception of voicing. *Journal of the Acoustical Society of America* 27, 761-764.

Docherty, G.J. and Ladd, D.R. (eds) 1992. *Papers in Laboratory Phonology II: Gesture, Segment, Prosody*. Cambridge: Cambridge University Press.

Fry, D.B. 1955. Duration and intensity as physical correlates of linguistic stress. *Journal of the Acoustical Society of America* 27, 765-768.

Fry, D.B. 1958. Experiments in the perception of stress. *Language and Speech* 1, 126-152.

Gussenhoven, C. and Rietveld, A.C.M. 1992. Intonation contours, prosodic structure and preboundary lengthening. *Journal of Phonetics* 20, 283-303.

Harrington, J., Cassidy, S., Fletcher, J. and McVeigh, A. 1993. The mu+ system for corpus based speech research. *Computer Speech and Language* 7, 305-331.

Harrington, J., Fletcher, J. and Beckman, M.E. 1996. Manner and place conflicts in the articulation of accent in Australian English. Paper presented at the 5th Conference in Laboratory Phonology, Northwestern University, 6-8 July 1996.

Harrington, J., Fletcher, J. and Roberts, C. 1995. Coarticulation and the accented/unaccented distinction: evidence from jaw movement data. *Journal of Phonetics* 23, 305-322.

Harris, K. 1978. Vowel duration change and its underlying physiological mechanisms. *Language and Speech* 21, 354-361.

House, A.S. and Fairbanks, G. 1953. The influence of consonantal environment upon the secondary acoustical characteristics of vowels. *Journal of the Acoustical Society of America* 25, 105-113.

Huss, V. 1978. English word stress in post-nuclear position. *Phonetica* 35, 86-105.

Kelso, J.A.S., Vatikiotis-Bateson, E., Saltzman, E.L. and Kay, B. 1985. A qualitative dynamic analysis of reiterant speech production: phase portraits, kinematics, and dynamic modeling. *Journal of the Acoustical Society of America* 77, 266-280.

Kent, R. and Netsell, R. 1971. Effects of stress contrasts on certain articulatory parameters. *Phonetica* 24, 23-44.

Kingston, J. and Beckman, M.E. (eds). 1990. *Papers in Laboratory Phonology I: Between the Grammar and the Physics of Speech.* Cambridge: Cambridge University Press.

Klatt, D.H. 1976. Linguistic uses of syllable duration in English: Acoustic and perceptual evidence. *Journal of the Acoustical Society of America* 59, 1208-1221.

Klatt, D.H. 1979. Synthesis by rule of segmental durations in English sentences. In B. Lindblom and S. Öhman (eds) *Frontiers of Human Communication Research.* New York: Academic Press, 287-299.

Kuehn, D.P. and Moll, K.L. 1976. A cineradiographic study of VC and CV articulatory velocities. *Journal of Phonetics* 4, 303-320.

Ladd, D.R. 1992. An introduction to intonational phonology. In Docherty and Ladd (eds), 321-334.

Ladd, D.R. 1996. *Intonational Phonology*. Cambridge: Cambridge University Press.

Laeufer, C. 1992. Patterns of voicing-conditioned vowel duration in French and English. *Journal of Phonetics* 20, 411-440.

Lehiste, I. 1972. The timing of utterances and linguistic boundaries. *Journal of the Acoustical Society of America* 51, 2018-2024.

Lehiste, I. 1975. The phonetic structure of paragraphs. In A. Cohen and S. Nooteboom (eds) *Structure and Process in Speech Perception.* Heidelberg: Springer, 195-203.

Levelt, W.J.M. 1992. Accessing words in speech production: stages, processes and representations. *Cognition* 42, 1-22.

Liberman, M. and Pierrehumbert, J. 1984. Intonational invariance under changes in pitch range and length. In M. Aronoff and R. Oerhle (eds) *Language Sound Structure.* Cambridge, MA: MIT Press, 157-233.

Lindblom, B. 1963. Spectrographic study of vowel reduction. *Journal of the Acoustical Society of America* 35, 1773-1781.

Lindblom, B. 1967. Vowel duration and a model of lip mandible coordination. *Speech Transmission Laboratory Quarterly Progress and Status Report* 4/1967, 1-29.

Lyberg, B. 1979. Final lengthening — partly a consequence of restrictions on the speed of fundamental frequency change? *Journal of Phonetics* 7, 187-196.

Nakatani, L.H. and Aston, C.H. 1978. Acoustic and linguistic factors in stress perception. Ms., Bell Laboratories.

Nakatani, L.H., O'Conner, K.D. and Aston, C.H. 1981. Prosodic aspects of American speech rhythm. *Phonetica* 38, 84-105.

Nittrouer, S., Munhall, K.G., Kelso, J.A.S., Tuller B. and Harris, K. 1988. Patterns of interarticulator phasing and their relation to linguistic structure. *Journal of the Acoustical Society of America* 84, 1653-1661.

Nooteboom, S.G. 1973. The perceptual reality of some prosodic durations. *Journal of Phonetics* 1, 25-45.

Öhman, S., Zetterlund, S., Nordstrand, L. and Engstrand, O. 1979. Predicting segment duration in terms of a gesture theory of speech production. *Proc. IXth International Congress of Phonetic Sciences*, vol. 2, 305-311.

Olive, J.P. 1977. Rule synthesis of English from diadic units. *Proc. 1977 IEEE International Conference on Acoustics, Speech and Signal Processing,* 568-570.

Ostry, D. and Munhall, K. 1985. Control of rate and duration of speech movements. *Journal of the Acoustical Society of America* 77, 640-648.

Pierrehumbert, J. 1980. *The Phonetics and Phonology of English Intonation.* Doctoral dissertation, Massachusetts Institute of Technology.

Pierrehumbert, J. This volume. Tonal elements and their alignment.

Pierrehumbert, J. and Steele, S. 1989. Categories of tonal alignment in English. *Phonetica* 46(4), 181-196.

Port, R.F. 1981. Linguistic timing factors in combination. *Journal of the Acoustical Society of America* 69, 262-274.

Saltzman, E. 1986. Task dynamic coordination of the speech articulators: a preliminary model. In H. Heuer and C. Fromm (eds) *Generation and Modulation of Action Patterns*. New York: Springer-Verlag, 129-144.

Saltzman, E. and K. Munhall. 1989. A dynamical approach to gestural patterning in speech production. *Ecological Psychology* 1, 333-382.

Silverman, K.E.A. and Pierrehumbert, J.B. 1990. The timing of prenuclear high accents in English. In Kingston and Beckman (eds), 72-106.

Stevens, K.N. and House, A.S. 1963. Perturbation of vowel articulations by consonantal context: an acoustical study. *Journal of Speech and Hearing Research* 6, 111-128.

Summers, V. 1987. Effects of stress and final-consonant voicing on vowel production: articulatory and acoustic analyses. *Journal of the Acoustical Society of America* 82, 847-863.

Summers, V. 1988. F1 structure provides information for final-consonant voicing. *Journal of the Acoustical Society of America* 84, 485-492.

Vanderslice, R. and Ladefoged, P. 1972. Binary suprasegmental features and transformational word-accentual rules. *Language* 48, 819-839.

Wightman, C.W. and Ostendorf, M. 1994. Automatic labeling of prosodic patterns. *IEEE Transactions on Speech and Audio Processing,* 469-481.

Dept. of Linguistics, Ohio State University, Columbus, OH, USA

STEFANIE SHATTUCK-HUFNAGEL

PHRASE-LEVEL PHONOLOGY IN SPEECH PRODUCTION PLANNING: EVIDENCE FOR THE ROLE OF PROSODIC STRUCTURE *

1. INTRODUCTION

It has long been recognized that speakers (and speech synthesizers) cannot produce well-formed multi-word utterances by simply stringing together word forms retrieved from long-term storage. Two kinds of phenomena compellingly illustrate the need for more complex processing for the production of fluent conversational speech: phrase-level prosody, such as intonation and timing, and the pervasive phonetic modification of word forms in context, such as distinctive feature assimilation and segmental reduction. Although both of these phenomena involve the phonetic shape of a word in an utterance, neither can be predicted from the lexical forms of the words in a sentence, and they have often been ascribed to higher level structures such as the phrase.

The higher-level phrasal structure that governs both prosodic patterns and phonetic modification has often been assumed to be surface-level syntactic structure (e.g. Mohanan, 1986, 1995 and references). Evidence for the view that syntax directly controls phrase-level phonology was found in acoustic-phonetic indications of the location of syntactic breaks by means such as pauses, in the spoken disambiguation of ambiguous syntactic structures (Lehiste 1973, Price et al. 1991), in evidence for final lengthening at the boundaries of syntactic constituents (Klatt 1976), and in the governance of sentence-level intonational prominence placement by syntactic structure (Bresnan 1971). In addition, certain prosodic phrasings are prohibited by syntactic structure, e.g.

Syntactic: *(Old men and women) (wear gloves)*

Prosodic: *Old men and women] [wear gloves*

**Old men] [and women wear gloves*

where the symbol] [indicates the boundary of a major prosodic constituent e.g. an intonational phrase. See Selkirk's contribution to this volume for

M. Horne (ed.), Prosody: Theory and Experiment, 201–229.

further discussion and Section 2.3 below for more on the meaning of 'prosodic phrasing'.

Although the evidence is strong that syntactic structure plays an important role in determining the intonational, durational and featural modification patterns of spoken utterances, a growing body of evidence supports two important further claims. First, syntactic structure is only one of many factors that influence these patterns, and second, these patterns as a whole can be most readily accounted for by postulating an alternative set of structures, i.e. the prosodic hierarchy of constituents and prominences postulated in current prosodic theories (Liberman and Prince 1977; Hayes 1984, Halle and Vergnaud 1987; Selkirk 1980, 1984, 1986, 1995; Ladd 1996; Beckman and Pierrehumbert 1986, Beckman and Edwards 1994). Proposed constituents in the hierarchy include intonational phrases, phonological phrases, accentual phrases, clitic groups, prosodic words, rhythmic stress feet, syllables and moras, etc.; proposed levels of prominence include nuclear pitch accent, prenuclear pitch accent, full vowels without pitch accent and reduced vowels. Shattuck-Hufnagel and Turk (1996) summarize several proposed versions of the prosodic hierarchy, and discuss some of the evidence that supports the hypothesis that these structures directly govern phrase-level phonological and phonetic phenomena. One major advantage of prosodic theories, and of the processing models that draw on those theories, is their ability to account for aspects of spoken utterances that are not easily dealt with by traditional morpho-syntactic structure. Examples include the tendency of language users to divide an utterance into phrases of approximately equal length (Gee and Grosjean 1983), patterns of pre-boundary lengthening in spoken utterances (Ferreira 1991), and the role of rhythm in the placement of phrase-level prominences within words (Shattuck-Hufnagel et al. 1994), discussed in Section 2 below. These and other findings lend support to the hypothesis that prosodic structure provides the framework that integrates a variety of factors to determine the specific phonological and phonetic shape of a particular utterance, and that the traditional morphosyntactic structure of sentence, including both syntactic and lexical specifications, are contributing factors to this prosodic structure rather than its sole and direct determinant.

1.1 What is prosodic structure and how is it determined for a particular utterance?

The prosodic constituent structure of an utterance has generally been proposed to be derived rather directly from the morphosyntactic constituent structure of the sentence (e.g. Selkirk 1984, Nespor and Vogel 1986), and there is no question that in many examples the correspondence between e.g. the intonational phrasing and the surface syntactic phrasing of an utterance is very close. For example, an utterance of the sentence *The mayor of Chicago won their support* is likely to be produced as a single intonational phrase spanning the whole sentence, or as two intonational phrases, corresponding to the subject noun phrase and verb phrase of the sentence (Selkirk 1984):

The mayor of Chicago] [won their support

However, other examples show that these correspondences are not complete, as in *Sesame Street is brought to you by the Children's Television workshop*. Among other possibilities, this sentence can be produced as two intonational phrases, with the subject noun phrase combined with just a fragment of the verb phrase,

Sesame Street is brought to you by] [the Children's Television Workshop

In this rendition, the first intonational phrase is a spoken constituent that is not a well-formed syntactic element. Other examples are found in sentences where the subject is a single pronoun (Gee and Grosjean 1983). A sentence like *He sat down at the table with the red checkered tablecloth*, if it is produced with two intonational phrases, may be divided into approximately equal pieces, such as

He sat down at the table] [*with the red checkered tablecloth*

rather than as

He] [*sat down at the table with the red checkered tablecloth*

although the latter would more directly reflect the location of the major syntactic constituent boundary between the subject noun phrase and the verb phrase.

These examples, and many others, illustrate the observation that the major prosodic break of an utterance is not always located at the site of the major syntactic break, and that prosodic constituent structure does not always line up in straightforward ways with syntactic constituent structure. (See Hale and Selkirk 1987, Selkirk 1984 and others for discussion of the proposal that only one edge of a prosodic constituent must line up with the edge of a syntactic constituent.)

Examples like this suggest that prosodic structure is influenced not only by syntactic structure but also by other considerations; some of the factors that have been suggested are speaking rate, number of words and syllables, semantic focus, discourse structure and rhythm. Such considerations lead naturally to what might be termed the Prosodic Planning Hypothesis: the view that prosodic structure governs the systematic changes in the phonetic shape of words that are often observed in fluent speech, including not only traditionally prosodic aspects such as patterns of duration, F0 and amplitude, but also segmental modification in context (i.e. the phonetic implementation of the abstract meaning-contrastive specifications of words (Nespor and Vogel 1986)), and may even provide the basic framework for the planning of an utterance, including the serial ordering of its words and segments (Shattuck-Hufnagel 1992). A corollary of this Prosodic Planning Hypothesis is that the morphosyntactic structure of a sentence affects the phonology only indirectly, insofar as it combines with other factors to determine the utterance-specific prosodic structure for a given rendition of that sentence (see Shattuck-Hufnagel and Turk 1996 and references therein).

This paper reviews three lines of evidence from studies done in our laboratory that support this view, and relates them to results from other laboratories that bear on the same claim. The first line of work investigates the phenomenon which has sometimes been called 'stress shift', e.g. when words like *JapaNESE* are produced with prominence on an early syllable in phrases like *JAPanese FOOD* (where upper case indicates a prominent syllable). We view this as the early occurrence of a phrase-level intonational prominence (pitch accent) within the word, rather than as the leftward shift of lexical stress, and postulate that placement of pitch accents within their words is governed to some extent by the prosodic constituent structure of the utterance (Section 2). This study was inspired in part by Bruce's (1977) investigation of the importance of the location of F0 excursions within words and syllables for the contrasting specification of different word accents in Swedish. The second set of investigations concerns the glottalization of word-initial vowels, as in *a, and, of, illegal, opportunity*, etc. We have confirmed and extended Pierrehumbert and Talkin's (1992) finding that such glottalization is also governed by prosodic structure, i.e. is more likely to occur at intonational phrase onsets and pitch-accented syllables (Section 3). The third line of evidence suggests the importance of phrase-level rhythm in speech production planning (Section 4). Taken together with emerging results from other laboratories, these findings raise the possibility that phrase-level prosody, in the form of both constituent structure and prominence patterns, plays an important role in the planning and production of utterances. The possibility that prosodic structure provides the framework of structural locations or 'slots' that guides the serial ordering of morphosyntactic components during the planning of a particular utterance is explored in a preliminary way in Section 5. The findings discussed in this paper raise a host of questions about the precise nature of the role that prosodic structure plays in speech planning representations, as well as about the precise form those representations take; some of these questions are summarized in Section 6.

2. INTONATION STRUCTURE DETERMINES 'STRESS SHIFT'

2.1 Apparent 'stress shift'

The intonational phrase (Pierrehumbert 1980, Beckman and Pierrehumbert 1986) is a constituent in many versions of the prosodic hierarchy. If we can find evidence that the intonational phrase structure of an utterance influences other aspects of its production, that evidence would support the hypothesis that prosodic constituent structure plays a role in speech production planning. Such evidence has emerged from a recent reanalysis of the phenomenon of 'stress shift'. This term has been used to describe the early occurrence of prominence in a word with late main lexical stress, as in *the MASsachusetts GOVernor* (where upper case represents the early prominence in *MASsachusetts*, apparently shifted from its citation location on the main-stress syllable *-chu-*, and also represents the expected main stress prominence in *GOVernor*).

Since Liberman's (1975) and Liberman and Prince's (1977) groundbreaking work on metrical phonology, the phenomenon of apparent stress shift has played an important role in discussion of the rhythmic structure of utterances, and has provided a key argument for the inclusion of prosodic structure in linguistic representations. It was described as the movement of the major prominence in a word from its main-stress syllable to an earlier full-vowel syllable. Words that can undergo this leftward shift of prominence have a full-vowel syllable preceding their main-stress syllable that can serve as a 'docking site' (Beckman and Edwards 1994) for the earlier prominence, as in *Cornell*, *Japanese*, *Mississippi* and *Massachusetts*. For example, when the word *Mississippi* is uttered in isolation, in citation form, its main prominence occurs on the third syllable, as in *MissisSIPpi*. But when *Mississippi* is combined with a following word that carries a large prominence on its first syllable, such as *LEGislator*, the main prominence in *Mississippi* seems to shift to the left, appearing on the first syllable, *MIS-*. That is, many speakers have the intuition that the main prominence on *Mississippi* will occur on the first syllable when this word occurs in phrases such as *MISsissippi LEGislator*.

In Generative Phonology, this phenomenon is described as the leftward movement of main lexical stress in order to avoid a stress clash with the stressed syllable of the following word (Liberman 1975, Liberman and Prince 1977, Prince 1983, Hayes 1984, Nespor and Vogel 1986, 1989, Selkirk 1984, among others), and the maintenance of a quasi-regular rhythmic pattern of prominent and less prominent syllables is believed to be a motivating factor (Hayes 1984). The effect is blocked by a phrase boundary, so that if the following clash context occurs in the next phrase, stress shift will not occur, e.g. *In MissisSIPpi] [LEGislators often FILlibuster.*

2.2 Early empirical tests of 'stress shift'

Although the intuitions behind this formulation of stress shift are strongly systematic for many native speakers of English (and of Dutch, see Gussenhoven 1991), early empirical tests of the predictions of the theory did not produce encouraging results. One problem was finding a reliable acoustic correlate of the shifted prominence. For example, Cooper and Eady (1986) looked at the durations and F0 patterns of the syllable that is predicted to receive the shifted stress when the word occurs in a clash context, compared with the same syllable in non-clash contexts (e.g. *Penn-* in *PENNsylvania RELatives* vs. in *PennsylVANia*). They were unable to find any systematic acoustic correlates of shifted stress in clash contexts. In the perceptual domain, Beckman and her colleagues (1987, 1990) looked at the location of perceived prominence in clash and non-clash contexts, and found that even non-clash contexts can reliably produce a perceived relocation of the major prominence, e.g. for the word *Chinese* in the phrase *Chinese antiques*, where prominence can occur not on the main stressed syllable *-nese* but on the earlier full-vowel syllable *Chi-*, i.e. *CHInese anTIQUES*. Why should an apparent stress shift be perceived in non-clash contexts, and why should it be so difficult to measure its correlate even in clash contexts?

2.3 'Stress shift' as Early Accent Placement within the word

A possible explanation of these findings is that, in many cases, apparent stress shift does not reflect the movement of lexical main stress to the earlier syllable, but is better described as the combination of two events: occurrence of a phrase-level intonational prominence (called a pitch accent, Bolinger 1958) on the earlier full-vowel syllable e.g. the *CHI* - of *Chinese*, and non-occurrence of a pitch accent on the later main-stress syllable *-nese*, e.g.

CHI	*nese*	*anTIQUES*
|	|	|
Pre-nuclear pitch accent	No pitch accent	Nuclear pitch accent

On this view, prominence on an earlier-than-expected syllable within the word is an intonational phenomenon, and reflects the tendency of speakers to place a phrase-level prominence (pitch accent) on a pre-main-stress full-vowel syllable within the accented word, in certain well-defined prosodic contexts. Beckman and Edwards (1994), Bolinger (1981), and Shattuck-Hufnagel (1989) have proposed that one such circumstance is when the word bears the first accent in an intonational phrase, and Gussenhoven (1991) and Monaghan (1990) have suggested that this might arise from a constraint on pitch accent deletion that obligatorily protects the first pitch-accentable syllable of a phrase.

Analysis of pitch accent placement within words and phrases in limited samples of read laboratory speech (Beckman et al. 1990, Horne 1990, Shattuck-Hufnagel 1991) supported the hypothesis that apparent stress shift is perceived when Early Pitch Accent within the word occurs. Subsequently, the existence of a corpus of prosodically labelled Boston University FM radio news speech (Ostendorf et al. 1995) provided the opportunity to test this hypothesis for a wider range of prosodic contexts. The speech in the BU FM corpus was produced by seven different speakers, with communicative intent, and most of the corpus was recorded in the studio during broadcast; additionally, four news stories were read aloud in the laboratory in radio news style by six of the seven speakers. The portion of the corpus that we analysed had been labelled for prosodic structure using the ToBI system (Silverman et al. 1992), so-named because it combines some of the Tonal elements proposed in Pierrehumbert (1980) and Beckman and Pierrehumbert (1986) as part of the intonational grammar of American English, and the between-word Break Indices suggested by Price et al. (1991) as an estimate of the depth of the prosodic boundary between each pair of words in a spoken utterance. (See Pitrelli et al. 1994 for an empirical evaluation of the reliability of this transcription system.) Among other things, ToBI labels specify the location of boundaries of Full Intonational Phrases and Intermediate Intonational Phrases and the location of pitch accents by syllable. (The contrast between Full and Intermediate Intonational Phrases was proposed by Beckman and Pierrehumbert (1986). An Intermediate Intonational Phrase is a string of words produced with a coherent intonation contour containing at least one pitch accent and a following phrase accent that controls the F0 from the final or nuclear pitch accent to the end of the phrase. The Intermediate Intonational Phrase is lower in the prosodic hierarchy than the Full Intonational Phrase, which Pierrehumbert (1980) defined with the same tonal requirements as the Intermediate Intonational Phrase, but with the added requirement of a Boundary Tone associated with the phrase boundary and realized on the final syllable of the phrase. A well-formed Full Intonational Phrase contains one or more Intermediate Intonational Phrases.) Shattuck-Hufnagel et al. (1994) examined the distribution of pitch accents within more than 400 shiftable words like *Massachusetts, Japanese* and *illegal* in a subset of the prosodically labelled FM radio news database. Results confirmed that a) the first or onset pitch accent of an Intermediate Intonational Phrase tends to occur on the earlier full-vowel syllable of a stress-shiftable (now Early Accentable) word, b) early accent within the word is more likely when its placement on the main-stress syllable would result in a clash with a pitch accent on the first syllable of the following word, and c) the nuclear or final accent of an Intermediate

Intonational Phrase occurs on the main stress syllable of the last accented word. In this sense, the phrase-onset prenuclear pitch accent and the final (nuclear) pitch accent of an Intermediate Intonation Phrase together bracket the bulk of the phrase, perhaps giving the listener a partial cue to the intonational phrase structure of the spoken utterance.

2.4 Additional findings re Early Accent Placement

This account of apparent stress shift as the Early Placement, within its word, of the first pitch accent of an intonational phrase is strengthened by another finding reported in Shattuck-Hufnagel et al. (1994). In the BU FM news corpus, if an Early-Accentable word carries all the pitch accents in an intonational phrase, that word has a tendency to be double-accented, with a prenuclear pitch accent on the earlier full vowel syllable and the nuclear pitch accent on the later main-stress syllable, as in *It was in MASsaCHUsetts*. This observation supports Bolinger's (1965) proposal that a well-formed intonational phrase has both an initial early pitch accent and a final late pitch accent. (See also the 'hat pattern' described by 't Hart and colleagues in 1975 and 1990.) It also provides an argument against the view that main stress must shift to the earlier full vowel syllable in order for a pitch accent to occur there, since, in a double-accented word, a pitch accent occurs on both the earlier full vowel syllable and the later main-stress syllable.

The early-accent view of apparent stress shift provides an account of Cooper and Eady's (1986) report that the duration and F0 of the presumably shifted-to syllable do not change systematically in stress clash contexts, where a shift of stress leftwards to the earlier syllable in the word was predicted to occur, to prevent a stress clash. This result may have been obtained because the target syllable was pitch accented in both the stress-clash and non-stress clash contexts, in parallel to Beckman et al.'s (1987, 1990) finding that *Chinese* was early-accented in both a clash context (*CHInese DRESser*) and a non-clash context (*CHInese anTIQUE*). In the non-clash condition, Cooper and Eady's target words like *Pennsylvania* may have been produced with both a phrase-onset pitch accent and a nuclear pitch accent (*PENNsylVANia*), and in the clash context with just a phrase-onset pitch accent (*PENNsylvania RELatives*). If this were the case, then no difference between the clash and non-clash cases would be predicted. When the initial syllable *Penn-* was compared for the clash vs. no clash conditions, measures may been indistinguishable because in both cases this syllable carried the initial pitch accent of the intonational phrase. (Just such a pattern of pitch accent locations was reported for a different set of clash vs non-clash utterances, such as *MISsissippi LEGislator* and *MISsisSIPpi,* in Shattuck-Hufnagel 1991).

The BU FM radio news data also showed that both intonational constituent structure and rhythm influence the occurrence of Early Pitch Accent within the word, just as earlier generative phonologists had observed in the stress shift framework (see especially Hayes 1984). That is, Early Accent was not only associated with the onset accent in a new phrase, but was also more likely to occur if the accented word was immediately followed by a pitch-accented syllable in the same intonational phrase, i.e. by a pitch accent clash context. An example from the BU corpus is *MASsachusetts HOSpitals*. Bolinger (1981) suggested that pitch accent rhythm, along with the rhythmic alternation between full vowels and reduced vowels, was responsible for the rhythmic patterns of spoken utterances, and Selkirk (1984) also suggests a role for pitch accent. She notes that stress shift on a target word requires a clash with a larger prominence on the first syllable of the following word (in her written examples, it appears that this is likely to be the nuclear pitch accent of the phrase) and postulates that the shift of main lexical stress to an earlier syllable of the target word is followed by reassignment of pitch accent to that syllable as well. The results for the FM radio news data (showing that potential pitch accent clash with the following word is avoided by Early Pitch Accent in the target word) support the view that pitch-accent rhythm plays a role in production planning.

2.6 Implications of Early Accent Placement for the active use of prosodic structure in phrase-level phonological planning.

The results of these studies of early prominence within the word offer substantial support for the claim that prosodic constituent structure plays an active role in the planning of pitch accent location within the word, at least for the languages that have been examined (English and Dutch.) That is, the location of an Early Accentable word in its Intermediate Intonational Phrase strongly influences the placement of pitch accents within the word: phrase-onset accents tend to occur on an early full-vowel syllable, and phrase-final accents on the main-stress syllable. This observation indicates that Intermediate Intonational Phrases are a part of the underlying representation that speakers refer to when they plan their speech. The results also show that speakers are reluctant to place pitch accents on two adjacent syllables, indicating a role for the rhythmic regularization of phrase-level prominences (see Section 4). These studies underline the importance of analysing the actual prosody used by speakers of particular utterances, since this may be a) different from theoretical predictions based on native speaker intuitions and preferences, b) highly variable from speaker to speaker, and c) variable from utterance to utterance by the same speaker. The findings confirm an emerging consensus that

– intonational phrases can have additional accents beyond the nuclear pitch accent (the tonic, in British tradition; see Halliday 1967),

– individual words can have more than one pitch accent,

– the placement of pitch accents within the word is systematic and therefore potentially informative about larger structures, and

– it is important to keep track of which syllables within the word are accented, as well as of which words in the phrase are accented.

2.7 *Unresolved questions re Early Accent Placement*

To date, published studies leave a number of holes in our understanding of the factors that govern pitch accent placement within the word. For example, there has not yet been a direct test of an alternative account in terms of surface syntactic structure, like that provided by Ferreira (1991) for pre-boundary lengthening, discussed above. Shattuck-Hufnagel et al.'s (1994) analysis did not directly compare the efficacy of intonational phrase structure vs. syntactic structure in accounting for Early Accent within the word, in part because it is not immediately obvious what level or type of syntactic constituent might be compared with the prosodic constituent. Another question of interest is whether the overall pattern of pitch accent placement within intonational phrases is better accounted for by a model of pervasive pitch accent placement on every full vowel, with subsequent deletion of alternating accents moving leftward from the nuclear pitch accent, and with obligatory sparing of the initial accent of the phrase (Gussenhoven 1991, Horne 1990, Monaghan 1990), or by a model of sparse pitch accent placement in the locations required by factors such as phrase onset marking, semantic focus and rhythm (Shattuck-Hufnagel et al. 1994). Related to this last question is the problem of lapses: i.e. whether pitch accent placement repairs lapses in rhythmic regularity as well as rhythmic clashes, as suggested (for stress shift) by Nespor and Vogel (1989) and Hayes (1984).

Despite these lacunae, the sum of the evidence supports the claim that the placement of early pitch accent in the word is sensitive to the position of that accent in the intonational phrase structure of the utterance, as well as to the location of nearby pitch accents. This evidence for the psychological effectiveness of prosodic structure is compatible with the findings that will be described in the next section, where again it appears that prosodic structure provides a reasonable account of what would otherwise be a set of seemingly non-systematic observations.

3. INTONATIONAL STRUCTURE DETERMINES THE GLOTTALIZATION OF WORD-INITIAL VOWELS

Further evidence for the claim that prosodic structure is represented during production planning is provided by the observation that at least some aspects of systematic variation in the phonetic shapes of words depends on their location in the prosodic constituents of each specific utterance. Such evidence has emerged from the study of variation in the sound source produced at the vocal folds in word-initial vowels, i.e. glottalization. (As we have noted for 'stress' and other terms used in prosody, terminology for voice quality is somewhat confusing. 'Glottalization' here refers to a vocal gesture which is perceived by the listener to involve the glottis, and which affects the wave form in various ways which are still being defined. (See Batliner et al. (1993), Dilley et al. (1996) and Hagen (1997) for discussion.)

3.1 In what contexts does glottalization of word-onset vowels occur?

Word-initial vowels, as in *Arlene*, *apple*, and *Alabama*, are commonly glottalized in American English, but not always. Early formulations sought to explain the observed distribution in terms of the immediately preceding segmental context (e.g. more common at a vowel-vowel hiatus as in *the apple*) or in terms of syntactic structure and word stress (Umeda 1978, Gimson 1980). However, these formulations left many examples unaccounted for. The development of an explicit grammar of prosodic structure over the past several decades, as well as a transcription system to capture some of the generally agreed-on prosodic aspects of spoken utterances (Silverman et al. 1992), has made it possible to test the hypothesis that prosodic structure provides a more complete account of the distribution of glottalization for word-initial vowels.

The hypothesis that glottalization is more likely for word-initial vowels that occur at prosodically significant locations, such as the onset of a new intonational phrase or a pitch accent, was first tested by Pierrehumbert and Talkin (1992). They looked at utterances containing word-initial vowels in a limited set of contexts, and found a strong effect of both intonational phrase boundary and pitch accent for their two speakers. These results were extended in Pierrehumbert (1994): similar effects were found in a different segmental and prosodic context, one which was no longer associated with low tones which might themselves elicit a creaky, glottalized source.

3.2 Glottalization of word-onset vowels in fluent communicative speech

The prosodically-labelled Boston University FM radio news speech (described above and in Ostendorf et al. 1995) provided the opportunity to test the hypothesis that glottalization occurs at prosodically significant locations for a wider range of speakers, segmental contexts and prosodic contexts. A subset of more that 3700 vowel-onset words were additionally labelled for whether or not the onset vowels with lexical stress were actually produced with full-vowel quality. (Roughly, a lexically-stressed syllable is regarded as one with full vowel quality in the lexical representation of its word; however, full-vowel syllables that follow the main-stress syllable, like *-chute* in *parachute* or *-o* in *logo*, are not generally regarded as lexically stressed.) We carried out this analysis because preliminary observation suggested that vowels marked with some degree of stress in the lexicon can be severely reduced in actual utterances. Finally, the target words were labelled for whether or not a glottalization occurred at each word-initial vowel. Although earlier studies had postulated that only lexically stressed vowels would be glottalized, Pierrehumbert and Talkin (1992) reported consistent glottalization of reduced vowels as well, so we examined both kinds of vowels when they occurred in word-initial position. The criterion for a positive judgment of glottalization required both perception of glottalized quality, and visual observation of irregular pitch period duration. The labelled subset of vowel-initial words were produced by five of the seven corpus speakers, two male and three female.

The results of this study (Dilley et al. 1994, Dilley and Shattuck-Hufnagel 1995, Dilley et al. 1996) confirmed Pierrehumbert and Talkin's (1992) original findings, i.e. an increased likelihood of glottalization for word-initial vowels at intonational phrase onsets and at pitch accents, now observed for additional speakers, a different speaking genre and a set of unconstrained segmental and prosodic contexts. To determine whether these effects were attributable to factors such as delayed offset of glottalization at the end of a preceding phrase or the startup of phonation after a pause, or whether they could be attributed to prosodic structure per se in the form of an intonational phrase boundary or pitch accent, we analysed separately the tokens that were NOT preceded by a pause longer than 50 milliseconds or a glottalized segment. This subset of tokens showed the same pattern of results, indicating that the presence of an intonational boundary or prominence itself, with or without a preceding pause or glottalized segment, increases the likelihood of glottalization for word-initial vowels, in a wide variety of prosodic and segmental contexts.

In addition, because of the size of the data set, we were able to report several new findings. For example, the rate of glottalization for reduced vowels was significantly greater for Full Intonational Phrases than for Intermediate Intonational Phrases, providing additional evidence for the distinction between these two levels of constituents (Beckman and Pierrehumbert 1986). Because the ToBI system for transcribing prosody provides pitch accent labels by individual syllable rather than simply for the word as a whole, we were also able to distinguish the effect of pitch accent on the target syllable vs. on a later syllable in the word. Results showed that an accent on the word-initial vowel itself was more likely to elicit glottalization than an accent on a syllable later in the word; the presence of a pitch accent later in the word increased the likelihood of glottalization of the initial vowel somewhat, but not as much.

3.3 Individual speaker variation in glottalization behavior

One of the most striking findings of the study was the degree and nature of the differences among the five individual speakers. Although the association of glottalization with intonational phrase onsets and pitch accents was confirmed for all speakers, the overall rates of glottalization were strikingly different across the five speakers, ranging from 13% to 44% of tokens. In addition, the preferred acoustic manifestation of glottalization was often strikingly different (Dilley and Shattuck-Hufnagel 1995). For example, one speaker often showed the kind of widely separated pulses, each characterized by exponential decay, that are associated with glottal stop; other speakers more often showed irregular pitch period duration in the first few recognizably vocalic pitch periods of the vowel itself; and yet another speaker often exhibited a different phenomenon (i.e. reduced amplitude in the absence of any noticeable durational irregularity) at the same locations (intonational phrase onsets and pitch accents) where he sometimes glottalized (see Hillenbrand and Houde (1996) and Pierrehumbert (1994) for further investigation of the acoustic correlates of perceived glottalization).

3.4 Preliminary results for phrase-final glottalization

The variety of audible acoustic phonetic patterns and visible wave-form patterns associated with the perception of glottalization suggests the importance of further study of the underlying mechanisms, to determine whether the striking differences observable in visual displays of the signal are associated with different physiological mechanisms, or reflect different degrees of attainment of a single glottal gesture or family of gestures. It would also be useful to determine the relationship between the glottalization we observed in word-initial vowels and that reported for final voiceless stops (e.g. for final /t/ and /p/ in American English (Pierrehumbert 1994) or in phrase-final creak (Henton and Bladon 1988, for several dialects of British English.) Kohler's (1994) extensive analysis of glottalization in German suggests there may be differences between languages, and it would be interesting to investigate whether these differences could be described in terms of differences in prosodic structure.

Another point of interest for the hypothesis that voice source variation is conditioned in part by intonational structure emerged from a pilot study of phrase final creak, which is characterized by irregular and/or widely spaced pulses in the sonorant segments at the end of a phrase. In a corpus of utterances drawn from a number of different sources and ToBI-labelled, a pilot study found almost no examples of phrase-final creak at the ends of Intermediate Intonational Phrases, but many occurrences at the ends of Full Intonational Phrases, possibly because F0 fell to lower levels in the final fall at the end of the larger constituent. This finding offers additional support for Beckman and Pierrehumbert's (1986) proposal for the distinction between the two levels of intonational phrases. It also suggests that, like pitch accent placement within the word (Section 2 above), the type of voice source variation we call glottalization often brackets the intonational phrase when segmental circumstances permit: glottalization of an onset vowel, and of the final portion of the phrase. Interestingly, we occasionally observed the glottalization of other word-onset sonorant segments, such as /w/ or /l/, when these segments occurred at the beginning of the phrase.

3.5 Other acoustic and articulatory correlates of prosodic constituent structure

The observation that the distribution of glottalization of word-initial vowels is sensitive to the presence and size of phrase-level prosodic constituent boundaries (Intermediate and Full Intonational Phrases), and also to the presence of phrase-level prosodic prominences (Pitch Accents) strengthens the hypothesis that the elements of the prosodic hierarchy play a role in the phrase-level representations that speakers generate when they plan utterances for production. Such results are compatible with an additional line of research at the Phonetics Laboratory at UCLA, which looks at patterns of articulation in different prosodic contexts. Using reiterant utterances, Fougeron and Keating (1997) have shown a pattern of articulatory strengthening for the initial consonants of prosodic constituents, as measured by amount of tougue contact with the palate during /n/. Their results show not only that the initial consonant is articulated with greater contact than constituent-internal consonants, but also that the degree of strengthening is greater for the initial consonant of constituents higher in the prosodic hierarchy, e.g. utterance > phrase > word. Durational results show a similar pattern.

Despite substantial differences between individual speakers, these effects are reliable and consistent, and have been replicated for French by Fougeron (1996). They indicate a substantive role for prosodic structure in the planning of articulatory gestures. A similar conclusion is suggested by reports that the degree of pre-boundary, i.e. final lengthening increases systematically as the level of the prosodic constituent rises in the prosodic hierarchy (Wightman et al. 1992, Ladd and Campbell 1991, Berkovits 1993). Findings of this kind raise the possibility that prosodic structure may serve as the overall planning framework for the phonological and phonetic shape of an utterance. In the next section we turn to a study which bears on this hypothesis.

4. EFFECTS OF RHYTHM ON SPEECH PRODUCTION PLANNING: PRELIMINARY STUDIES

In addition to intonational constituent structure, another aspect of prosody that may play a role in production planning is rhythm. This term, like the related term 'stress', often goes undefined in theoretical discussions. It has at least two possible interpretations: a) regular alternations between greater and lesser prominence, and b) the pattern of greater and lesser prominences in an utterance, whether regular in alternation or irregular. The second meaning can be illustrated by the irregularly timed strong taps or beeps of a Morse code message. Although it is sometimes difficult to be absolutely certain which meaning an investigator is using, most discussions of rhythm in spoken utterances appear to be concerned with the first interpretation. That is, they ask whether speech tends to be organized in a metrically regular way. Empirical studies of spoken isochrony have found it difficult to establish such regularity (Lehiste 1973, Fletcher 1991, Wenk and Wioland 1982, Dauer 1983), although others suggest a role for regular timing, possibly at the level of the foot. (Conceptions of the relevant foot vary; see Fant, Kruckenberg and Nord 1991, Scott 1982, Abercrombie 1965, Thompson 1980.) It would appear to be intuitively reasonable that at least some aspect of the rhythmic organization of an utterance can cross word boundaries, i.e. concatenate words into feet, and perhaps even concatenate fragments of adjacent words into constituents. This can be seen in the following examples, where the symbol | divides successive feet:

| JACK had | TOLD me | ALL a- | -BOUT it | (from Lehiste 1973a)

and

| Your maCHINES | will soon PASS | at my DESK | toDAY | (Fourakis and Monahan 1988)

and

I | KNOW that on the | NIGHT be- | FORE there was a | LOT of expec- | TAtion and ex- | -CITE ment on my | PART | (Thompson 1980)

The theoretical status of these cross-word-boundary constituents is far from clear; see Abercrombie (1965) for some discussion).

4.1 Prosodic prominence and the definition of prosodic constituents

One aspect of the pattern of prominence placement in spoken utterances has to do with its role in defining constituents. Theories of metrical phonology have traditionally dealt with an explanation of 'stress' patterns, whether at the word level or at the phrase level, and recent linguistic discussions of this issue often postulate that prominence is an important aspect of the definition of constituents. For example, Halle and Idsardi (1995) discuss their metrical theory of bracketed grids in terms of Liberman's (1975) insight that "stress is a phonetic means for marking various kinds of groupings of linguistic elements". A related view, which emphasizes that each type of prosodic constituent has a most-prominent element or head, is described in Beckman and Edwards (1990, 1994). For example, the metrical foot has a full-vowel (lexically-stressed) syllable as its head; the Intermediate Intonational Phrase has the Nuclear Pitch Accented syllable as its head; etc. This formulation does not require a rhythmic regularity in the occurrence of prominent elements, but shares with Liberman the view that prominence marks structural groupings, focussing on the relationship between prosodic constituent structure and rhythmic prominence.

4.2 Regular patterns of prominence in perception and production

On the other hand, theories of prosodic prominence placement sometimes postulate a tendency toward regular rhythm at the phrase level. For example, Hayes (1984) discusses the possibility that speakers of American English prefer to place phrase-level prominences at regular intervals of e.g. 4 syllables, avoiding structures with intervals of 3 or 5 syllables. Nespor and Vogel's (1989) discussion of avoidance of both clashes and lapses of rhythmic stress embodies a similar assumption. This view focuses on the regular temporal relationship between successive rhythmic prominences, or isochrony, as reflected in the intuitions of native speakers. More quantitative studies of listeners' perception of rhythmic regularity in spontaneously spoken utterances (e.g. Couper-Kuhlen 1993; Dilley and Shattuck-Hufnagel 1996; Dilley 1997) suggest that some but not all portions of utterances can be reliably labelled by listeners as rhythmically regular. It would be particularly useful to have a clearer understanding of two aspects of speech rhythm: (1) which (if any) linguistic elements are associated with perceived beats, and (2) which acoustic parameters signal the occurrence of the beat. If the perception of rhythm involves a construction or interpretation by the listener, rather than a simple registration of regular acoustic alternation, the answers to these questions may be difficult to find (see Lehiste (1977) and Dilley (Forthcoming) for discussion). As we have noted, two types of linguistic contrasts are often proposed as the determinants of speech rhythm: the contrast between lexically-stressed and reduced syllables, and the contrast between pitch accented and non-pitch-accented syllables. Empirical analyses of rhythm at the lexical level, i.e. looking at the timing of sequential alternations between lexically stressed and unstressed syllables, have shown little evidence for a tendency toward isochrony in terms of absolute duration, even in languages that are sometimes described as isochronous. Few parallel empirical studies have focussed explicitly on pitch accent rhythm, although it is sometimes difficult to be sure precisely what kind of prominence is being studied, because the distinction between phrase-level and lexical-level rhythmic prominence patterns is not expressly considered. (In metrical phonology frameworks, grid-based representations with greater or lesser numbers of gridmarks in the column above each syllable, and tree-based representations with alternating strong-weak branches at different levels in the tree, are formal mechanisms for expressing several different levels of rhythm, where prominence at higher levels builds on prominence at lower levels. But the phonetic correlates of these patterns have not been easy to discover.)

It is possible that the rhythmic regularity so characteristic of the intuitions of native speakers of American English about utterances in their language is not defined in the domain of absolute time. However, there is quantitative evidence that speakers favor rhythmic regularity in the sense of alternations between elements with greater and lesser degrees of prominence, even if not at the level of regular temporal duration. For example, the finding (described in Section 2) that early accent placement in the word is influenced by pitch accent clash avoidance (Shattuck-Hufnagel et al. 1994) suggests a role for rhythm, in the sense that speakers of American English avoid placing pitch accents too close together. Nespor and Vogel (1989) also discuss the repair of both clashes (potential occurrences of prominences too close together) and lapses (potential occurrences of prominences too far apart.) This hypothesis is further supported by the observation that words with adjacent full vowels (e.g. *maintain*, *obese*, *employee*) behave irregularly with respect to both main-stress specification in the lexicon and pitch accent location in labelled utterances (Shattuck-Hufnagel 1995). If most lexically-full-vowel syllables are candidates for pitch accent placement, as Bolinger (1965, 1981), Gussenhoven (1991), Beckman and Edwards (1994) and others have suggested, and speakers prefer not to locate pitch accents on adjacent syllables, then adjacent-full-vowel words present a potential challenge to the rhythmic system, because they provide potential sites for adjacent pitch accents in the same word and therefore raise the possibility of a pitch-accent clash.

4.3 Are utterances with irregular patterns of prominence harder to produce?

These findings and discussions raise an interesting question about the possible role of rhythm in the planning of spoken utterances. Investigation of this issue is handicapped by our current lack of a theory of the phonology of rhythm: what are its elements, and how can they be combined to form distinct rhythmic patterns? However, it is possible to ask whether an utterance perceived to be produced with regular rhythm is harder or easier to produce than one with irregular rhythm, if one can systematically construct pairs of sentences which can be characterized in this way.

The hypothesis that representations of rhythm play a role in speech production planning has been examined in our laboratory in work on error elicitation in various prosodic contexts on speech error rates. In pilot experiments to address this issue, we asked speakers to produce two kinds of sentences: those designed to elicit a metrically regular production, characterized by (or at least easily permitting) a regular and predictable alternation between strong and weak syllables, and those designed to elicit a production which is metrically irregular, characterized by sequences of full-vowel syllables separated by different numbers of weaker-vowel syllables, such as monosyllabic function words and reduced-vowel syllables of content words. The contrasting sentences of a pair contain the same number of syllables (and a subsequence of the same words), but differ in the rest of their phonological content in a way which makes it likely that speakers will produce different patterns of strong vs. weak syllables, at least according to the intuitions of the stimulus designers. For example, the sentence *I asked her to slip me a sip on the spot* is designed to elicit rhythmically regular utterances, such as

I ASKED her to SLIP me a SIP on the SPOT.

(In these representations, upper case indicates rhythmic prominence.) In contrast, the sentence *I asked my guest to slip me a sip real fast* is designed to elicit rhythmically irregular utterances, such as

I ASKED my GUEST to SLIP me a SIP REAL FAST.

Speakers were asked to produce the sentences both in normal form and using reiterant imitation, in which the normal utterance is imitated as closely as possible with a string of fluently integrated repetitions of the single syllable *ma*. Three renditions of the normal and reiterant versions were elicited for each of 30 such pairs, in two sessions of 30 sentences each, counterbalanced for presentation of intended-rhythmic vs. intended-non-rhythmic members of each pair.

Pilot results for the reiterant versions produced by five subjects suggest that sentences that were designed to be rhythmically irregular evoke more errors than do sentences designed to be rhythmically regular. In addition, speakers judged the designed-to-be irregular sentences to be more difficult to produce.

Thus, initial indications are that, at least for this task, sentences which intuitions suggest will be produced with regular rhythm are measurably easier for speakers to imitate in reiterant form than sentences which intuitions suggest will be produced with irregular rhythm. Additional analyses will be required to deal with an unexpected phenomenon: for some of the sentences designed to elicit rhythmically irregular utterances, speakers were able to impose a rhythmic regularity at the phrasal level despite the irregularity at the level of alternation between intuitively stronger and weaker syllables. For example, the sentence *We dressed our kids in black and white in back of Johnny's restaurant* was designed to be produced with a regular alternation between weak and strong syllables. As expected, renditions of this sentence by the 5 pilot speakers are perceived as rhythmically regular, as expected:

> *We DRESSED our KIDS in BLACK and WHITE in BACK of JOHN-ny's REStaurant.*

The contrasting sentence *We took five rolls of black and white in back of John's restaurant* was designed to elicit rhythmically irregular productions, such as

> We TOOK FIVE ROLLS of BLACK and WHITE in BACK of JOHN'S REStaurant.

However, speakers produced this sentence in such a way as to permit a perceptible rhythmic regularity, apparently weakening *five* and adjusting *John's* in some way to produce a relatively regular alternation between stronger and weaker elements:

> *We TOOK five ROLLS of BLACK and WHITE in BACK of JOH-OHN'S REStaurant.*

This apparent distinction between rhythm defined as the potential for a natural alternation between stronger and weaker syllables, as in *DRESSED our KIDS,* and rhythm defined at the level of alternation between syllables with greater and lesser phrase-level rhythmic prominence, as in *TOOK five ROLLS*, is somewhat reminiscent of the description in Bolinger's (1981) paper "Two kinds of vowels, two kinds of rhythm". Bolinger postulates that the rhythm of full vs reduced vowels supports an additional, higher-level rhythm, which is defined by an alternation between pitch-accented vs. non-pitch accented full-vowels. Although it is not clear that the phrase-level rhythm which we observed anecdotally corresponds to a pattern of pitch-accented vs. non-pitch accented full vowels, since these utterances have not yet been prosodically labelled, our observation does lend support to the notion of a phrase-level rhythm that builds on the rhythm defined by stronger vs. weaker vowels or syllables. It also highlights the importance of determining the prosodic pattern that is actually used by the speaker of stimulus utterances, rather than assuming that one's intuitions as a native speaker will predict speaker behavior (see Shattuck-Hufnagel (1995) and Shattuck-Hufnagel and Turk (1996) for extensive discussions of this point.) We are pursuing the question of how perceived rhythmic regularity affects both error rates and perceived difficulty of production using the method developed by Dilley and Shattuck-Hufnagel 1996; Dilley 1997, Forthcoming) for labelling the perceived rhythmic structure of an utterance, and the ToBI transcription system to locate the pitch accents. Although these investigations are at a very preliminary stage, when combined with the evidence for rhythmic constraints on pitch accent placement discussed above in Section 2 they indicate that it may be important to understand the role of rhythm in speech production planning.

5. THE ROLE OF PROSODIC STRUCTURE IN SPEECH PRODUCTION PLANNING: THE PROSODIC PLANNING HYPOTHESIS REVISITED

If prosodic structure plays an active role in speaking, as the previously described results suggest it does, then it is natural to ask exactly what this role is. One hypothesis (Shattuck-Hufnagel and Turk 1996) is that, when an utterance is being planned, the developing prosodic structure is the locus of integration for a number of different factors that influence the shape of an utterance. These might include syntactic structure, semantics, discourse structure, pragmatics and emotion, as well as performance constraints which are not related to the specification of the message, such as speaking rate, preferred limits on number of elements per phrase, preferred patterns of alternation in rhythmic strength, memory constraints, etc. (Nespor and Vogel 1983). Furthermore, it may be that the resulting prosodic structure of an utterance provides a kind of performance framework, into which the abstractly-specified elements of the segments, words and phrases of the morphosyntactic representation must be fitted in order to receive specifications for the aspects of their duration, amplitude, F0 etc. that are related to phrase-level (as opposed to lexical-level) phonology and phonetics. This is reminiscent of Mohanan's (1986) post-lexical phonology module, although that formulation did not invoke prosodic phonology per se; see Inkeles and Zec (1990) for a lively discussion of the relationship between lexical and postlexical phonological phenomena, and Mohanan (1995) for an update of his views.

5.1 A production planning framework separate from its contents

The Prosodic Planning Hypothesis, that prosodic structure serves as the planning framework for phonological and phonetic planning of a specific utterance of a morphosyntactically specified phrase or sentence, fits comfortably with the claim described in Section 1 in the discussion of the computation of intonational, timing and segmental-modification patterns that depend on higher-level structure. In other words, it is compatible with the view that an utterance-specific prosodic framework, separate from the surface syntactic structure but influenced by it, is developed for each utterance during the planning process, and that this framework is what determines the eventual phonetic shape of the utterance. Although not always couched in terms of prosodic structure, such a content-independent but utterance-specific frame has long been proposed on the basis of constraints and patterns observed in speech errors (Garrett 1975, 1980, 1982; Shattuck-Hufnagel 1979, 1983, 1987, 1992). Lashley (1951) pointed out that the occurrence of exchange errors such as *to cage Pevin* for *to page Kevin* suggests that the word *Kevin* is actively being processed well before the moment of its scheduled occurrence in the utterance; otherwise, its /k/ would not be available to replace the initial /p/ in the target word *page*. Shattuck-Hufnagel (1979) extended this line of thinking by noting that the symmetrical occurrence of the two exchanged error segments in each other's locations suggests that the speaker computes the location-defining framework not only before the articulation of the first target word but also independent of the framework's contents. If there were no such framework, then the segment displaced by the first part of the exchange error could occur in some random later location – but this has not been commonly observed. Apparently, the framework maintains the downstream 'slot' even though its target segment has already been used to fill an earlier slot, and the element displaced for the earlier slot becomes the most likely candidate to fill the later, downstream slot.

Production planning models in this vein postulate a set of stored elements, a planning frame of larger structures, and a mechanism for integrating the elements into the frame. Shattuck-Hufnagel (1992) raises the possibility that the planning framework for the serial ordering of individual phonemic segments is related to the prosodic structure of an utterance. Levelt (1989), Roeloffs (1992) and others have also formulated phonological encoding models which invoke a prosodic frame, and Dell (1986) has suggested that the planning framework is organized into canonical syllables. Initial formulations of the phonological encoding process have focussed largely on modelling the production of single word utterances, where phrase-level prosodic structure is not particularly complex. However, as planning models begin to address multiword and multiphrase utterances with more complex prosody, it will become increasingly important to deal with the two aspects of phrasal phonology discussed at the beginning of this paper: phrase-level prosody, such as intonation and timing, and phonetic modifications both within words and across word boundaries. Development of models that can address these questions will allow us to evaluate the hypothesis that the planning structures that govern the serial ordering of segmental elements in spoken utterances also govern prosodic parameters such as intonation and timing as well as contextual constraints on segmental modification.

5.2 Evidence that the planning framework is represented separately from its contents

Whatever the precise nature of the planning structure, there is growing experimental evidence for the claim that speakers represent such an utterance-specific framework separately from its segmental contents. For example, a number of priming studies have shown that the pronunciation of a word is facilitated by previous exposure to a word that is similar in CV shape, even if the segmental content that makes up the CV shape is different (Sevald and Dell 1994, Meijer 1994). This is presumably because the construction of a frame is made easier by the prior processing of a similar frame. In addition, evidence from phoneme-level speech errors shows that speakers tend to add a segment to the onset consonant of a word if a nearby word also has a CC onset, again suggesting a role for the word frame independent of its segmental contents (Stemberger 1990). Although these studies deal with single-word utterances rather than the complexities of fluent conversational speech, they give empirical support to the claim that a sentence frame independent of its contents plays a role in production processing, and prosodic structure is a natural candidate for this structural frame. One can speculate that serial ordering errors occur during the integration of morphosyntactic components into the developing prosodic frame, which can then be used to compute aspects of the phonetic shape of the utterance which are not specified by morphosyntactic information. Investigation of this possibility will be facilitated by the collection of a large number of such errors in tape-recorded form, so that the prosody of the utterance, the error (and the error correction, if any) can be analysed (see Cutler 1983). Such a resource will be provided by the MIT Digitized Error Corpus, now under construction in our laboratory. In addition, the experiment on rhythmic regularity described in Section 4 above will measure the effect of rhythm on segmental serial ordering errors. If a connection can be established between prosodic factors such as rhythm and the likelihood of serial ordering errors, it will strengthen the hypothesis that prosodic structure plays a role in the serial ordering process. In the final section of the paper, we relate this issue to a number of other questions about the role of prosody in speech production planning.

6. UNRESOLVED QUESTIONS

This paper has summarized some aspects of the growing body of evidence which shows that prosody plays an important role in speech production planning. The increasing specificity in statements of prosodic theory makes it possible to develop increasingly explicit predictions about the effects of postulated prosodic structure on measureable aspects of the phonetic shape of spoken utterances. This makes it likely that the next decade will see the emergence of a comprehensive theory of prosodic structure, of the factors that influence a speaker's choice among the possible well-formed prosodic structures for a particular utterance of a given sentence, and of the phonetic consequences of those choices.

A number of themes emerging from current research suggest lines that this future research might take. One is the need for direct comparison between predictions made by surface syntactic structure vs. predictions made by prosodic structure about phonetic modification, in contexts where these predictions are maximally different. (For an alternative approach in which the grammar defines the intonational constituents, see Steedman 1991, Prevost and Steedman 1994, but see also Beckman 1996). To put this issue in more general terms, a better understanding of the relationship between prosodic structure and the various factors that influence it would be a great advance. Another useful direction would be the exploration of the striking individual speaker differences that have emerged from current studies, both in the kinds of structures that are evidenced in measurable dimensions of the speech signal, and in the nature of the acoustic and articulatory correlates that speakers make use of for the same prosodic structures. Differences among languages, both in prosodic structures and in the production processing mechanisms that make use of them, are another fertile area for study. For example, many investigators have reported that segmental interaction errors in American English, such as *He'll get his tar kowed away* for *...car towed away*, occur preferentially between word-onset consonants (Shattuck-Hufnagel 1992 and references). However, Igoa et al. (1987) report no such asymmetry in favor of word-onset consonants in a corpus of errors in Spanish, where the syllable provides a better account of the position similarity constraints on patterns of segmental interaction errors. What does this kind of difference suggest about either the processing representations that underlie segmental interaction errors, or about the prosodic structures of the language? Finally, it would be of great interest to have a better understanding of the relationship among various aspects of prosodic structure, such as rhythm, intonation and constituents that govern phonological processes. For example, Jun (1993) has argued that the intonational constituents of Korean are equivalent to the constituents whose boundaries determine whether a phonological rule will apply or not. Hayes and Lahiri (1991) reach similar conclusions about Bengali. Nespor and Vogel

(1986) integrated intonational phrases into their hierarchy of prosodic constituents, and concentrated their attention on evidence that the domains of segmental modification rules could offer in support of their proposed hierarchy of constituents.

The question of the relationship among constituent structures that govern such apparently disparate domains as timing, F0 contours, the location of prominences and segmental modifications as well as serial ordering processes and errors will be a challenging but important one in the next decade. Despite (or perhaps because of) the large gaps in our understanding of what prosodic structure is, what determines its shape for a given utterance and how it is manifested in the signal, it appears that the study of effects of prosodic constituent structure and prominence on the acoustic-phonetic shape of utterances will provide a rich source of new information on how speakers represent the information they need during sentence processing for speech production.

NOTES

* Work supported by NIH (RO1 DC 02125), NSF (IRI-8905249), and NSF-ARPA (IRI-9314967).

REFERENCES

Abercrombie, D. 1965. Syllable quantity and enclitics in English. *Studies in Phonetics and Linguistics*. Oxford: Oxford University Press, 26-34.

Batliner, A., Burger, S., Johne, B. and Kießling, A. 1993. MÜSLI: A classification scheme for laryngealizations. *Proc. ESCA Workshop on Prosody, Working Papers* (Dept. of Linguistics, Lund Univ.) 41, 176-179.

Beckman, M. 1996. The parsing of prosody. *Language and Cognitive Processes* 11, 17-67.

Beckman, M., DeJong, K. and Edwards, J. 1987. The surface phonology of stress clash in English. Presented at the Annual Meeting of the Linguistic Society of America.

Beckman, M. and Edwards, J. 1990. Lengthenings and shortenings and the nature of prosodic constituency. In J. Kingston and M.E. Beckman (eds), *Papers in Laboratory Phonology I: Between the Grammar and the Physics of Speech*. Cambridge: Cambridge University Press, 152-178.

Beckman, M. and Edwards, J. 1994. Articulatory evidence for differentiating stress categories. In P. Keating (ed), *Papers in Laboratory Phonology III: Phonological Structure and Phonetic Form*. Cambridge: Cambridge University Press, 7-33.

Beckman, M. and Pierrehumbert, J.B. 1986. Intonational structure in Japanese and English. *Phonology Yearbook* 3, 255-309.

Beckman, M., Swora, M., Rauschenberg, J. and DeJong, K. 1990. Stress shift, stress clash and polysyllabic shortening in a prosodically annotated discourse. *Proc. International Conference on Spoken Language Processing*, (Kobe), vol. 1, 5-8.

Berkovits, R. 1993. Utterance-final lengthening and the duration of final stop closures. *Journal of Phonetics* 21, 479-489.

Bolinger, D. 1958. A theory of pitch accents in English. *Word 14,* 109-149.

Bolinger, D. 1965. Pitch accent and sentence rhythm. In I. Abe and T. Kanekiyo (eds), *Forms of English: Accent, Morpheme, Order.* Tokyo: Hokuou, 139-180.

Bolinger, D. 1981. *Two Kinds of Vowels, Two Kinds of Rhythm.* Ms., Distributed by the Indiana University Linguistics Club, Bloomington, Indiana.

Bresnan, J. 1971. Sentence stress and syntactic transformations. *Language* 47, 257-281.

Bruce, G. 1977. *Swedish Word Accents in Sentence Perspective.* Lund: Gleerups.

Cooper, W. and Eady, J. 1986. Metrical phonology in speech production. *Journal of Memory and Language* 25, 369-384.

Couper-Kuhlen, E. 1993. *English Speech Rhythm.* Amsterdam: John Benjamins.

Cutler, A. 1983. Speakers' conceptions of the function of prosody. In A. Cutler and D.R. Ladd (eds), *Prosody: Models and Measurements.* New York: Springer, 79-92.

Dauer, R. 1983. Stress-timing and syllable-timing reanalyzed. *Journal of Phonetics* 11, 51-62.

Dell, G. 1986. A spreading activation theory of retrieval in sentence production. *Psychological Review* 93, 283-321.

Dilley, L. 1997. Some factors influencing duration between syllables judged perceptually isochronous. *Journal of the Acoustical Society of America* 102(5) Part 2, 3205-6.

Dilley, L. Forthcoming. *A rhythm transcription system for American English speech.* Ms., Speech Group, MIT.

Dilley, L. and Shattuck-Hufnagel, S. 1995. Variability in glottalization of word onset vowels in American English. *Proc. XIIIth International Congress of Phonetic Sciences* (Stockholm), vol. 4, 586-589.

Dilley, L. and Shattuck-Hufnagel, S. 1996. Rhythm and intonational phrase structure influences on the placement of pitch accents in American English. *Journal of the Acoustical Society of America* 100(4) Part 2, 2826.

Dilley, L., Shattuck-Hufnagel, S. and Ostendorf, M. 1994. Prosodic constraints on glottalization of vowel-initial syllables in American English. *Journal of the Acoustical Society of America* 95 (5 pt 2), 2978.

Dilley, L., Shattuck-Hufnagel, S. and Ostendorf, M. 1996. Glottalization of word-initial vowels as a function of prosodic structure. *Journal of Phonetics* 24, 423-444.

Fant, G., Kruckenberg, A. and Nord, L. 1991. Stress patterns and rhythm in the reading of prose and poetry with analogies to music performance. In J. Sundberg, L. Nord and R. Carlson (eds), *Music, Language, Speech and Brain.* Houndmills: Macmillan, 380-407.

Ferreira, F. 1991. Creation of prosody during sentence production. *Psychological Review* 100, 233-253.

Fletcher, J. 1991. Rhythm and final lengthening in French. *Journal of Phonetics* 19, 193-212.

Fougeron, C. 1996. Articulation of French nasal segments depending on their prosodic position. Presented at the January meeting of the Linguistic Society of America, San Diego.

Fougeron, C. and Keating, P. 1997. Demarcating prosodic groups with articulation. *Journal of the Acoustical Society of America* 97 (5, pt 2), 3384.

Garrett, M.F. 1975. The analysis of sentence production. In G. Bower (ed), *Psychology of Learning and Motivation* 9. New York: Academic Press, 133-175.

Garrett, M.F. 1980. The limits of accommodation: Arguments for independent processing levels in sentence production. In V.A. Fromkin (ed), *Errors in Linguistic Performance: Slips of the Tongue, Ear, Pen and Hand.* New York: Academic Press, 263-272.

Garrett, M.F. 1982. Production of speech: observations from normal and pathological language use. In A.W. Ellis (ed), *Normality and Pathology in Cognitive Functions.* London: Academic Press, 19-76.

Gee, J. and Grosjean, F. 1983. Performance structures: a psycholinguistic and linguistic appraisal. *Cognitive Psychology* 15, 411-458.

Gimson, A. 1980. *An Introduction to the Pronunciation of English.* 3rd edition, London: Edward Arnold.

Gussenhoven, C. 1991. The English rhythm rule as an accent deletion rule. *Phonology* 8, 1-35.

Hagen, A. 1997. *Linguistic Functions of Glottalizations and Their Language Specific Use in English and German.* M.A. Thesis, Friedrich-Alexander Universität Erlangen-Nürnberg and MIT.

Hale, K. and Selkirk, E. 1987. Government and tonal phrasing in Papago. *Phonology* 4, 151-183.

Halle, M. and Idsardi, W. 1995. General properties of stress and metrical structure. In J. Goldsmith (ed), *The Handbook of Phonological Theory.* Oxford: Blackwell, 403-443.

Halle, M. and Vergnaud, J-R. 1987. *An Essay on Stress.* Cambridge, Mass.: MIT Press.

Halliday, M.A.K. 1967. *Intonation and Grammar in British English.* The Hague: Mouton.

't Hart, J. and Collier, R. 1975. Integrating different levels of intonation analysis. *Journal of Phonetics* 3, 235-255.

't Hart, J., Collier, R. and Cohen, A. 1990. *A Perceptual Study of Intonation.* Cambridge: Cambridge University Press.

Hayes, B. 1984. The phonology of rhythm in English. *Linguistic Inquiry* 15, 33-74.

Hayes, B. and Lahiri, A. 1991. Bengali intonational phonology. *Natural Language and Linguistic Theory* 9, 47-96.

Henton, C. and Bladon, A. 1988. Creak as a sociophonetic marker. In L.M. Hyman and C. N. Li (eds), *Language, Speech and Mind: Studies in Honour of Victoria A. Fromkin.* London: Routledge, 3-29.

Hillenbrand, J. and Houde, R. 1996. The role of F0 and amplitude in the perception of intervocalic glottal stops. *Journal of Speech and Hearing Research* 39, 1182-1190.

Horne, M. 1990. Empirical evidence for a deletion formulation of the rhythm rule in English. *Linguistics* 28, 959-981.

Houde, R. and Hillenbrand, J. 1994. The role of voice pitch in the perception of glottal stops. *Journal of the Acoustical Society of America* 95 (S1), 2892.

Igoa., del Viso, S. and Garcia-Albea, J. 1987. A corpus of spontaneous slips of the tongue in Spanish. Technical Report for Research Grant 2225-83 from Comisión Asesora de Investigación Científica y Técnica, Spanish Ministry of Education.

Inkeles, S. and Zec, D. (eds) 1990. *The Phonology-Syntax Connection.* Chicago: University of Chicago Press.

Jun, S-A. 1993. *The Phonetics and Phonology of Korean Prosody.* PhD thesis, Ohio State University, Department of Linguistics.

Klatt, D.H. 1976. Linguistic uses of segmental duration in English: Acoustic and perceptual evidence. *Journal of the Acoustical Society of America* 59, 1208-1221.

Kohler, K. 1994. Glottal stops and glottalization in German. *Phonetica* 51, 38-51.

Ladd, D.R. 1996. *Intonational Phonology.* Cambridge: Cambridge University Press.

Ladd, R. and Campbell, N. 1991. Theories of prosodic structure: Evidence from syllable duration. *Proc. XIIth International Congress of Phonetic Scienc*es (Aix-en-Provence), vol. 2, 290-293.

Lashley, K. 1951. The problem of serial order in behavior. In L.A. Jefferess (ed), *Cerebral Mechanisms in Behavior.* New York: Wiley, 112-136.

Lehiste, I. 1973. Phonetic disambiguation of syntactic ambiguity. *Glossa 7,* 107-121.

Lehiste, I. 1977. Isochrony reconsidered. *Journal of Phonetics* 5, 253-263.

Levelt, W. 1989. *Speaking. From Intention to Articulation.* Cambridge, MA: MIT Press.

Liberman, M. 1975. *The Intonational System of English.* PhD thesis, MIT Department of Linguistics.

Liberman, M. and Prince, A. 1977. On stress and linguistic rhythm. *Linguistic Inquiry* 8, 249-336.

Meijer, P. 1994. *Phonological Encoding: The Role of Suprasegmental Structures.* The Hague: CIP-Gegevens Koninklijke Bibliotheek.

Mohanan, K. 1986. *The Theory of Lexical Phonology.* Dordrecht: Reidel.

Mohanan, K. 1995. The organization of the grammar. In J Goldsmith (ed), *The Handbook of Phonological Theory.* Oxford: Blackwell, 24-69.

Monaghan, A. 1990. Rhythm and stress shift in speech synthesis. *Computer Speech and Language* 4, 71-78.

Nespor, M. and Vogel, I. 1986. *Prosodic Phonology.* Dordrecht: Foris.

Nespor, M. and Vogel, I. 1989. On clashes and lapses. *Phonology* 6, 69-116.

Ostendorf, M., Price, P. and Shattuck-Hufnagel, S. 1995. *The Boston University Radio News Corpus.* Boston University ECS Technical Report ECS-95-001.

Pierrehumbert, J.B. 1980. *The Phonology and Phonetics of English Intonation.* PhD thesis, MIT Department of Linguistics. Distributed by the Indiana University Linguistics Club, Bloomington, Indiana.

Pierrehumbert, J.B. 1994. Knowledge of variation. *Papers from the 30th Regional Meeting of the Chicago Linguistic Society (CLS 30).* University of Chicago, 232-256.

Pierrehumbert, J.B. and Hirshberg, J. 1990. The meaning of intonational contours in the interpretation of discourse. In P. Cohen, J. Morgan and M. Pollack (eds), *Intentions in Communication.* Cambridge, Mass: MIT Press, 112-136.

Pierrehumbert, J.B. and Talkin, D. 1992. Lenition of /h/ and glottal stop. In G.J. Docherty and D.R. Ladd (eds), *Papers in Laboratory Phonology III: Gesture, Segment, Prosody.* Cambridge: Cambridge University Press, 90-116.

Pitrelli, J., Beckman, M., and Hirschberg, J. 1994. Evaluation of prosodic transcription reliability in the ToBI framework. *Proc. International Conference on Spoken Language Processing,* (Yokohama), vol. 1, 123-126.

Prevost, S. and Steedman, M. 1994. Specifying intonation from context for speech synthesis. *Speech Communication* 15, 139-153.

Price, P., Ostendorf, M., Shattuck-Hufnagel, S. and Fong, C. 1991. The use of prosody in syntactic disambiguation. *Journal of the Acoustical Society of America* 90, 2956-2970.

Prince, A. 1983. Relating to the grid. *Linguistic Inquiry* 14, 19-100.

Roeloffs, A. 1992. A speaking-activation theory of lemma retrieval in speaking. *Cognition* 42, 107-142.

Scott, D. 1982. Duration as a cue to the perception of a phrase boundary. *Journal of the Acoustical Society of America* 71, 996-1007.

Selkirk, E.O. 1980. The role of prosodic categories in English word stress. *Linguistic Inquiry* 11, 563-605.

Selkirk, E.O. 1984. *Phonology and Syntax.* Cambridge: MIT Press.

Selkirk, E.O. 1986. On derived domains in sentence phonology. *Phonology Yearbook* 3, 371-405.

Selkirk, E.O. 1995. The prosodic structure of function words. In J. Beckman, L. Walsh Dickey and S. Urbanczyk (eds), *Papers in Optimality Theory.* Univ. of Mass. *Occasional Papers in Linguistics* 18, 439-470.

Sevald, C. and Dell, G. 1994. The sequential cuing effect in speech production. *Cognition* 53, 91-127.

Silverman, K., Beckman, M., Pitrelli, J., Ostendorf, M., Wightman, C., Price, P., Pierrehumbert, J. and Hirschberg, J. 1992. ToBI: a standard for labeling English prosody. *Proc. International Conference on Spoken Language Processing* (Banff), vol. 2, 867-870.

Shattuck-Hufnagel, S. 1979. Speech errors as evidence for a serial order mechanism in sentence production. In W.E. Cooper and E.C.T. Walker (eds), *Sentence Processing.* Hillsdale, N.J.: Erlbaum, 295-342.

Shattuck-Hufnagel, S. 1983. Sublexical units and suprasegmental structure in speech production planning. In P.F. MacNeilage (ed), *The Production of Speech.* New York: Springer, 109-136.

Shattuck-Hufnagel, S. 1987. The role of word onset consonants in speech production planning: New evidence from speech error patterns. In E. Keller and M. Gopnik (eds), *Motor and Sensory Processing in Language.* Hillsdale, N.J.: Erlbaum, 17-51.

Shattuck-Hufnagel, S. 1989. Stress shift as the placement of phrase-level pitch markers. *Journal of the Acoustical Society of America* 86, S493.

Shattuck-Hufnagel, S. 1991. Acoustic correlates of stress shift. *Proc. XII Congress of Phonetic Sciences* (Aix-en-Provence), vol. 4, 266-269.

Shattuck-Hufnagel, S. 1992. The role of word structure in segmental serial ordering. *Cognition* 42, 213-259.

Shattuck-Hufnagel, S. 1995. Pitch accent patterns in adjacent-stress vs. alternating stress words in American English. *Proc. XIIIth International Congress of Phonetic Sciences* (Stockholm), vol. 3, 656-659.

Shattuck-Hufnagel, S., Ostendorf, M. and Ross, K. 1994. Stress shift and early pitch accent placement in lexical items in American English. *Journal of Phonetics* 22, 357-388.

Shattuck-Hufnagel, S. and Turk, A. 1996. A prosody tutorial for investigators of auditory sentence processing. *Journal of Psycholinguistic Research* 25, 193-247.

Silverman, K., Beckman, M., Pierrehumbert, J., Ostendorf, M., Wightman, C., Price, P., and Hirschberg, J. 1992. ToBI: A standard scheme for labeling prosody. *Proc. Second Intl. Conf. on Spoken Language Processing* (Banff), 867-870.

Steedman, M. 1991. Structure and intonation. *Language* 68, 260-296.

Stemberger, J. 1990. Wordshape errors in language production. *Cognition* 35, 123-157.

Thompson, H. 1980. *Stress and salience in English.* CSL-80-8, Palo Alto: Xerox PARC.

Umeda, N. 1978. Occurrence of glottal stops in fluent speech. *Journal of the Acoustical Society of America* 64, 1582-1592.

Wenk, B. and Wioland, F. 1982. Is French really syllable-timed? *Journal of Phonetics* 10, 193-216.

Wightman, C., Shattuck-Hufnagel, S., Ostendorf, M. and Price, P. 1992. Segmental durations in the vicinity of prosodic phrase boundaries. *Journal of the Acoustical Society of America* 91, 1707-1717.

Speech Communication Group, Research Laboratory of Electronics, Massachusetts Institute of Technology, Cambridge, MA, USA

ELISABETH SELKIRK

THE INTERACTION OF CONSTRAINTS ON PROSODIC PHRASING

1. INTRODUCTION

The notion that the prosodic phrasing structure of a sentence plays a crucial role in organizing the segmental, tonal and prominence structures of a sentence's phonological representation and its phonetic implementation as well is quite widely assumed in work in both phonology and phonetics. It is also quite widely assumed that this prosodic phrasing structure is independent of, but related to, the syntactic and/or information structure of a sentence. Yet no consensus has emerged within the various traditions of research on prosodic phrasing concerning the nature of the relation between prosodic phrasing and these other distinct types of grammatical representation. Certain approaches foreground the role for syntactic constraints on prosodic phrasing[1], others the role for constraints appealing to aspects of information structure[2]. There are, moreover, properly phonological constraints on prosodic phrasing which ignore these interface representations[3]. An adequate theory has to recognize the full diversity of constraints on prosodic phrasing, and in addition, make explicit the manner in which these constraints interact[4].

The study of constraint interaction is at the heart of work in optimality theory (Prince and Smolensky 1993). Optimality theory sees a grammar as a set of ranked constraints on output representations. The constraints are hypothesized to be universal; it is in the ranking of the constraints with respect to each other that languages are claimed to differ. Central to optimality theory is the notion that constraints are in fact violable, but only minimally, and only under pressure from some higher ranked constraint. The claim is that the (grammatical) output representation generated on the basis of an input representation is not necessarily well-formed, in the sense of respecting all constraints, but rather is the *optimal* output representation possible, the one that best satisfies the constraint hierarchy. It is predicted, then, that a constraint may be disobeyed in some grammatical surface representation, but at the same time, that this constraint violation should arise only in order to satisfy some higher ranked constraint.

The aim of this paper is to illustrate potential gains to be made by understanding the functioning of constraints on prosodic phrasing from an

M. Horne (ed.), Prosody: Theory and Experiment, 231–261.

optimality theoretic point of view. First, using data from studies of sentence phonology in various Bantu languages, I would like to show that optimality theoretic constraint rankings provide the basis for an insightful typology of crosslinguistic variation in prosodic phrasing: the rather different patterns of prosodic phrasing attested in different languages can be understood as the consequence of different rankings of the same, putatively universal, constraints. Here the presentation is largely based on work by Truckenbrodt (1995). Then I would like to air some preliminary hypotheses about the hierarchy of constraints on prosodic phrasing in English, looking at the interaction of focus- and syntax-based constraints with each other and with constraints that are properly phonological in character. We will see that including an optimality theoretic ranking as part of the grammar of constraints extends the empirical coverage of a constraint system in highly desirable ways.

2

A number of illuminating indepth studies of the sentence phonology of Bantu languages have appeared in the last decade or so. These include works which explicitly investigate the patterns of prosodic phrasing that are revealed in tonal and segmental phenomena. Based in part on the thoroughgoing account by Kisseberth and Abasheikh (1974) of the distribution of contrastive and rule-governed vowel length in ChiMwiini[5], in Selkirk (1986) I proposed that the syntax-prosodic structure relation is characterized by a set of interface constraints which require that the edge of every constituent of a designated type in the surface syntactic structure of a sentence coincide with (= align with) the edge of a prosodic constituent of a designated type in prosodic structure. In ChiMwiini, the distribution of vowel length was argued to be defined with respect to the major phonological phrase (MaP), and the distribution of major phrases in a sentence was argued to be determined by a constraint aligning the right edge of a maximal projection (XP) in syntactic structure with a MaP edge in prosodic structure[6]. In the McCarthy and Prince (1993) generalized alignment format, this constraint would be expressed as follows[7]:

(1) **Align$_R$ XP**

Align (XP, R; MaP, R)

"The right edge of any XP in syntactic structure must be aligned with the right edge[8] of a MaP in prosodic structure."

To see the effects of this constraint in ChiMwiini, consider the following examples with input syntactic structure (a) and output major phrase organization (b). In (2) the subject NP is separated off in a distinct MaP from material in the following verb phrase, as shown by the presence of vowel length in the output: underlying and rule-governed vowel length is realized on the surface only in the penultimate or antepenultimate position of the major phrase.

(2) a. [[mwa:na [wa [si:mba]$_{NP}$]$_{PP}$]$_{NP}$] [ni [si:mba]$_{NP}$]$_{VP}$]$_{S}$

b. $_{MaP}$(mwana wa si:mba)$_{MaP}$ $_{MaP}$(ni si:mba)$_{MaP}$

‘A child of a lion is a lion.’

Note the disappearance of pre-pre-antepenultimate underlying length in *mwa:na* in (b), but the maintaining of underlying length in the penultimate syllable of *si:mba.* The major phrasing in (b) can be ascribed to Align$_{R}$ XP, which calls for a MaP edge to coincide with the right edge of the subject NP. Next, in example (3) the verb has two complements.

(3) a. [[panzi:ze$_{V}$ [cho:mbo]$_{NP}$ [mwa:mba]$_{PP}$]$_{VP}$]$_{S}$

b. (panzize cho:mbo)$_{MaP}$ (mwa:mba)$_{MaP}$

‘He ran the vessel onto the rock.’

The first complement to the verb is joined with the verb in a major phrase, as shown by the absence of surface length in the second vowel in *panzi:ze*, which appears in a preantepenultimate position in the MaP. That *cho:mbo* retains its vowel length shows it is not in the same MaP with the following complement. The systematic presence of MaP breaks at the right (or left) edges of maximal projections, such as seen in examples (2) and (3) from ChiMwiini, are what will be called Align XP effects in what follows.

Investigations of sentence phonology in other Bantu languages have also detected what can be analyzed as Align XP effects. These languages include Chaga (McHugh 1987, 1990), Kimatuumbi (Odden 1987, 1990, Truckenbrodt 1995), Kikuyu (Clements and Ford 1981) and Kinyambo (Bickmore 1989, 1990). In Chaga, for example, McHugh (1987) argues that various tonal phenomena take the major phrase as their domain. On the basis of such tonal evidence, the verb phrase in (4a) would be phrased in output structure as (4b), with the direct object NP separated into a distinct MaP from the preceding indirect object.

(4) a. [[á̲muénenga]$_{V}$ [prayáni]$_{NP}$ [mbúru]$_{NP}$] $_{VP}$

b. (amú̲enénga prayá̲ní̲)$_{MaP}$ (mburû)$_{MaP}$

‘(S)he has given Brian a goat.’

This breaking up of the verb phrase constituent in Chaga is a characteristic Align XP effect, as we saw above. There are other Bantu languages, however, in which the effects of Align XP are not discerned. These include Digo (Kisseberth 1984), Chichewa (Kanerva 1989, 1990) and Chizigula (Kenstowicz and Kisseberth 1990). In Chizigula, for example, what is at issue is the behavior of various phrasal tone-spreading phenomena as well as the migration of a lexical floating high tone from its word of origin to the penult (presumably most prominent) position of the phrase containing that word. In (5) we can see an underlying floating high tone migrate from the first to the second complement following the verb, indicating that both complements are contained within the same major phrase in Chizigula.

(5) a. $[[\text{nambikila}]_{\text{V}}\ [\overset{\text{H}}{\underline{\text{m}}}\text{vyele}]_{\text{NP}}\ [\text{nyama}]_{\text{NP}}]_{\text{VP}}$

b. $(\text{nambilika}\ \underline{\text{m}}\text{vyele nyáma})_{\text{MaP}}$

'I am cooking the woman meat.'

If Align XP were respected here, there would be no tonal migration possible from the first to the second complement, which would be in a separate MaP. So the phrasing data from languages like Chizigula could be taken as problematic for claims that Align XP is a universal constraint, at play in all languages. But a deeper, richer, account of cross-linguistic variation which assumes the universality of constraints is available given the optimality theoretic hypothesis that cross-linguistic differences are to be traced to differences in the relative ranking of constraints.

2.1

Truckenbrodt (1995) argues for a second class of constraints on the morphosyntax-prosodic structure interface, the Wrap constraints, which require that the terminal material of a morphosyntactic constituent type α be included ('wrapped') within a prosodic structure constituent type β. The function of the Wrap constraint is cohesional: a morphosyntactic constituent whose contents are separated into distinct prosodic constituents constitutes a violation of the constraint. The constraint Wrap XP proposed by Truckenbrodt has the formulation in (6).

(6) **Wrap XP**

Wrap (XP; MaP)

"The elements of an input morphosyntactic constituent of type XP must be contained within a prosodic constituent of type MaP in output representation."

Wrap XP and Align XP are inherently in conflict with each other – the satisfaction of one entails the violation of the other. We can see this in the tableau in (7). The input syntactic structure sits at the top of the lefthand column and below it are listed two candidate output representations, one consisting of two major phrases, with a break falling between the complements of VP, and the other consisting of a MaP which contains the entire VP.

(7)

$[[\text{verb}]\ [\text{noun}]_{NP}\ [\text{noun}]_{NP}]_{VP}$	Align $_R$ XP	Wrap XP
a. $(\text{ verb noun })_{MaP}(\text{ noun })_{MaP}$		*
b. $(\text{ verb noun noun })_{MaP}$	*	

Candidate (a) satisfies Align_R XP, in that it contains MaP edges coinciding with all right edges of syntactic XP-level phrases in the input, but it violates Wrap XP, since the entire VP is not contained within a single MaP. (A violation is indicated by a '*' mark in the appropriate cell.) Candidate (b) incurs one violation of Align_R XP, in that the VP-medial NP edge in the input does not coincide with a MaP in the output, but (b) does satisfy Wrap XP since there is a single MaP containing the entire VP. Truckenbrodt proposes that in the languages like ChiMwiini or Chaga which show Align XP effects (call them Type A languages) the constraint Align XP is higher ranked than the constraint Wrap XP, and so is responsible for the violation of Wrap XP in these languages: Align XP >> Wrap XP. On the other hand, for languages like Chizigula in which the effects of Align XP are not typically observed (call them Type B languages), it is proposed that Wrap XP dominates Align XP in the constraint hierarchy: Wrap XP >> Align XP.

Let us look at just how the desired results are obtained, using optimality theory tableaux. In the tableau in (8), Align_R XP is ranked higher than Wrap XP, as indicated by the solid line between the constraint columns. It is this ranking that is responsible for selecting candidate (a) as the optimal, grammatical, output candidate out of all the candidates generable on the basis of the input representation[9].

(8)

$[[\text{verb}]\ [\text{noun}]_{NP}\ [\text{noun}]_{NP}]_{VP}$	Align $_R$ XP	Wrap XP
☞a. (verb noun $)_{MaP}$(noun $)_{MaP}$		*
b. (verb noun noun $)_{MaP}$	*!	

Candidates (a) and (b) both violate some constraint, but candidate (b) violates a higher ranked constraint, so it is candidate (a) which is the better candidate, the *optimal* one. The fatal higher violation is indicated by the exclamation mark. In tableau (9) the ranking of the two constraints is reversed. This is the ranking posited for the type B languages showing no Align XP effects in the basic cases. Now candidate (b) turns out to be optimal.

(9)

$[[\text{verb}]\ [\text{noun}]_{NP}\ [\text{noun}]_{NP}]_{VP}$	Wrap XP	Align$_R$ XP
a. (verb noun $)_{MaP}$(noun $)_{MaP}$	*!	
☞b. (verb noun noun $)_{MaP}$		*

Candidate (a) violates the higher ranked constraint, fatally, and therefore the constraint violation exhibited by (b) is consistent with its optimal status. This, then, is an example of the sort of typology made available in optimality theory, assuming a universal constraint repertoire. Possible cross-linguistic differences are just those that would derive from differences in constraint ranking.

It might seem that an alternative conception would be to say simply that in the Type A languages the grammar includes the constraint Align XP, and in the Type B languages it does not. But it turns out that this alternative is empirically inadequate, precisely in the simplistic presumption that in the Type A languages Align XP is always at play and in the type B languages never at play. The factual situation is more complicated, and an optimality theoretic approach is required in order to model it. It turns out that under the appropriate circumstances the effects of a subordinate, lower-ranked, constraint may emerge, even though in most circumstances this constraint is violated in the language. Truckenbrodt (1995) argues that this is the case in Chichewa, whose phrasing is the subject of a study by Kanerva (1989, 1990).

Chichewa may on the face of it appear to be a language in which Align XP has no effect. Indeed, this is what Kanerva (1989) concluded, based on an extensive investigation of the sentence phonology of Chichewa. For example, he shows that a verb phrase with multiple complements is prosodically phrased without any breaks within it, at least in conditions where no VP-internal consituents are in Focus:

(10) Chichewa: Neutral Case

a. [[anaménya]$_{V}$ [nyu^{m}bá]$_{NP}$ [^{n}dí [mwalá]$_{NP}$]$_{PP}$]$_{VP}$

b. (anaményá nyu^{m}bá ^{n}dí mwáála)$_{MaP}$

'He hit the house with a rock.'

(The assumptions about analysis into major phrases are based on the distribution of vowel length – which is always, and only, found in the penult syllable of a MaP – and on various tonal phenomena.) Truckenbrodt (1995) argues, on the other hand, that in the grammar of Chichewa the constraint Align$_{R}$ XP is present, but subordinate to Wrap XP in the constraint hierarchy. This lower rank has the consequence that Align$_{R}$ XP is violated in this case:

(11) Chichewa: Neutral Case

[[anaménya]$_{V}$ [nyu^{m}bá]$_{NP}$ [^{n}dí mwalá]$_{PP}$]	Wrap XP	Align XP
☞a. (anaményá nyu^{m}bá ^{n}dí mwáála)$_{MaP}$		*
b. (anaményá nyuú^{m}ba)$_{MaP}$(^{n}dí mwáála)$_{MaP}$	*!	

A particularly important aspect of Kanerva's study of Chichewa is his systematic examination of the phonological effects of Focus. Using responses to *wh-* questions, sentences with focus-sensitive particles and sentences with constructions calling for contrastive focus, Kanerva found that the right edge of a Focus constituent systematically coincided with the edge of a MaP. For example, the sentence we have been considering would be differently phrased with the direct object in Focus.

(12) Chichewa: Direct Object Focus

a. [[anaménya]$_{V}$ [nyu^{m}bá]$_{NP\text{-}FOCUS}$ [^{n}dí mwalá]$_{PP}$]

b. (anaményá nyuú^{m}ba)$_{MaP}$ (^{n}dí mwáála)$_{MaP}$

'He hit the HOUSE with a rock'

Let us assume that such Focus effects are due to a constraint Align Focus, as has been suggested by a number of authors (Pierrehumbert and Beckman 1988, Kanerva 1989, Vogel and Kenesei 1990a-b, Hayes and Lahiri 1991, Jun 1993, among others), and that in Chichewa it is the Align$_R$ XP version that is at play:

(13) **Align$_R$ Focus**[10]

Align (Focus, R; MaP, R)

"Align the right edge of a Focus constituent in informational or syntactic structure with the right edge of a major phrase (MaP) in the phonological structure."

Truckenbrodt (1995) points out that the presence of these Focus effects in Chichewa shows that the constraint Align$_R$ Focus is higher ranked than Wrap XP, which is violated in this example. (14) is a tableau showing the effects of this ranking.

(14)

[[anaménya]$_V$ [**nyu^{m}bá**]$_{NP\text{-}FOCUS}$ [^{n}dí mwalá]$_{PP}$]	Align$_R$ Focus	Wrap XP
a. (anaményá **nyu^{m}bá** ^{n}dí mwáála)$_{MaP}$	*!	
☞b. (anaményá **nyuú^{m}ba**)$_{MaP}$(^{n}dí mwáála)$_{MaP}$		*

Even though it respects Wrap XP, candidate a, with the violation of the higher ranking Align$_R$ Focus, is rejected, in favor of candidate b, which instead satisfies Align$_R$ Focus and thus violates Wrap XP.

Since constraint ranking is transitive, the ranking Align$_R$ Focus >> Wrap XP plus the ranking Wrap XP >> Align$_R$ XP gives the composite ranking (15):

(15) Align$_R$ Focus >> Wrap XP >> Align$_R$ XP

It is precisely this ranking, and, specifically, the inclusion of Align XP in the constraint hierarchy of Chichewa, that permits an explanation for a fact for which Kanerva had no account. The fact is that when the verb is focussed in two-complement VP's like those we have been considering, there appears a MaP break not only at the right of the focussed verb but also at the right of the first complement.

(16) Chichewa: Verb Focus Case

a. [[anaménya]$_{\text{V-FOCUS}}$ [nyu^{m}bá]$_{\text{NP}}$ [^{n}dí mwalá]$_{\text{PP}}$]

b. (anaméenya)$_{\text{MaP}}$ (nyuú^{m}ba)$_{\text{MaP}}$ (^{n}dí mwáála)$_{\text{MaP}}$

'He HIT the house with a rock'

The MaP at the right edge of the verb is to be expected, given Align$_{R}$ Focus. But why should a MaP also appear after the first complement, which is not itself focussed? As Truckenbrodt points out, the presence of this seemingly unexpected phrase break after the direct object is a direct consequence of assuming the constraints and the ranking in (15). This is shown in tableau (17)[11]:

(17)

[[anaménya]$_{\text{V-FOCUS}}$ [nyu^{m}bá]$_{\text{NP}}$ [^{n}dí mwalá]$_{\text{PP}}$]	Align$_{R}$ Focus	Wrap XP	Align$_{R}$ XP
☞ a.(anaméenya)$_{\text{MaP}}$(nyuú^{m}ba)$_{\text{MaP}}$(^{n}dí mwáála)$_{\text{MaP}}$		*	
b. (anaméenyáa)$_{\text{MaP}}$ (nyu^{m}bá ^{n}dí mwáála)$_{\text{MaP}}$		*	*!
c. (anaményá nyu^{m}bá ^{n}dí mwáála)$_{\text{MaP}}$	*!		*

(The shading in the tableau appears in cells to the right of (i.e. below) a cell containing a fatal violation (*!), in order to show that any further, lower, constraint violations are irrelevant to the determination of the optimal candidate.) In the tableau we see that candidate (a), which has a phrase break after the direct object in addition to a break after the focussed verb, is optimal. This is because satisfaction of the highest ranked Align$_{R}$ Focus constraint forces the violation of Wrap XP. Given that Wrap XP must be violated in any candidate satisfying Align$_{R}$ Focus, as seen in candidates a and b, it is no longer determinant in evaluating the relevant candidates, and in that case the effects of the lower ranked Align$_{R}$ XP are allowed to emerge. It is the candidate which satisfies Align$_{R}$ XP – with the presence of a phrase break after the direct object – that is the optimal output for the input with verb in Focus. So, Truckenbrodt concludes, we must assume that both Wrap XP and Align$_{R}$ XP are at play in the grammar of Chichewa.[12]

2.2

Grammaticized in the form of constraints, the cohesional function of Wrap XP and the demarcative function of Align XP are at odds with each other. It is not possible that in a given prosodic structure the elements of the same maximal projection in syntactic structure be included in the same major phrase, as called for by Wrap XP, and at the same time that the edges of maximal projections within that higher phrase be denoted by major phrase breaks, as called for by Align XP. When it comes to the syntax-prosodic structure interface, languages must opt for a dominant cohesional strategy or a dominant demarcative strategy, as represented by the relative rankings of Wrap XP and Align XP. The constraint Align Focus is also demarcative, and thus at odds with Wrap XP. A language like Chichewa shows that even where a dominant cohesional strategy is chosen, through the ranking Wrap XP >> Align XP, the demarcative strategy takes precedence when it comes to Align Focus, which dominates Wrap XP in the ranking Align Focus >> Wrap XP.

In other realms, cohesional and demarcative strategies are not necessarily opposed. Phonological and phonetic phenomena which are demarcative in the sense of being defined with respect to the edges of prosodic units, coexist happily in the same phrase with other phonological or phonetic phenomena which are defined across the span of the phrase. To take an example from Bruce and his co-workers' work on Swedish phrasal prosody (e.g. Bruce et al. 1991, 1993), the presence of a phrase accent at the edge of a prosodic phrase (construed to be demarcative) in no way precludes the presence of a downstepping pattern across the phrase (construed to be cohesional). It is in characterizing the interface between prosodic structure and other grammatical representations – morphosyntactic or informational – that the cohesional and demarcative functions come into conflict and where a language must apparently 'decide', through constraint ranking, which of the functions will predominate.

3.

The facts that we will consider in the following sections give support for hypothesizing (i) that the constraints $Align_R$ XP, Wrap XP and $Align_R$ Focus form part of an account of major phonological phrasing in English, and (ii) that phonological constraints on phrasing interact with these interface constraints in providing a full account of English major phrasing. Some of these notions have surfaced in one way or another in previous work, though the manner in which major phrasing in English relates to syntactic structure or information structure has not yet been the subject of any comprehensive investigation[13]. The remarks that follow are rather speculative, pointing to directions for further research, rather than firm results acquired on the basis of systematic study. They are intended to show the gains that are potentially to be made by understanding the grammar of constraints on phrasing in optimality theoretic terms.

3.1

The hallmark of the English major phrase (aka intermediate phrase), according to Beckman and Pierrehumbert (1986), is the presence of the phrase accent, a Low or High peripheral tone located at the phrase's right edge, and in what follows we will use the presence or absence of this peripheral tone to diagnose the presence of major phrase breaks[14]. A first observation about phonological phrasing in English is that a fluent, but not particularly rapid rendering of a verb phrase with multiple complements and no VP-internal Focus does not require the presence of a major phrase break within it. For example, the sentences in (18) may all be pronounced with one major phrase. Which is to say that no peripheral phrase accents are necessary between the successive complements.

(18) Available in neutral infomational contexts:

a. (She lóaned her róllerblades to Róbin.)$_{MaP}$
b. (She púshed Sám's bóat into the wáter.)$_{MaP}$
c. (She gáve Zóe a báckrub.)$_{MaP}$
d. (She sént her sincere regréts to Luís)$_{MaP}$

(The acute accent in these sentences stands for a High pitch accent on the relevant syllable, and the presence of a right MaP bracket should be taken as indicating the presence of a L phrase accent, unless otherwise stated.) Let us consider a circumstance in which such utterances, with accents on all the content words and no VP-internal focus, would be appropriate. B comes across a notebook of A's which has a number of one sentence journal entries each describing an event of some day. C asks B to report what A has written, and B utters the sentences in (18). With these utterances, B has made no presumptions about A's point of view, nor has B, as speaker, imposed any particular perspective on the information the utterance contains. The utterances lack a focus or topic structure, and all information is treated as new. These are 'neutral' pronunications of such sentences. The fact that all these elements of the VP may be included in the same MaP in this case might seem to suggest that Wrap XP dominates Align XP in the grammar of English, as in Chichewa. But if that were so, then there would be no explanation for the fact that there are variants of the utterances in (18) which are produceable in the same 'neutral' circumstances and which *do* show a MaP break after the first complement:

(19) Also available in neutral informational contexts:

a. (She lóaned her róllerblades)$_{MaP}$ (to Róbin.)$_{MaP}$
b. (She púshed Sám's bóat)$_{MaP}$ (into the wáter.)$_{MaP}$
c. (She gáve Zóe)$_{MaP}$ (a báckrub.)$_{MaP}$
d. (She sént her sincere regréts)$_{MaP}$ (to Luís)$_{MaP}$

This is the pattern of phrasing that would be expected with the opposite ordering of Wrap XP and Align$_R$ XP in the constraint hierarchy. Making the universalist assumption that Wrap XP and Align XP are indeed part of the constraint grammar of English, it must be concluded that they are not ranked with respect to each other, i.e. that they occupy the *same* rank in the English constraint hierarchy, and moreover that no other higher ranked constraint rules in favor of one or the other. As we see in the tableau in (20), when Wrap XP and Align$_R$ XP are same-ranked (in the notation: separated only by a dotted line), candidates (a) and (b) show the same number and rank of constraint violations, and so are equally optimal.

(20)

[she [lóaned] [her róllerblades]$_{NP}$ [to Róbin]$_{PP}$]$_{VP}$	Wrap XP	Align$_R$ XP
☞a. (she lóaned her róllerblades to Róbin)$_{MaP}$		*
☞b. (she lóaned her róllerblades)$_{MaP}$(to Róbin)$_{MaP}$	*	

Assuming the existence of Wrap XP and Align XP, then, this pattern of variability in the pronunciation of a sentence with the same informational structure requires us to assume no domination of one by the other. On this account, English is a language whch simply fails to choose between a cohesional and a demarcative strategy for phrasing.

There is a further fact which requires us to elaborate this explanation for the attested variation between (18) and (19). This is the apparent unacceptability – in the 'neutral' informational context under discussion – of the further pattern of phrasing seen in (21), where the verb, the first complement and the second complement all form major phrases on their own. This sort of pronunciation is possible in English, but seems restricted to the case where the verb is focussed (which we will discuss below).

(21) Unavailable when verb is not focussed:

a. (She lóaned)$_{MaP}$ (her róllerblades)$_{MaP}$ (to Róbin.)$_{MaP}$
b. (She púshed)$_{MaP}$ (Sám's bóat)$_{MaP}$ (into the wáter.)$_{MaP}$
c. (She gáve)$_{MaP}$ (Zóe)$_{MaP}$ (a báckrub.)$_{MaP}$
d. (She sént)$_{MaP}$ (her sincere regréts)$_{MaP}$ (to Luís)$_{MaP}$

So the grammar must exclude this pronunciation for inputs where the verb is not focussed. It is worth noting at this point that the further logically possible pattern of phrasing in (22), where a phrase break occurs after the verb but not after the first complement, seems not to be available at all.

(22) a. *(She lóaned)$_{MaP}$ (her róllerblades to Róbin.)$_{MaP}$
b. *(She púshed)$_{MaP}$ (Sám's bóat into the wáter.)$_{MaP}$
c. *(She gáve)$_{MaP}$ (Zóe a báckrub.)$_{MaP}$
d. *(She sént)$_{MaP}$ (her sincere regréts to Luís)$_{MaP}$

Consider now the more complete tableau (23), which includes in (c) a candidate from (21) and in (d) t7he corresponding candidate from (22):

(23)

[she [lóaned] [her róllerblades]$_{NP}$ [to Róbin]$_{PP}$]$_{VP}$	Wrap XP	Align XP
☞ a. (she lóaned her róllerblades to Róbin)$_{MaP}$		*
☞ b. (she lóaned her róllerblades)$_{MaP}$(to Róbin)$_{MaP}$	*	
🍎 c. (she lóaned)$_{MaP}$ (her róllerblades)$_{MaP}$(to Róbin)$_{MaP}$	*	
d. (she lóaned)$_{MaP}$ (her róllerblades to Róbin)$_{MaP}$	*	*!

Note that candidate (d) is correctly rejected as nonoptimal. It violates both Wrap XP and Align XP. Having these two violations is fatal, given that the alternative candidates incur only one violation with respect to this constraint pair. The problem here has to do with candidate (c), which is wrongly considered to be just as optimal as candidates (a) and (b). (The 'bad apple', with a bite taken out of it, indicates a candidate wrongly considered to be optimal.) It incurs just one violation, of Wrap XP. Recall from the discussion of Chichewa that it is necessary to assume that Wrap XP does not care *how many* major phrases a particular XP is parcelled out into; it is violated whenever it is the case the elements of a particular XP are not all wrapped up in a single MaP. Thus candidate (c), like candidate (b) incurs just one violation of Wrap XP. In both cases the violation is incurred by the VP, which fails to be included in a single MaP in prosodic structure. So the nonoptimality of candidate (c) in these informational circumstances must be the consequence of some other constraint(s) in the grammar of English.

3.2

It has been suggested that constraints on the minimum and maximum size of prosodic constituents are part of the universal repertoire. These size constraints assess the wellformedness of a constituent of a particular level of prosodic structure C^i in terms of the number of constituents of a particular lower level C^{i-1} that it contains. Perhaps what rules out the excessive phrasing of candidate (c) in (23) are phonological constraints on the size of the major phrase. Such constraints would assess each MaP for its composition in terms of constituents at the next lower level of phrasing, which I assume to be the minor phrase (aka accentual phrase). I take the defining feature of the minor/accentual phrase to be that it contains an accent.[15] We might suppose that there are two size constraints on major phrases, one that calls for an MaP to consist of *at least* two minor/accentual phrases (call it Binary Minimum (MaP), or $\text{BinMin}_{\text{MaP}}$), and another that calls for a MaP to consist of *no more* than two minor/accentual phrases (call it Binary Maximum(MaP) or $\text{BinMax}_{\text{MaP}}$). In what follows, we will simply assume a single binarity constraint, Binary(MaP), given in (24iii), since there is no evidence from the English data to be considered that allows one to ascertain a distinct role for BinMin(MaP) and BinMax(MaP).

(24) (i) **Binary Maximum(MaP)**
A major phrase may consist of at most two minor/accentual phrases.

(ii) **Binary Minimum(MaP)**
A major phrase must consist of at least two minor/accentual phrases.

(iii) **Binary(MaP)**
A major phrase consists of just two minor/accentual phrases.

We will assume that in the unmarked case the accent count of an utterance reflects the number of minor or accentual phrases into which the utterance is organized.

If we add the binary size constraint to the grammar of English phrasing that we are constructing, then we can account for the unacceptability of candidate (c) alongside the availability of both (a) and (b). We see this in tableau (25):

(25)

[she [lóaned] [her róllerblades]$_{NP}$ [to Róbin]$_{PP}$]	Wrap XP	Align$_R$ XP	BinMaP
☞ a. (she lóaned her róllerblades to Róbin)$_{MaP}$		*	*
☞ b. (she lóaned her róllerblades)$_{MaP}$ (to Róbin)$_{MaP}$	*		*
c. (she lóaned)$_{MaP}$ (her róllerblades)$_{MaP}$ (to Róbin)$_{MaP}$	*		**!*
d. (she lóaned)$_{MaP}$ (her róllerblades to Róbin)$_{MaP}$	*	*!	*

(The vertical dotted lines indicate that the constraints flanking the lines are same-ranked with respect to each other; the vertical solid line between the interface constraints and the size constraint indicates that the interface constraints are higher ranked than the size constraint.) Candidates (a) and (b) are still both optimal given this expanded constraint hierarchy. (a) incurs a violation of Align XP, and in so doing incurs a violation of Bin(MaP)) as well; (b) incurs a violation of Wrap XP, and in addition a violation of Bin(MaP). Given the same ranking of the members of Wrap XP and Align XP, the marks against these candidates are equivalent. As for (d), its nonoptimality in this candidate set can be traced to its two violations of the interface constraints. But the second violation of the interface constraint pair in (d) would be fatal only if the interface constraint pair were ranked above the size constraint. If the interface constraints and the size constraint were same-ranked, it is the cumulative three violations of constraints in d) (as against the two violations in (a) and (b)) that would be determinate. Candidate (c) comes out much worse than either (a) or (b), and this is because of its three violations of Bin(MaP). Note that these results would be obtained whether or not Binary(MaP) were ranked below the interface constraints. If the constraints were all same-ranked, the constraint violations in each line would simply be summed up, but it would still be the case that candidate (c) was not optimal, since its total number of constraint violations would be greater than those of candidates (a) and (b), which have an equivalent number of total violations. The higher rank of the interface constraints, assumed in the tableau but not crucial to the present argument, will be argued for in the next section.

Considered from the point of view of the size constraint alone, (d) is as optimal as candidates (a) and (b). Like (a) and (b) it incurs just one size constraint violation. This means that an alternative conception of English phrasing based on the notion that phonological size constraints alone account for the phrasing patterns is not adequate. And we saw above that the interface constraints Wrap XP and Align XP are not on their own adequate to provide an account of the facts of major phrasing in English either. The hypothesis entertained here is that the English constraint hierarchy includes constraints of both general types, and that they together provide the basis of the phrasing patterns seen thus far.

It is not possible to explore much further here the consequences of assuming a role for the constraint Bin(MaP) in accounting for English phrasing patterns. But it should be pointed out that this constraint predicts that small changes in the phonological composition of a sentence can have significant phrasing consequences. For example, suppose that the indirect object of the example sentence discussed above was the phrase *Róbin's síster* , with two accented words, rather than the the singly accented *Róbin.* This would give the sentence *She lóaned her róllerblades to Róbin's síster*. As the tableau in (26) shows, the constraint system now predicts that there is only one optimal candidate, the one with an analysis into two perfectly binary major phrases:

(26)

[she [lóaned] [her róllerblades]$_{NP}$ [to Róbin's síster]$_{PP}$]	Wrap XP	Align XP	Bin(MaP)
a. (she lóaned her róllerblades to Róbin's síster)$_{MaP}$		*	*!
☞ b. (she lóaned her róllerblades) (to Róbin's síster)$_{MaP}$	*		
c. (she lóaned) (her róllerblades) (to Róbin's síster)$_{MaP}$	*		*!*
d. (she lóaned) (her róllerblades to Róbin's síster)$_{MaP}$	*	*!	* *

Candidate (b) violates Wrap XP, but incurs no violation at all of Bin(MaP). It is a better candidate than (a) or any of the others, all of which violate some size constraint in addition to incurring at least one violation of the interface constraints. As the sole optimal candidate, (b) should be the only available pronunciation of the input given. Intuitions suggest that (b) is highly preferred at the fluent, not particularly rapid speech rate at which the other examples above were assessed. This is a promising result, though one, like the others, which would need to be solidified on the basis of non-intuition-based investigation.

3.3

Let us consider next cases where one of the constituents of the complex VPs that we have been examining is a Focus. I believe English is like Chichewa in aligning the edge of a major phrase with the right edge of a Focus constituent (cf. Vogel and Kenesei 1990a-b). This effect emerges in varying the position of Focus in sentences having the verb phrase structure of those examined above, as we see in (27):

(27)

(i) [Verb XP YP_{FOC} $]_{VP}$:	She [lóaned [her róllerblades] [to Róbin] $_{FOC}]_{VP}$
a.	(She lóaned her róllerblades to **Róbin**)$_{MaP}$
b.	(She lóaned her róllerblades)$_{MaP}$ (to **Róbin**)$_{MaP}$
c.	*(She lóaned)$_{MaP}$(her róllerblades)$_{MaP}$(to **Róbin**)$_{MaP}$
d.	*(She lóaned)$_{MaP}$ (her róllerblades to **Róbin**)$_{MaP}$
(ii) [Verb XP_{FOC} YP$]_{VP}$:	She [lóaned [her róllerblades]$_{FOC}$ [to Róbin]$_{VP}$
a.	*(She lóaned her **róllerblades** to Róbin)$_{MaP}$
b.	(She lóaned her **róllerblades**)$_{MaP}$ (to Róbin)$_{MaP}$
c.	*(She lóaned)$_{MaP}$(her **róllerblades**)$_{MaP}$(to Róbin)$_{MaP}$
d.	*(She lóaned)$_{MaP}$ (her **róllerblades** to Róbin)$_{MaP}$
(iii) [$Verb_{Foc}$ XP YP$]_{VP}$:	She [lóaned$_{Foc}$ [her róllerblades] [to Róbin]]$_{VP}$
a.	*(She **lóaned** her róllerblades to Róbin)$_{MaP}$
b.	*(She **lóaned** her róllerblades)$_{MaP}$ (to Róbin)$_{MaP}$
c.	(She **lóaned**)$_{MaP}$(her róllerblades)$_{MaP}$(to Róbin)$_{MaP}$
d.	*(She **lóaned**)$_{MaP}$ (her róllerblades to Róbin)$_{MaP}$

In the representations (a-d) under each different Focus case are the grammaticality judgments assigned to each of the relevant candidates. The available phrasings can be checked out by considering the discourse contexts where they would be felicitously produced. Suppose the same scenario as above, where B has read A's journal entry, e.g. *I have loaned my rollerblades to Robin*, and is reporting it to C. What's different in these cases is that B is imposing an informational structure on the utterance, expressed as a Focus on one or the other of the constituents. What's the same is that all the information is 'new', and hence the pitch accenting is as above. Case (i) involves Focus of the indirect object. Suppose that B is surprised that Robin, of all people, would be the person to whom A would lend her rollerblades. B might then utter the sentence with *Robin* in Focus, as in (27i-a) or (27i-b). The available phrasings in this case actually turn out to be identical to those of (25ab), where *Robin* is not in Focus[16]. This is predicted, since in the case of (27i-ab) the MaP edge required by Align Focus is sentence-final, where a MaP edge is required in any case. Consider next case (ii), where the direct object is in Focus. We have the same journal-entry reporting scenario as above, except that what B finds surprising is that A would lend Robin her precious rollerblades, of all things. This surprise is articulated in B's report to C by a Focus on *rollerblades*. As indicated in (27ii-b), a phrase break follows the Focus in this case. It is important to underline that *Robin* is new in this discourse, and so is accented; the major phrase following the major phrase which ends in the Focus thus contains an accent itself. The significance of this fact will be discussed below. Finally, consider case (iii), where the verb is in Focus. We have the same journal-reading scenario, with a slightly different twist. C mentions the rumor that A has been getting rid of all her things, and asks B if there is any evidence of that in the journal. B replies with (27iii-c), putting a Focus on *loan*, and wondering if this might be the evidence C is looking for. As above, what follows the verb is informationally new, and hence accented.

What is particularly interesting about this last case with Focus on the verb is the apparent need to locate a MaP break both after the focussed verb and after the direct object as well. The pronunciation in (27iii-c) is preferred to the pronunciation in (27iii-d)[17], which also respects the Align$_R$ Focus constraint. My hypothesis here is that the explanation for the phrasing pattern in (27iii) involves ranking Align$_R$ Focus above the morphosyntactic interface constraints in English, as in Chichewa.

(28) Align$_R$ Focus >> Wrap XP, Align$_R$ XP

The tableau in (29) shows that candidates (a) and (b) are both non-optimal because of the violation of high-ranked Align$_R$ Focus that they each incur.

(29)

[she [lóaned]$_{\text{V-FOC}}$ [her róllerblades]$_{\text{NP}}$ [to Róbin]$_{\text{PP}}$]	Align$_R$ Focus	Wrap XP	Align$_R$ XP
a. (she **lóaned** her róllerblades to Róbin)$_{\text{MaP}}$	*!		*
b. (she **lóaned** her róllerblades)$_{\text{MaP}}$(to Róbin)$_{\text{MaP}}$	*!	*	
☞c. (she **lóaned**)$_{\text{MaP}}$ (her róllerblades)$_{\text{MaP}}$(to Róbin)$_{\text{MaP}}$		*	
d. (she **lóaned**)$_{\text{MaP}}$ (her róllerblades to Róbin)$_{\text{MaP}}$		*	*!

The remaining candidates, which respect Align$_R$ Focus, both incur a violation of Wrap XP, and the choice between the two therefore falls to Align$_R$ XP. Since Align$_R$ XP is not violated in candidate (c), (c) is optimal. Thus far, then, the explanation for the multiple phrasing under verb Focus within a VP with two complements parallels that for the Chichewa case, despite the fact that Align$_R$ XP and Wrap XP are same-ranked in English.

Note now that the result that candidate (c) is optimal in case (27iii) will stand only if the Wrap XP-Align XP pair are ranked higher than Bin(MaP), the binary size constraint on major phrases I suggested is playing a role in English. This is because, as we saw above, candidate (c) incurs three violations of Bin(MaP) while candidate (d) incurs just one violation of Bin(MaP). If the size constraint were at the same rank as the morphosyntactic interface constraints, it is (d) that would be ruled optimal. We see this in (30), where the bad apple indicates that the candidate is wrongly ruled optimal.

(30)

[she [lóaned]$_{FOC}$ [her róllerblades]$_{NP}$ [to Róbin]$_{PP}$]	Align Foc	Wrap XP	Align XP	Bin (MaP)
a. (she **lóaned** her róllerblades to Róbin)$_{MaP}$	*!		*	*
b. (she **lóaned** her róllerblades)(to Róbin)$_{MaP}$	*!	*		*
c. (she **lóaned**) (her róllerblades)(to Róbin)$_{MaP}$		*		***!
🍎 d. (she **lóaned**)(her róllerblades to Róbin)$_{MaP}$		*	*	*

This then is an argument in favor of ranking the Wrap XP-Align XP pair above Bin(MaP), as in (31):

(31) Align$_R$ Focus >> Wrap XP, Align$_R$ XP >> Bin(MaP)

With Bin(MaP) ranked lower, the result that (c) is the optimal candidate is preserved. We see this in tableau (32):

(32)

[she [lóaned]$_{FOC}$ [her róllerblades]$_{NP}$ [to Róbin]$_{PP}$]	Align Foc	Wrap XP	Align XP	Bin (MaP)
a. (she **lóaned** her róllerblades to Róbin)$_{MaP}$	*!		*	*
b. (she **lóaned** her róllerblades)(to Róbin)$_{MaP}$	*!	*		*
☞ c. (she **lóaned**) (her róllerblades)(to Róbin)$_{MaP}$		*		***
d. (she **lóaned**) (her róllerblades to Róbin)$_{MaP}$		*	*!	*

Given this ranking, the greater number of violations of Bin(MaP) in candidate (c) are irrelevant; it is the violation of the two higher-ranked morphosyntactic interface constraints in (d) that is fatal.

Finally, let us assure ourselves that the proper patterns of phrasing for the other Focus cases in (27) are obtained with this constraint hierarchy. Tableau (33) shows that in the case of direct object Focus, it is only candidate (b) that is ruled optimal.

(33)

[she [[lóaned] [her róllerblades]$_{NP\text{-}FCO}$ [to Róbin]$_{PP}$]	Align Foc	Wrap XP	Align XP	Bim (MaP)
a. (she lóaned her **róllerblades** to Róbin)$_{MaP}$	*!		*	*
☞b. (she lóaned her **róllerblades**)(to Róbin)$_{MaP}$		*		*
c. (she lóaned) (her **róllerblades**)(to Róbin)$_{MaP}$		*		**!*
d. (she lóaned)(her **róllerblades** to Róbin)$_{MaP}$	*!	*	*	*

The two candidates that respect Align Focus, (b) and (c), both violate Wrap XP; they are therefore distinguished by their number of Bin(MaP) violations, and it is (b) that wins, as desired. In the tableau in (34), we have the case of the sentence-final indirect object focus.

(34)

[she [[lóaned] [her róllerblades]$_{NP}$ [to Róbin$_{FOC}$]$_{PP}$]	Align Foc	Wrap XP	Align XP	Bin(MaP)
☞a. (she lóaned her róllerblades to **Róbin**)$_{MaP}$			*	*
☞b. (she lóaned her róllerblades)(to **Róbin**)$_{MaP}$		*		*
c. (she lóaned) (her róllerblades) (to **Róbin**)$_{MaP}$		*		**!*
d. (she lóaned) (her róllerblades to **Róbin**)$_{MaP}$		*	*!	*

All candidates in this set respect the Align$_R$ Focus constraint. Candidate (d) fatally violates both Wrap XP and Align$_R$ XP, leaving it to the Bin(MaP) to determine which of the other candidates are optimal. As before, (c) is ruled out due to the number of violations of Bin(MaP), and so it is (a) and (b), which each violate Bin(MaP) once, that are again both optimal.

To sum up this section, then, the data paradigm above in (27) gives a first round of evidence in favor of the presence in the grammar of English of a constraint aligning the edge of a MaP with the right edge of a Focus constituent, i.e. Align_{R}Focus. That constraint plays a crucial role in explaining the patterns available for the verb Focus and direct object Focus cases. The paradigm also gives further evidence in favor of the presence of Align_{R}XP in the grammar, in the preference of (c) over (d) in the verb Focus case. It is of interest, then to observe that these alignment constraints are often enough violated in utterances of English. The next section examines such cases.

3.4

The grammaticality judgments reflected in the data in (27) and in (18)-(21) crucially involved utterances where the informational content was all new, and as a consequence the content words were all pitch accented. The patterns of phrasing attested would not necessarily be the same under different assignments of pitch accent. Consider now, by contrast, utterances containing a nonfinal Focus which is followed by no accented material. Such an utterance would be naturally produced in the following dialogue, where the material following the focussed verb *lóaned* in B's response is old information:

(35) A. That Roberta! What a cheapskate! She rénted$_{FOC}$ her róllerblades to Róbin!
B. That's not true. She lóaned$_{FOC}$ her rollerblades to Robin.

In B's response, both *rollerblades* and *Robin* would lack an accent. The standard assumption is that the major phrasing (aka intermediate phrasing) of such an utterance would be as in (36)[18]:

(36) (She **lóaned** her rollerblades to Robin)$_{MaP}$

Here no MaP edge appears at the right edge of the Focus, rather the first phrase break following the Focus coincides with the end of the sentence. This is not the phrasing predicted by the constraint hierarchy in (31), which as it stands calls for the inappropriate phrasing in (37).

(37) (She **lóaned**)$_{MaP}$ (her rollerblades)$_{MaP}$ (to Robin)$_{MaP}$

I want to suggest that the choice of the phrasing in (36) as optimal over that in (37) is the consequence of a high-ranking phonological constraint on minor (accentual) phrases which requires that any minor phrase include at least one pitch accent:

(38) **MiPAccent**

A minor phonological phrase (MiP) contains at least one accent.[19]

Since any major phrase must consist of at least one minor phrase[20], this constraint has the effect of a constraint requiring that any major phrase contain at least one accent. Such a constraint on major (aka intermediate) phrases has been presumed in descriptive generalizations about the properties of English prosodic phrasing (e.g. Pierrehumbert and Beckman 1988) and is elevated to the status of law in the ToBI guidelines (Beckman and Ayers 1994) for transcribing English intonation, which disallow transcriptions where a sequence of intermediate (aka major) phrase breaks flank a string containing no accent.

We will instead want to understand this to be a constraint on minor/accentual phrases if we are to maintain our account of the role for size constraints on major phrasing that was sketched above. That account assumed that each accent in the utterance corresponded to a minor/accentual phrase. It is the constraint MiPAccent that guarantees that result.

It is enough to assume that the properly phonological constraint MiPAccent is higher ranked than Align$_R$Focus, an interface constraint, to ensure that the phrasing attested in (36) is indeed that which would be ruled optimal by the grammar. The tableau in (39) provides the argument for this MiPAccent >> Align$_R$Focus ranking. Assume that in the set of candidates considered here each MaP dominates just one minor phrase: $_{MaP}(_{MiP}(....)_{MiP})_{MaP}$.

(39)

[she [lóaned]$_{V\text{-FOCUS}}$ [her rollerblades]$_{NP}$ [to Robin]$_{PP}$]	MiPAccent	Align Focus
☞a. (she **lóaned** her rollerblades to Robin)$_{MaP}$		*
b. (she **lóaned** her rollerblades)$_{MaP}$(to Robin)$_{MaP}$	*!	*
c. (she **lóaned**)$_{MaP}$ (her rollerblades)$_{MaP}$(to Robin)$_{MaP}$	*!*	
d. (she **lóaned**)$_{MaP}$ (her rollerblades to Robin)$_{MaP}$	*!	

In candidate (c), the two MaPs which follow the MaP break at the right edge of Focus each lack an accent, which means that their component MiP's each lack an accent. Each MiP in MaP thus incurs a violation of MiPAccent. Just one violation of MiPAccent is enough to be fatal, since there is another candidate, (a), which shows no violation at all of that constraint. Thus candidates (b) and (d) are nonoptimal since in each there is one accentless MaP (dominating an accentless MiP) in the representation. We see then that candidate (a) is ruled optimal, even though it violates Align$_R$Focus, because the other candidates all show violations of the higher ranked MiPAccent. Here the high rank of the phonological constraint MiPAccent is achieving just the desired effect.

We are not at the end of our account, however, for a further set of candidates must be ruled out, ones which satisfy MiPAccent through the insertion (epenthesis) of a pitch accent. I am assuming that pitch accents are morphemes in English and thus form part of the surface syntactic representation (cf. Gussenhoven 1983, 1984, Selkirk 1984, 1995b, Pierrehumbert and Hirschberg 1990) that is in an interface relation with the phonology. The epenthesis of a pitch accent into the phonological representation (or the deletion of a pitch accent present in the interface syntactic representation) would incur violations of the faithfulness constraints that regulate the relation between interface representations in optimality theory. In the correspondence theory of faithfulness (McCarthy and Prince 1995), there are DEP constraints which penalize epenthesis and MAX constraints which penalize deletion. The intuitions captured are that, ideally, the output is strictly dependent on the input (hence DEP), and that, ideally, the input is maximized in the output (hence MAX). Correspondence theory can also hold of the relation between interface representations. A constraint DEP(Accent), which rules out the presence of any pitch accent in the phonological output that is not in the corresponding interface representation, will complete our account .

(40) **DEP (Accent)**

> An accent in the output representation must have a corresponding accent in the interface representation.

Ranking DEP(Accent) above Align$_R$Focus will guarantee that the candidate we saw above in (39a) is indeed optimal, in comparison with candidates that insert accents in order to satisfy the high-ranked MaP. This can be seen in the tableau in (41):

(41)

[she [lóaned]$_{V\text{-}FOCUS}$ [her rollerblades]$_{NP}$ [to Robin]$_{PP}$]	MiP Accent	DEP (Accent)	Align Focus
☞a. (she **lóaned** her rollerblades to Robin)$_{MaP}$			*
e. (she **lóaned**)$_{MaP}$ (her róllerblades)$_{MaP}$(to Róbin)$_{MaP}$		*! *	
f. (she **lóaned**)$_{MaP}$ (her róllerblades to Róbin)$_{MaP}$		*!*	
g. (she **lóaned**)$_{MaP}$ (her róllerblades to Robin)$_{MaP}$		*!	
h. (she **lóaned**)$_{MaP}$ (her rollerblades to Róbin)$_{MaP}$		*!	

The violations of the higher ranked DEP(Accent) are fatal, and therefore responsible for the choice of (a) as optimal; candidate (a) retains the lack of input accenting on the verbal complements, in satisfaction of DEP(Accent). Looking at tableaux (39) and (41) together, we see that the violation of Align$_R$Focus in the optimal (a) is a consequence of the higher ranking of both MiPAccent and DEP(Accent).

To summarize, we have now augmented the partial constraint ranking in (31) with the addition of MiPAccent, a constraint that is properly phonological and holds of the output representation itself, and DEP(Accent), a faithfulness constraint that regulates the relation between the input and output representations. It is proposed that these two are higher ranked than the interface constraint Align$_R$ Focus, and by transitivity also higher than the interface constraints Align Focus dominates – Wrap XP and Align$_R$ XP – as well as the lower ranked phonological constraints on major phrase size, Bin(MaP). Together, the final constraint ranking is hypothesized to be as in (42)[21]:

(42) MiPAccent, DEP(Accent) >> Align$_R$Focus >> Wrap XP , Align$_R$XP >> Bin(MaP)

This constraint hierarchy predicts that violations of Align$_R$Focus and Align$_R$XP will be incurred in optimal output candidates which lack the accents to support the phrasing to the right of the MaP break that these constraints call for. Here is a case where the demarcative function satisfied by respect of the alignment constraints is subordinated to the demands of surface phonological wellformedness and faithfulness to underlying phonological form. The optimality theoretic approach allows us to understand that the constraints Align$_R$Focus and Align$_R$XP are indeed at play in the grammar of English, but not satisfied in every surface representation, precisely when some other (higher-ranked) constraint of the language would be violated.

3.5

The preceding sections lay out a complex hypothesis about the nature of the grammar of constraints on major phrasing in English. Most of the constraints that play a role in the constraint hierarchy proposed have either been independently motivated for English or other languages, or are an instance of a type of constraint that has been independently motivated. The task undertaken here has been to confront a small array of data on English phrasing in order to arrive at an initial hypothesis about how such constraints might be interacting in the grammar of English. In so doing the intent has been to illustrate the workings of a constraint-based grammar conceived in optimality theoretic terms and the sorts of insights to be gained within that framework. The larger goal will be to expand the base of factual generalizations concerning major phrasing in English, ideally through the use of natural speech corpora and/or experimental procedures, and to test the hypothesis embodied in (42) against that base. As this hypothesis suggests, any such investigation will have to control for the morphosyntactic and informational structure of the input sentences as well as for properly phonological factors such as prosodic constituent size and accentuation. Both interface constraints and properly phonological constraints on output representation appear to have an influence on major phrasing in English, and an understanding of the role of each constraint cannot be reached without an understanding of its relation to the others.

NOTES

[1] For example, Selkirk 1986, Nespor and Vogel 1986, Chen 1987, Inkelas and Zec 1990, 1996.

[2] For example, Bruce 1977, Beckman and Pierrehumbert 1986, Pierrehumbert and Beckman 1988.

[3] These include the constraints on prosodic domination which are collectively referred to as the strict layer hypothesis (see, e.g. Selkirk 1995a), constraints on the size of prosodic phrases (e.g. Gee and Grosjean 1983, Selkirk and Tateishi 1988, Dresher 1994, Ghini 1993, Helsloot 1995, Delais-Roussarie 1996), and constraints on their tonal composition, including the requirement that a peripheral tone be present, that a pitch accent be present, or that the lexical tones present be arrayed in a certain fashion. In the latter case, representative references would be too numerous to cite.

[4] Selkirk and Tateishi 1988, Kanerva 1989, Selkirk 1989, Vogel and Kenesei 1990a-b, Hayes and Lahiri 1991, and Jun 1993, for example, have pointed to the diverse types of constraints on phrasing. It is only in the context of optimality theory, however, that an understanding of the interaction of the various constraint types has become an explicit topic of investigation, as in the work of Nagahara (1994), Helsloot (1995), Selkirk (1995a), Truckenbrodt (1995), Delais-Roussarie (1996), for example.

[5] ChiMwiini is a language closely related to Swahili that is spoken in the town of Brava in Somalia.

[6] There is considerable diversity in the terms used to refer to prosodic constituents larger than the word. The terms *major (phonological) phrase, intermediate phrase* and even simply *phonological phrase* have been used by one or another scholar to designate the same level of prosodic structure. Since there are no phonological properties which universally characterize this level of phrasing, a nomenclature which evokes instead the level of syntactic constituency to which this prosodic phrasing is related in the statement of the interface constraint has a desirable mnemonic effect. Hence the term *major* phonological phrase for the level of prosodic phrasing aligned with syntactic *maximal* projections.

[7] Constraints of this type involve universal quantification over the first category and existential quantification over the second, thus they read: "For any XP in the input representation, align its edge (R,L) with the edge (R,L) of some MaP in the output representation".

[8] The requirement that the *right* edge of some MaP be aligned with any XP will typically have the same effects as the requirement that a left edge be aligned, or simply that there be a MaP 'break' at that location. This is because, in the unmarked case at least, the presence of a right MaP bracket implies the presence of an immediately following left MaP bracket, since the utterance is parsed as a sequence of MaP's. See Selkirk (1995a) for a proposal concerning an optimality theoretic treatment of the basic constraints on prosodic structure that are often subsumed under the rubric of the Strict Layer Hypothesis.

[9] Tableau (8) displays only the most relevant two of the potentially large range of candidates that are generable on the basis of the input representation. In principle the constraint system must evaluate all such candidates.

[10] Truckenbrodt (1995) suggests an interesting alternative conception of the correlation between Focus and phrasing which does without the constraint Align Focus which directly relates Focus and phrasing. His suggestion is that the interface of Focus and phonological representation involves a constraint calling for the Focus to correspond to a major phrase prominence. The alignment of phrase edges with Focus would be due to a constraint, widely attested in phonology, calling for a prominence of a constituent to be located at the edge of a constituent. The presence of prominence on Focus would thus

induce the presence of the phrase break. This proposal is a very appealing one. I am not adopting it here, though, in order to simplify this introductory discussion of constraint interaction.

[11] In the row corresponding to candidate (c), which fatally violates the highest-ranked $Align_R$Focus constraint, there is a cell to the right of the one containing the fatal violation that is shaded. This shading is a notational device that serves to highlight the fact that any further constraint violations of constraints below the one incurring the fatal violation are irrelevant.

[12] One might want to ask at this point whether Wrap XP and Align XP are the only members of their respective families relevant to sentence phonology. At this point it would seem most fruitful to assume there might be yet further Align and Wrap constraints, which might play a role in characterizing the distribution of intonational phrases or minor phrases, for example. This is an interesting area for research.

[13] Though there have been steps in the direction of an account of English major/intermediate phrasing made by Bing (1979), Gussenhoven (1984), Beckman and Pierrehumbert (1986), Ladd (1986), Vogel and Kenesei (1990a,b), Selkirk (1995a) among others.

[14] There are other reflexes of major phrasing in English. On the basis of material from an FM radio news corpus, Shattuck-Hufnagel et al. (1994) argue that the intermediate phrase (aka major phrase) is the domain with respect to which the phenomenon of early accent, also referred to as rhythmic inversion, is defined. Similarly, Vogel and Kenesei (1990a) had argued that the phonological phrase was the domain of rhythmic inversion in English. Selkirk (1995a) argues that the location of nonaccented function words at the right edge of a MaP is responsible for their appearing there in strong, unreduced form.

[15] Beckman and Pierrehumbert (1986) and Pierrehumbert (1993) reject the notion that the accentual/minor phrase plays a role in English, based on the assumption that an accentual phrase may never have any more than one accent. There exist cases of single words bearing two accents in English. For example, an emphatic rendering can give the pronunciation *Càlifórnia.* Since such words do not divide up into two accentual phrases, or two prosodic words, for that matter, Beckman and Pierrehumbert consider them to be evidence that the accentual/minor phrase plays no role in English. Given an optimality theoretic perspective, according to which constraints are violable, this is not an argument. A constraint against more than one accent within an accentual phrase could be violated, and these double-accented words are precisely cases where such a violation would be forced. Suppose that the double accents are part of the input, and that their presence in the output is ensured by a faithfulness constraint which disallows deletion of input accents. Suppose moreover that there were some constraint(s) which prevented noncompound

words in the input from being divided up into two prosodic words in the output (see Selkirk 1995a). All these constraints have independent motivation in the grammar of English. If these constraints were to outrank any constraint(s) calling for accentual phrases to have at most one accent, then it would be predicted that double accenting within accentual phrases would be allowed in just these cases.

[16] This is not to say that the pronunciations in the two cases would be fully identical. In the case where *Robin* is in Focus there is likely to be a greater pitch prominence on the constituent.

[17] The avoidance of a phrasing like that in (27iii-d) in English, where the verb is followed by a phrase break and two complements to the verb are joined in a single prosodic phrase, has been quite generally assumed in earlier work, including Bing 1979, Selkirk 1984 and Hirst 1993.

[18] Beckman and Pierrehumbert (1986) and Beckman and Ayers (1994) analyze sentences like (36), with a nuclear pitch accent at some distance from the end of the sentence, as a single intermediate phrase (aka major phrase). (The sentence is also analyzed as a single intonational phrase. The intonational phrase is one up from major phrase in the prosodic hierarchy, according to Selkirk (1986), Beckman and Pierrehumbert 1986).) This analysis seems right. Note that an alternative analysis, according to which the sentence has a nested MaP structure as in (i) cannot be maintained.

H* L- L- L%

$_{IP}(_{MaP}(_{MaP}($She **loaned** $)_{MaP}$ her rollerblades to Robin $)_{MaP})_{IP}$

This analyses would predict a fall from the H* pitch accent to the L- phrase tone within the verb. But this is not attested. Rather, it is the syllable *her*, one after the accented syllable, which carries the target low tone following the H*. This is the position that a L- would take if it spread leftward from the MaP edge at the end of the sentence, as is proposed in Pierrehumbert (1980). The unattested MaP structure in (i) must therefore be ruled out. We assume the nonoptimality of the nested MaP structure in (i) to be due to the high rank of a constraint NonRec(MaP) from the family of NonRecursivity constraints (Selkirk 1995a), which rules out such self-embedding of prosodic constituents from the same level in the prosodic hierarchy.

[19] This constraint will in the end probably not have this formulation. An alternative conception would be to see the constraint as one calling for the most prominent syllable of any MiP to be marked by the presence of an associated pitch accent. This conceptualization is more firmly grounded crosslinguistically.

[20] This follows from the very nature of the prosodic hierarchy, according to which a constituent at one level dominates, or is headed by, a minimum of one constituent of the next level down (cf. Selkirk 1995a).

[21] The comma in (26) and the dotted line between MaPAccent and DEP(Accent) in the tableau indicate that these two constraints are not crucially ranked with respect to each other. However, their position both to the left of Align$_R$Focus indicates they are both higher ranked than it.

REFERENCES

Beckman, J., Dickey, L.W. and Urbanczyk, S. (eds), 1995. *Papers in Optimality Theory.* Univ. of Mass. *Occasional Papers in Linguistics* 18.

Beckman, M. and Ayers, G. 1994. *Guidelines for ToBI Transcription.* Department of Linguistics, Ohio State University.

Beckman, M. and Pierrehumbert, J. 1986. Intonational structure in English and Japanese. *Phonology* 3, 255-309.

Bickmore, L. 1989. *Kinyambo Prosody.* Doctoral Dissertation, University of California at Los Angeles.

Bickmore, L. 1990. Branching nodes and prosodic categories. In S. Inkelas and D. Zec (eds), 1-17.

Bing, J. 1979. *Aspects of English Prosody.* Doctoral dissertation, University of Massachusetts at Amherst.

Bruce, G. 1977. *Swedish Word Accents in Sentence Perspective.* Lund: Gleerup.

Bruce, G., Granström, B., Gustafson, K. and House, D. 1991. Prosodic phrasing in Swedish. *Working Papers* 38, 5-17. Department of Linguistics, Lund University.

Bruce, G., Granström, B., Gustafson, K. and House, D. 1993. Prosodic modelling of phrasing in Swedish. In *Proc. ESCA Workshop on Prosody. Working Papers* (Dept. of Linguistics, Lund University) 41, 180-183.

Bruce, G., Granström, B., and House, D. 1991. Strategies for prosodic phrasing in Swedish. *Proc. XIIth International Congress of Phonetic Sciences* (Aix-en-Provence), 182-185.

Chen, M. 1987. The syntax of Xiamen tone sandhi. *Phonology* 4, 109-150.

Clements, G.N. and Ford, K. 1981. On the phonological status of downstep in Kikuyu. In D. Goyvaerts (ed), *Phonology in the 80's.* Ghent: Story-Scientia, 309-357.

Delais-Roussarie, E. 1996. Phonological phrasing and accentuation in French. Manuscript, University of Massachusetts at Amherst.

Dresher, B.E. 1994. The prosodic basis of the Tiberian Hebrew system of accents. *Language* 70, 1-52.

Gee, J. and Grosjean, F. 1983. Performance structures: a psycholinguistic and linguistic appraisal. *Cognitive Psychology* 15, 411-458.

Ghini, M. 1993. *Phonological Phrase Formation in Italian.* Master's thesis, University of Toronto.

Gussenhoven, C. 1983. Focus, mode and the nucleus. *Journal of Linguistics*, 19, 377-417.

Gussenhoven, C. 1984. *On the Grammar and Semantics of Sentence Accents.* Dordrecht: Foris.

Hayes, B. and Lahiri, A. 1991. Bengali intonational phonology. *Natural Language and Linguistic Theory* 9, 47-96.

Helsloot, K. 1995. *Metrical Prosody: A Template-and-Constraint Approach to Phonological phrasing in Italian.* Doctoral dissertation, University of Amsterdam.

Hirst, D. 1993. Detaching intonational phrases from syntactic structure. *Linguistic Inquiry* 24, 781-788.

Inkelas, S. and Zec, D. (eds), 1990. *The Phonology-Syntax Connection.* Chicago: University of Chicago Press.

Inkelas, S. and Zec, D. 1996. Syntax-phonology interface. In J. Goldsmith (ed), *The Handbook of Phonological Theory*. Oxford: Blackwell, 535-549.

Jun, S-A. 1993. *The Phonetics and Phonology of Korean Prosody*. Doctoral dissertation, Ohio State University.

Kanerva, J. 1989. *Focus and Phrasing in Chichewa Phonology*. Doctoral dissertation, Stanford University.

Kanerva, J. 1990. Focusing on phonological phrases in Chichewa. In S. Inkelas and D. Zec (eds), 145-162.

Kenstowicz, M. and Kisseberth, C. 1990. Chizigula tonology: the word and beyond. In S. Inkelas and D. Zec (eds), 163-194.

Kisseberth, C. 1984. Digo Tonology. In G.N. Clements and J. Goldsmith (eds), *Autosegmental Studies in Bantu Tone* . Dordrecht: Foris, 105-182.

Kisseberth, C. and Abasheikh, M. 1974. Vowel length in ChiMwi:ni. A case study of the role of grammar in phonology. In M. LaGaly, A. Bruck and R. Fox (eds), *CLS 10: Parasession on Natural Phonology*. Chicago: Chicago Linguistic Society, 193-209.

Ladd, D.R. 1986. Intonational phrasing: The case for recursive prosodic structure. *Phonology* 3, 311-340.

McCarthy, J.J. and Prince, A. 1993. Generalized alignment. In G. Booij and J. van Marle (eds), *Yearbook of Morphology*. Dordrecht: Kluwer, 79-153.

McCarthy, J.J. and Prince, A. 1995. Faithfulness and reduplicative identity. In J.N. Beckman, L.W. Dickey and S. Urbanczyk (eds), 249-384.

McHugh, B. 1987. Syntactic structure, empty categories and phrasal phonology in Chaga. In D. Odden (ed), *Current Approaches to African Linguistics* 4. Dordrecht: Foris, 247-65.

McHugh, B. 1990. *Cyclicity in the Phrasal Phonology of Kivunjo Chaga*. Doctoral dissertation, University of California at Los Angeles.

Nagahara, H. 1994. *Phonological Phrasing in Japanese*. Doctoral dissertation, UCLA.

Nespor, M. and Vogel, I. 1986. *Prosodic Phonology*. Dordrecht: Foris.

Odden, D. 1987. Kimatuumbi phrasal phonology. *Phonology* 4, 13-36.

Odden, D. 1990. Syntax, lexical rules, and postlexical rules in Kimatuumbi. In S. Inkelas and D. Zec (eds), 259-278.

Pierrehumbert, J. 1980. *The Phonetics and Phonology of English Intonation*. Doctoral dissertation, MIT.

Pierrehumbert, J. 1993. Alignment and prosodic heads. *Proc. Eastern States Conference on Formal Linguistics,* vol. 10. Department of Linguistics, Ohio State University, 268-286.

Pierrehumbert, J. and Beckman, M. 1988. *Japanese Tone Structure*. Linguistic Inquiry Monograph Series No. 15. Cambridge, MA: MIT Press.

Pierrehumbert, J. and Hirschberg, J. 1990. The meaning of intonational contours in the interpretation of discourse. In P.R. Cohen, J. Morgan and M.E. Pollock (eds), *Intentions in Communication*. Cambridge MA: MIT Press, 271-311.

Prince, A. and Smolensky, P. 1993. *Optimality theory: Constraint interaction in generative grammar*. Ms., Rutgers University and Johns Hopkins University.

Selkirk, E. 1984. *Phonology and Syntax: The Relation between Sound and Structure*. Cambridge, MA: MIT Press.

Selkirk, E. 1986. On derived domains in sentence phonology. *Phonology* 3, 371-405.

Selkirk, E. 1989. Modularity in the syntax-phonology mapping-- the case of Chinese dialects. Paper presented at GLOW 12, Utrecht.

Selkirk, E. 1995a. The prosodic structure of function words. In J.N. Beckman, L.W. Dickey and S. Urbanczyk (eds), *Papers in Optimality Theory*. University of Massachusetts, Amherst, GLSA, 439-470. Also appears in J.L. Morgan and K. Demuth (eds), *Signal to Syntax: Bootstrapping from Speech to Grammar in Early Acquisition*. Mahwah, NJ: Lawrence Erlbaum Associates, 1996, 187-214.

Selkirk, E. 1995b. Sentence prosody: intonation, stress and phrasing. In J. Goldsmith (ed), *The Handbook of Phonological Theory*. Oxford:Blackwell, 550-569.

Selkirk; E. and Tateishi, K. 1988. Minor phrase formation in Japanese. In L. Macleod et al. (eds), *Papers from CLS* 24, 316-336. Chicago: Chicago Linguistic Society.

Shattuck-Hufnagel, S., Ostendorf, M. and Ross, K. 1994. Stress shift and early pitch accent placement in lexical items in American English. *Journal of Phonetics* 22, 357-388.

Truckenbrodt, H. 1995. *Phonological Phrases: their Relation to Syntax, Prominence and Focus*. Doctoral dissertation, MIT.

Vogel, I. and Kenesei, I. 1990a. Focus and phonological structure. Paper presented at GLOW 13, Cambridge.

Vogel, I. and Kenesei, I. 1990b. Syntax and semantics in phonology. In S. Inkelas and D. Zec (eds), 339-364.

Dept. of Linguistics, University of Massachusetts at Amherst, Amherst, MA, USA

MARI OSTENDORF

PROSODIC BOUNDARY DETECTION

1. INTRODUCTION

Prosodic constituent structure, or the perceived grouping of words in speech, plays a role in human speech communication in virtually every language. Speakers use prosodic phrasing to contribute meaning to and sometimes disambiguate the sequence of words that comprise an utterance by highlighting its information structure. From a speech analysis perspective, prosodic phrase structure provides the link that seems to most effectively explain continuously varying acoustic correlates (pauses, F0 patterns, duration lengthening, etc.) in terms of the word sequence of an utterance (syntactic, semantic and discourse structure). Just as both speakers and listeners use prosodic phrases in human speech communication, so computational models of prosodic phrase structure can be useful both for communicating meaning in synthesized speech and for extracting meaning in automatic speech understanding. In fact, prosodic phrase structure is probably even more important for computer speech processing than for humans, because computers have a much less detailed semantic representation and less extensive knowledge of the world than humans and thus word sequences tend to be more often ambiguous in computer language processing than for human listeners.

As both speech and language processing technology have advanced dramatically over the past decade, speech systems are better positioned to take advantage of the information coded in prosodic patterns. For example, in speech understanding, the role of prosody in reducing syntactic ambiguity becomes increasingly important as language processing systems handle more syntactic constructions and as speech recognition systems entertain more word hypotheses in systems with 20k or larger vocabularies. As speech understanding technology matures and human-computer dialog becomes a reality, a new need for prosody has arisen in the problem of determining the dialog state. Finally, in speech translation, the role of prosodic pattern recognition is at least as important as in speech understanding, since phenomena like emphasis may be coded in the prosodic patterns of one language but in the choice of wording in another.

Recognition of prosodic phrase structure is probably the first and most important problem in computational modeling of prosody for automatic speech

M. Horne (ed.), Prosody: Theory and Experiment, 263–279.

processing, and certainly has had the biggest impact on computer speech processing to date. Thus, the goal of this chapter is to survey various methods for detecting prosodic phrases and using phrase information in speech processing applications, taking as a premise the goal of detecting symbolic linguistically-based prosodic phrase structure (as opposed to some automatically learned acoustic unit)[1]. The problem of detecting prosodic constituent structure is not straightforward, both because of differences in theoretical representations and because the suprasegmental nature poses a challenge to computational modeling. In addition, depending on the particular application that the phrase detection algorithm is designed for, the cues available for use in detection may differ. As a result, there is an array of possible choices among different prosodic phrase recognition algorithms, with no one solution being appropriate for all applications or theories. This chapter will outline the various choices and trade-offs, acknowledging that phrase detection is still an active area of research.

In order to motivate some of the biggest differences in the various phrase detection algorithms, the chapter begins in Section 1.2 with a discussion of the theoretical issues that impact algorithm design. Next, in Section 1.3, the various alternatives for phrase recognition are outlined, including the mathematical modeling framework, the features included, and the options for allowing ambiguity. The particular choice of algorithm is influenced to some extent by the application, whether automatic labeling of phrases in speech where the word transcription is known or recognizing phrases in unrestricted speech for either transcription or understanding purposes, as discussed in Section 1.4. Finally, the chapter concludes in Section 1.5 with a discussion of open questions in computational modeling of prosodic phrase boundaries.

2 IMPLICATIONS OF LINGUISTIC THEORY

Unlike in speech recognition, where the definition of a word is widely accepted for many languages and the acoustic cues to a word are locally situated in time, prosody recognition is faced with controversy in the definition of the basic symbolic units and the problem of acoustic correlates that span multiple time scales and cue multiple types of information. For example, a change in F0 may be a cue to to a new topic (phrase-level pitch range), a change in focus (syllable-level pitch accent), the end of a phrase or speaker turn (phrase boundary tone), and/or simply the presence of a voiceless obstruent (segmental effects). The fact that the same acoustic measurements are associated with cues to multiple types of information structure associated with multiple time scales impacts the computational modeling assumptions, as will be discussed in Section 1.3. It also underscores the importance of having a symbolic intermediate level of representation, such as phrase constituent structure. Just as phones or other sub-word units are used as an intermediate level between acoustic features and words in large vocabulary continuous speech recognition, there is a practical need for intermediate prosodic units to allow for generalization over the many possible combinations of prosodic patterns that can occur. Unfortunately, there is no clear theoretical consensus on the appropriate units, in particular the number of levels of phrase constituents and the grammar of these constituents are issues of debate. However, most theories allow for more than one level of constituent structure and assume a strict layer hypothesis (edges of high level constituents coincide with edges of lower level constituents), so the computational models discussed here will handle at least these properties.

Accepting the notion of an embedded hierarchy of phrase constituents, there is one important theoretical issue that separates two classes of computational models, which we will refer to as superposition vs. linear-sequence models. Superposition models typically view intonational phrase structure as a sum of the contributions from different levels, including one or more phrase contours, accent components and possibly segmental perturbations. The best known example of this general class is the Fujisaki model (Hirose and Fujisaki 1982, Fujisaki and Kawai 1988) which has been applied to a number of languages, but any theory built on the notion of declination could fall into this category. Under the superposition theory, phrase boundary recognition requires the model to represent whole phrases and generally focuses on intonation patterns to the exclusion of other cues such as duration.

Linear-sequence theories, on the other hand, treat a phrase as a sequence of events with acoustic cues that are localized in time, i.e. do not span the entire phrase. In this class is the model of a phrase as a sequence of pitch accents followed by a phrase boundary marker, where examples include Bruce's model of Swedish intonation (Bruce 1977) and the ToBI labeling system (Silverman et al. 1992; Pitrelli et al. 1994) built on the theory of intonation developed by Pierrehumbert and colleagues (Pierrehumbert 1980, Beckman and Pierrehumbert 1986). Although linear-sequence models are generally developed to describe intonation, the local nature of the acoustic cues makes it easy to accommodate observed duration facts about phrasing in a computational model; detection involves features in a local window and can therefore include a variety of cues in addition to F0 features. An important issue for computational modeling under a linear-sequence theory is the domain of the phrase marker, e.g. the last syllable of the phrase vs. the last word vs. all syllables after the last (nuclear) accent for phrase-final position. As Beckman (1995) points out, the linear-sequence theories generally accept superposition at some level (e.g. to model segmental effects), and often include phrase-level phenomena such as pitch range though not as an additive component. Phrase-level phenomena are generally incorporated in linear-sequence phrase detection algorithms through feature normalization.

Within these two broad theoretical frameworks, superposition or sequential, computational models can be posed as otherwise theory-neutral in most instances. In other words, the same model can be generally configured so as to be consistent with a variety of linguistic theories. The details of the particular linguistic theory are then implemented in terms of grammatical constraints, definition and span of the symbolic units, specification of feature normalization, and structural constraints such as parameter tying.

Another factor that might impact the choice of computational models is the fact that prosodic "parses" themselves can be ambiguous (Beckman 1996). For example, in the ToBI system, it can be difficult to distinguish between the (H* L- H*) and (H* L+H*) combinations, which correspond to the presence vs. absence of an intermediate phrase boundary. This and the fact that some acoustic cues may not be reliably present (e.g. phrase onset glottalization (Dilley et al. 1996) suggest that it is useful to allow "soft" or "fuzzy" decisions in the detection output where there is ambiguity in the acoustic evidence. Probabilistic approaches to computational modeling are well suited to allowing soft decisions; the output can simply be the likelihood of the possible events (Ostendorf 1998). However, other methods of representing uncertainty are also possible, including simply allowing a number of candidate boundary locations (Nakai et al. 1995). In addition to representing ambiguity as to whether a prosodic phrase boundary is present, there is a need to represent the fact that placement of phrase boundaries is optional at many types of syntactic constituent boundaries. Again, probabilistic approaches to representing the relationship between prosody and syntax can handle this fact.

1.3 COMPUTATIONAL MODELING

A variety of algorithms have been proposed for analyzing prosodic contours and recognizing prosodic phrase boundaries. In the following survey of techniques, we take as a premise that only computational models which can be automatically trained and include some notion of uncertainty are of interest. This includes a large number of approaches – including probabilistic, neural network and template matching methods – but excludes rule-based approaches. Since so much is still not understood about prosodic structure and the associated acoustic correlates, automatic training methods are essential for handling our lack of knowledge about specific cues and how they interact, i.e. the variability that is not explained by current linguistic theory. As mentioned above, models that handle uncertainty are also useful simply because there is inherently some ambiguity in current representation systems of prosodic structure. The emphasis or bias here will be towards probabilistic models, because of the well understood training and recognition methodologies associated with this approach and because it has been the most successful for speech recognition.

As mentioned earlier, there are several design choices that can be made in a prosodic phrase recognition algorithm. Depending on the linguistic theory that one ascribes to, superposition vs. linear-sequence modeling, the model can represent whole contours or phrase edge phenomena. For the linear-sequence case, models can be further divided in terms of whether the acoustic observations are variable-length sequences of frame-based features, i.e. computed every 5-10 ms, or fixed-length transformations of the acoustic cues in a region. For both superposition and linear-sequence approaches, as in all recognition problems, the accuracy is improved if the prior probabilities of the possible events are accounted for, which raises the question of how to represent this probability and whether to incorporate the information available from the hypothesized word sequence associated with the acoustic features. Finally, there is the issue of how to represent hierarchical structure in a model and the impact on the problem of phrase detection. Each of these four issues will be discussed in further detail below.

The two main classes of computational models are distinguished by whether they model whole phrase patterns vs. sequences of events within a phrase. The whole phrase models use superposition to capture different levels of prosodic information and focus on the intonation contour alone. Whole phrases are usually recognized using a minimum mean squared error criterion, either to one of a specified set of phrase templates (Nakai et al. 1995) or via analysis-by-synthesis techniques (Geoffrois 1993), for the case of the Fujisaki model. An alternative technique, which uses neural networks to generate the sub-components (Morlec et al. 1996), has been used for synthesis but could easily be used in recognition with either of these approaches. Linear-sequence models represent a series of events by concatenating separate models for each event according to a grammar of the allowable sequences. This sequence of models is associated with the acoustic observations, which are not limited to the intonation contour, via a time segmentation of the phrase into regions associated with the time scale of the events (e.g. syllables). Linear sequence models may represent the observations directly (as do whole phrase models) or use some transformation of the observations, as will be discussed further shortly. With both the superposition and linear-sequence models, it is possible to ignore the detailed structure within a phrase when detecting phrase boundaries. For the superposition model, the more detailed levels of the intonation contour or of the phrase "template" are omitted with the hope that the mean-squared error associated with these events contributes less than would a misalignment of the template. For the linear-sequence model, detection can be based on the local cues associated with phrase edges using the same technology as is used in word spotting (Butzberger 1989). When it is practical to estimate models for the details of a phrase, better performance is possible because it reduces the chances of confusion between accent lending vs. boundary-marking pitch movements and duration lengthening. Of course, the improved performance comes at a higher cost. Probably more important than the extra computational cost, the more detailed phrase models also require a more detailed prosodic labeling of the training corpus.

The linear-sequence approaches can be further grouped in two sub-classes: the frame-based approach, which models the complete sequence of acoustic features associated with a region, and the transformation approach, which uses a vector that is some function of local F0 patterns and other cues within the region. To understand this distinction, consider the observation sequence for an event such as a syllable: $(y_1, ..., y_T)$ where each yt is a vector of cues such as F0 and energy in different frequency bands and $T$is the length of the syllable. The frame-based approach uses the observations yt directly, either with a minimum mean-squared error criterion or with conditional distributions $p(y_1, ..., y_T)| a)$ as in a probabilistic approach (a is an index for a particular prosodic event). The transformation approach maps the sequence of observations in a region to a fixed-length feature vector, say $Z = F(y_1, ..., y_T)$,and uses this vector in finding the "score" of an event a, such as the posterior probability $p(a \mid Z)$. Examples of vector elements z_i might include the average normalized duration of phones in the syllable, the length of a pause following the word, an indicator of whether the syllable is stressed, and the regression coefficients of a polynomial fit to the F0 contour. The posterior probability might be computed as a function of standard conditional distributions via Bayes rule, or using non-linear methods such as decision trees and neural networks.

Many frame-based approaches are based on hidden Markov models (HMMs) (Butzberger 1989, Jensen et al. 1993), although the superposition methods described above would also be classed as frame-based models. Frame-based methods work well for representing F0 contours, but have the disadvantage that duration cues, known to be very important, are effectively ignored. In addition, HMMs are known to be weak in modeling smoothly varying features because of their independence assumptions, so derivative features are used to provide some means of capturing the correlation of successive features. Segmental models (Ostendorf et al. 1996) provide a mechanism for using frame-based features that overcome some of these weaknesses, and in particular allow for an explicit model of duration. A segment model that represents syllable-level intonation patterns using a dynamical system model of feature correlation is described in Ostendorf and Ross (1997). An alternative to the segment model that allows the use of cues such as duration is the transformation approach, as illustrated by the different solutions proposed in Wightman and Ostendorf (1994), ten Bosch (1993), Campbell (1994), Kompe et al. (1994), Kompe et al. (1995). An advantage of the feature transformation approach is that it allows a much more general feature set and more flexible scoring functions that can better account for the interdependence of different cues, e.g. see Bruce et al. (1993), Horne et al. (1995). Although the feature transformation approach yields good performance for automatic labeling given a known word sequence, it has the disadvantage relative to frame-based approaches that it is theoretically problematic to make direct comparisons of hypotheses with different segmentations (see Ostendorf et al. (1996) for a discussion of this problem), which is a critical part of the recognition problem for unknown word sequences. Recent work in speech recognition provides a possible solution to this problem through the use of "anti-model" likelihood compensation (Glass et al. 1996), and it would be interesting to apply this extension to the prosodic phrase recognition problem.

In order to achieve lower error rates, it is useful to model the sequences of symbolic events as well as their acoustic realization. This can mean a prosodic grammar specified by the particular linguistic theory of the phonotactics of the symbolic units and/or the empirically estimated likelihood of sequences of the prosodic events. Although even very simple models are useful, it is generally better to account for the phrasing likelihood of a sentence as a whole, in which case the probability of the different phrasings must be represented in terms of a variable-length process because sentences vary in length. A simple and effective sequence model is a bigram model, which represents the probability of a sequence of prosodic labels as a Markov chain

$$p(a_1, \ldots, a_M) = \mathrm{p}(a_1) \prod_{i=2}^{M} \mathrm{p}(a_i | a_{i-1}) \tag{1.1}$$

as in Wightman and Ostendorf (1994), Kompe et al. (1994). If the symbols a_i are binary-valued (± break), then this is simply a phrase length distribution model. If the symbols include accent labels, then the model can capture effects such as the interaction between accent placement and phrase boundaries (Ostendorf and Ross 1997). If the sequences can be described by a finite-state grammar, then the grammar can be compiled into the form of Equation 1.1 with some label transitions having zero probability. Alternatively, the probability of the sequence can be described as the product of the probabilities of the rules used to generate it. Markov assumptions are advantageous because they allow an efficient optimal recognition search via dynamic programming. Linguistic theories that represent infinite recursion can be handled using a stochastic context-free grammar (Lari and Young 1990) and a stack recognition search strategy (Gopalakrishnan et al. 1995), though there have been no prosodic phrase recognition experiments in this framework.

Currently, algorithms for predicting phrase structure from text appear to have similar performance to algorithms for predicting phrase structure from acoustic observations (see Wightman and Ostendorf (1994), Ostendorf and

Veilleux (1994) for rough comparisons on the same corpus), which suggests that it may be useful to take advantage of known or hypothesized word sequences to determine the relative likelihood of a phrase boundary. In other words, Equation 1.1 can be modified to

$$\begin{aligned} p(a_1,\ldots,a_M|\mathcal{W}) &= p(a_1|\mathcal{W})\prod_{i=2}^{M} p(a_i|a_{i-1},\mathcal{W}) \\ &= p(a_1|f(\phi,\mathcal{W}))\prod_{i=2}^{M} p(a_i|f(a_{i-1},\mathcal{W})), \qquad (1.2) \end{aligned}$$

where W is the word sequence and the function $f(a,W)$ is introduced to extract key elements of the word sequence and reduce the number of parameters needed in the model (Ostendorf and Ross 1997). Since the relative frequency of phrase boundaries seems to be somewhat dependent on speaking style, it is not a good idea to include word sequence conditioning unless the sequence model is appropriate for the style being recognized.

In the above discussions, we have treated phrase boundary detection as involving a flat sequence of constituents, ignoring the fact that most theories represent a hierarchy of phrase constituents. If the hierarchy is simply a grammatical constraint and there are no interactions between the acoustic cues at the different levels, then the hierarchy can be accounted for in the label sequence model (Equation 1.1 or 1.2) and there is relatively little increase in complexity. The recognition search, typically a dynamic programming algorithm to jointly find prosodic labels and their time of occurrence, depends in complexity on the size of the basic units (frames, syllables or phrases) and the order of the Markov sequence model. For the case where there are interactions across levels, as in a superposition model or a linear-sequence model that requires pitch range normalization, both the recognition search space and the number of free parameters in the model can increase substantially. For example, the recognition search might require a multi-level dynamic programming algorithm (Ostendorf and Ross 1997). To reduce the potentially large recognition cost, a first-pass search based on simple feature analysis can reduce the possible set of phrase boundary candidates for subsequent more detailed phrase boundary recognition. The time-aligned output of a word recognition system (lattice or single word string) can also be used to restrict the candidate phrase ending times. To reduce the number of free parameters of the model, which would be infeasible to estimate reliably otherwise, it is essential to use the results of linguistic studies to either factor the model into different components and/or specify a parameter tying configuration. Superposition naturally factors the model into different components; the target for a region is the sum of the phrase, accent and segment level targets. In a sequential model, the target might be given by a model whose phrase-level parameters are shared by all other models used in that phrase, as in Ostendorf and Ross (1997).

1.4 APPLICATIONS AND IMPLICATIONS FOR PROSODIC PHRASE DETECTION

As mentioned earlier, prosodic phrase detection has a variety of applications and the particular application, as well as the particular architecture associated with the application, have important implications for the type of phrase detection algorithm that will be most effective. Key issues are whether an explicit decision vs. a score is required, whether phone duration information is available for use in recognition, and how language cues are integrated with acoustic cues. To illustrate this point, three applications will be discussed in this section: automatic prosodic labeling, constraining the speech recognition and language processing search process, and prosodic scoring of linguistic hypotheses.

Perhaps the most extensive use of prosodic phrase recognition is for automatic prosodic labeling, where the goal is to make an explicit decision about where the prosodic phrase boundaries are. Campbell (1997) has argued for automatic prosodic labeling as being critical to improving speech synthesis systems. In addition, automatic prosodic labeling will make it possible to study much larger speech corpora, and thereby improve our understanding of the mapping between syntax and prosodic phrase structure and the role of other factors, such as focus structure, that might affect the speaker's choice of prosodic phrase structure. An important aspect of automatic labeling, as opposed to other applications of prosodic phrase recognition, is that the word sequence associated with an utterance is known. In this case, it is possible to use forced alignment with a speech recognition system to obtain reasonably reliable phone and word alignments. Duration cues – appropriately normalized for inherent phone durations, contextual effects and speaking rate – can be very important for improving phrase detection accuracy (Campbell 1993, 1994; Wightman and Ostendorf 1994). In fact, duration alone leads to reasonably good prosodic recognition results for phrase boundaries (Wightman et al. 1990). In addition, the alignment of speech segments to words and syllables (or other sub-word units) can help eliminate false detections, since boundary tones are typically associated with the final syllable in a word. Since the time alignment is fixed, the variable segmentation problem raised for the transformational feature approach to phrase recognition is no longer an issue, so this more flexible approach for incorporating different prosodic cues may be the most effective solution. The known word sequence can also be used to improve the estimate of the prior probability of a particular prosodic label sequence, although with the caveat mentioned above that one would want to use this information only if the word conditioning model was trained on a similar style.

Another application of prosodic phrase recognition is as an information source used to constrain the search space for a speech recognition or understanding system. In this case, one could use either an explicit (Bear and Price 1990) or a fuzzy (boundary likelihood) (Bakenecker et al. 1994, Batliner et al. 1996) decision as to what type of prosodic phrase boundary occurs at a particular point in time, or a collection of hypothesized phrasings for the utterance (Nakai et al. 1995). (Where the boundary likelihood is used, generally an empirically determined threshold is incorporated to determine the cut-off point for labeling a hypothesized word juncture as a boundary, so effectively an explicit decision is used.) In a one-pass speech recognition search algorithm, the prosodic phrase boundary detection algorithm would be conducted before word recognition and would not have access to duration cues, with the exception of the impoverished cues associated with simple measures of syllable rate. Because the associated phrase detection algorithms are not sufficiently reliable, there has been little success reported for the one-pass, pre-word-recognition approach. If the constraints are applied to the word recognition search in conjunction with parsing and in a later stage of a multi-pass search where duration information and hypothesized word times are available from an earlier pass, then the results can be quite successful. Bakenecker et al. (1994) report a 70% reduction in the number of parse derivations with no loss in recognition performance due to use of recognized prosodic phrase boundaries. Batliner et al. (1996) report a 46% reduction in parser runtime by employing prosodic constraints on a small word graph (5 hypotheses/word), and find that parsing without prosodic constraints is unfeasible for larger word graphs (10 hypotheses/word). In most of these examples, the detected prosodic phrase boundaries are based on acoustic cues alone since they are used to eliminate candidate linguistic hypotheses. However, it is possible to integrate language cues with phrase boundary detection and obtain improved performance at a somewhat higher computational cost (Batliner et al. 1996).

Finally, detected prosodic phrase boundaries can be used to score linguistic hypotheses in terms of the consistency between the acoustic cues to prosodic phrasing and the expected phrasing associated with the candidate syntactic/semantic structure of the hypothesized word sequence. Hypothesis scoring does not offer the computation reduction advantages of the constraint approach described above, but the score is more useful for the problem of resolving ambiguity of multiple valid interpretations. In this application, prosodic phrase boundaries can be detected explicitly or their likelihood can be used in computing the score. If the phrase breaks are detected explicitly, one can compute a distance (or correlation) between detected phrase breaks and the predicted phrasing given the hypothesized syntactic structure (Ostendorf et al. 1993, Hunt 1997). However, since the goal is to obtain a score and not to impose constraints, better results are obtained by using the phrase break likelihood (Ostendorf et al. 1993; Veilleux and Ostendorf 1993a) or an equivalent soft decision in a distance-based approach. Two different types of likelihood scoring methods have been proposed. One approach (Veilleux and Ostendorf 1993a, Veilleux and Ostendorf 1993b) defines the score to be the likelihood of the i-th linguistic hypothesis W_i given the acoustic evidence

$$y_1^T = [y_1, \ldots, y_T],$$

which is computed in terms of the relative likelihood of the sequence of prosodic events $a_1^m = [a_1, \ldots, a_m]$:

$$p(W_i \mid y_1^T) = K_y p(W_i) p(y_1^T \mid W_i) = K_y p(W_i) \sum_{a_1^m} p(y_1^T \mid a_1^m) p(a_1^m \mid W_i) \tag{1.3}$$

where K_y is a normalizing constant that can be ignored in score maximization if a frame-based model of the acoustic cues is used. Assumptions that make computation of this probability practical include conditional independence of regions of y_t given the prosodic sequence a_1^m and Markov assumptions about the sequence a_1^m. The prosodic likelihood of each hypothesis is then combined with scores from other knowledge sources, for example in the N-best rescoring paradigm (Ostendorf et al. 1991). A very different approach is to use the prosodic phrase boundary likelihoods more directly in the decision process using semantic classification trees (Nöth et al. 1996). Here, feature vectors that include both syntactic/semantic information and prosodic phrase break likelihood $p(a \mid Z)$ are input to a classification tree, which is designed to predict the semantic interpretation of an utterance in an information query domain. For both of these approaches, it is assumed that syntactic/semantic hypotheses are available, so a prior pass of recognition must have been run. Thus, phone and word alignments are available for feature processing for the detection algorithm. Language cues are integrated with the acoustic cues, either as a component of the hypothesis score using a prosody-syntax likelihood $p(a^m{}_1 \mid W_i)$ or as an input to the classification tree.

1.5 OPEN QUESTIONS

In summary, there are currently several options available for prosodic phrase detection, which differ in terms of the acoustic cues that they can accommodate and the use of a hypothesized word sequence as well as in the theoretical basis and amount of detail of the intonation contour incorporated into the model. Which algorithm will be most effective will depend on the target application for phrase detection and computational resources available. However, the relative trade-offs of the different approaches are likely to change with advances in computational modeling and linguistic inquiry. For example, in speech recognition scoring (vs. labeling or search constraint) applications, frame-based methods currently offer the advantage over feature-transformation methods of a more theoretically-based framework. However, advances in using segmental features in recognition may address this problem (Glass et al. 1996), suggesting that the more flexible classifiers in the feature-transformation approach will lead to higher accuracy. In addition, as linguistics research provides a better understanding of the various cues associated with prosodic structure – beyond the F0, energy and duration cues used now, as suggested in Shattuck-Hufnagel (This volume) – the best model for representing the corresponding feature set will surely change. As computational resources continue to improve, more complex models that represent both phrase and accent structure will probably become more attractive than simple phrase boundary spotting approaches. These changes are important areas for future research, but it is likely that their impact on phrase detection will be somewhat evolutionary. Perhaps more significant changes will result from addressing the bigger questions of how to model spontaneous speech and how to account for the interaction between prosodic constituent structure and discourse factors.

In order for prosodic phrase detection to be more broadly useful, it is critical that algorithms be extended to handle spontaneous speech. In theory, the algorithms described above could all be applied to both spontaneous and read speech, but almost all work has focused on read speech. Preliminary experiments using a posterior distribution approach (Wightman and Ostendorf 1994) showed that error rates almost doubled in moving from read radio news speech to the goal-directed spontaneous speech in the ATIS corpus, from 8.4% to 14.6% error for detecting major prosodic phrase boundaries (Macannuco 1994). Yet, the speech in the ATIS corpus (Dahl et al. 1994) is relatively "careful" and easier to analyze than spontaneous unrestricted dialogs as in, for example, the Switchboard corpus (Godfrey et al. 1992). Thus, the algorithms described here will probably need substantial refinement to work successfully on spontaneous speech in general. At a minimum, the "knowledge" that is incorporated via the choice of features provided to the classifier must be revised to incorporate our growing understanding of the differences between read and spontaneous speech. For example, Bruce (1995) has observed longer phrases in read speech and differences in the location of the increase in F0 that is typically associated with a topic shift, and Hirschberg (1995) has observed that rising boundary tones are much more frequent in spontaneous speech than in read speech. Furthermore, several researchers have noted that pauses are a reliable indicator of prosodic phrase boundaries in read speech, but they are often associated with hesitations that occur within a phrase in spontaneous speech – a problem that resulted in the introduction of the "p" break diacritic in the ToBI labeling system (Pitrelli et al. 1994). Thus, phrase boundary recognition in spontaneous speech will probably require new feature processing and categorical representations of phrasing, as well as advances in computational modeling.

A second area that current phrase recognition algorithms do not address is the interaction between phrase structure and discourse structure in terms of their effect on the acoustic cues to each. Consider some of the acoustic (prosodic) cues to discourse structure that are summarized in Hirschberg (1993). Changes in F0 range are leveraged in most superposition models for phrase detection, but such changes have been noted by several researchers as signaling topic changes. Pause duration is generally pointed to as a cue for both major prosodic phrase boundaries (Wightman et al. 1992, Horne et al. 1995) and discourse structure. The tendency of speakers to put accents early and late in the phrase (Shattuck-Hufnagel et al. 1994) interacts with and is therefore a cue to phrase structure, but speakers also tend to accent words because they represent focused or new information. (Focus structure may also interact more directly with prosodic phrase structure (Selkirk, This Volume).) Given the overlap in the types of cues to prosodic phrase boundaries and discourse segmentation, it is not surprising that Mast et al. (1996) successfully detect dialog act boundaries using an algorithm that is similar to one for detecting prosodic phrase boundaries (Kompe et al. 1995). Although the integration of discourse processing and phrase detection poses practical problems for computational modeling because of the longer time scale of discourse events, it should be an important goal of future research in phrase modeling.

As speech technology improves, there is an increasing number of speech processing applications that could clearly benefit from prosodic information, which is apparent in the growing interest in prosodic phrase recognition. These applications – from human-computer dialog processing to speech translation – push both technical and theoretical questions in prosodic phrase modeling by providing the opportunity and the need to look at a wider variety and larger collection of speech data. In addition to providing a fertile ground for future research by raising such questions, the applications and the technology behind the various components of such speech systems may also contribute to their answers. Already there is a substantial impact of speech recognition technology in the prosodic phrase recognition algorithms described here. Furthermore, the existence of automatic speech understanding systems may make it possible to inject some control in experiments eliciting spontaneous speech in a dialog context. On the flip side, it may also be that technology developed for prosodic phrase recognition will eventually benefit speech processing technologies more generally. Although the problem of representing acoustic events at multiple time scales is critical to prosodic modeling, it also plays a role in speech recognition at some level and therefore advances in computational models of prosodic constituent hierarchies and associated feature normalization may impact speech processing more broadly than simply improving the bottom-line recognition accuracy.

ACKNOWLEDGMENTS

Partial funding for this work came from the American National Science Foundation, grant no. IRI-8905249. Any opinions, findings and conclusions expressed in this material are those of the author and do not necessarily reflect the views of NSF.

NOTES

[1] It has been argued that a linguistically-based symbolic representation of prosodic structure is not practical for use in computer speech processing because of the high cost of hand labeling prosodic structure, and encouraging results have been obtained for automatically learned units (Hunt 1997). However, the relatively good performance of automatic prosodic labeling algorithms suggests that large hand labeling efforts may not be necessary, and linguistically-based units are likely to be more robust across different speech styles and tasks, just as phone-based units appear to be more useful than automatically learned units in speech recognition.

REFERENCES

Bakenecker, G., Block, U., Batliner, A., Kompe, R., Nöth, E. and Regel-Brietzmann, P. 1994. Improving parsing by incorporating 'prosodic clause boundaries' into a grammar. *Proc. International Conference on Spoken Language Processing* (Yokohama), vol. 3, 1115-1118.

Batliner, A., Feldhaus, A., Geißler, S., Kiss, T., Kompe, R. and Nöth, E. 1996. Prosody, empty categories and parsing – a success story. *Proc. International Conference on Spoken Language Processing* (Philadelphia) vol. 2, 1169-1172.

Bear, J. and Price, P. J. 1990. Prosody, syntax and parsing. *Proc. 28th Annual Meeting, Association for Computational Linguistics*, 17-22.

Beckman, M. and Pierrehumbert, J. 1986. Intonational structure in Japanese and English. *Phonology Yearbook* 3, 255-309.

Beckman, M. 1995. Local shapes and global trends. *Proc. XIIIth International Congress of Phonetic Sciences* (Stockholm) vol. 2, 100-107.

Beckman, M. 1996. The parsing of prosody, *Language and Cognitive Processes* II, 17-67.

Bruce, G. 1977. *Swedish Word Accents in Sentence Perspective.* Lund: Gleerups.

Bruce, G. 1995. Modelling Swedish intonation for read and spontaneous speech. *Proc. XIIIth International Congress of Phonetic Sciences* (Stockholm) vol. 2, 28-35.

Bruce, G., Granström, B., Gustafson, K., and House, D. 1993. Prosodic modelling of phrasing in Swedish. *Proc. ESCA Workshop on Prosody, Working Papers* 41 (Dept. of Linguistics and Phonetics, U. of Lund), 180-183.

Butzberger, J. 1989. *Statistical Methods for Intonation Pattern Recognition.* Boston University M.S. Thesis.

Campbell, W. N. 1993. Automatic detection of prosodic boundaries in speech. *Speech Communication* 13, 343-354.

Campbell, W.N. 1994. Combining the use of duration and F0 in an automatic analysis of dialogue prosody. *Proc. International Conference on Spoken Language Processing* (Yokohama) vol. 3, 1111-1114.

Campbell, W.N. 1997. Synthesizing spontaneous speech. In Y. Sagisaka, N. Campbell and N. Higuchi (eds.), *Computing Prosody*. New York: Springer, 165-186.

Dahl, D. et al. 1994. Expanding the scope of the ATIS task: the ATIS-3 corpus. *Proc. ARPA Workshop on Human Language Technology*, 43-48.

Dilley, L., Shattuck-Hufnagel, S. and Ostendorf, M. 1996. Glottalization of vowel-initial syllables as a function of prosodic structure. *Journal of Phonetics*, 24, 423-444.

Fujisaki, H. and Kawai, H. 1988. Realization of linguistic information in the voice fundamental frequency contour of the spoken Japanese. *Proc. International Conference on Acoustics, Speech and Signal Processing*, 663-666.

Geoffrois, E. 1993. A pitch contour analysis guided by prosodic event detection. *Proc. Eurospeech* (Berlin), vol. 2, 793-796.

Glass, J., Chang, J. and McCandless, M. 1996. A probabilistic framework for feature-based speech recognition. *Proc. International Conference on Spoken Language Processing* (Philadelphia), vol. 4, 2277-2280.

Godfrey, J., Holliman E., and McDaniel, J. 1992. Switchboard: Telephone speech corpus for research and development. *Proc. International Conference on Acoustics, Speech and Signal Processing*, vol. 1, 517-520.

Gopalakrishnan, P., Bahl, L. and Mercer, R. 1995. A tree-search strategy for large vocabulary continuous speech recognition. *Proc. International Conference on Acoustics, Speech and Signal Processing*, vol. 1, 572-575.

Hirose, K. and Fujisaki, H. 1982. Analysis and synthesis of voice fundamental frequency contours of spoken sentences. *Proc. International Conference on Acoustics, Speech and Signal Processing*, 950-953.

Hirschberg, J. 1993. Studies of intonation and discourse. *Proceedings ESCA Workshop on Prosody, Working Papers* 41, (Dept. of Linguistics and Phonetics, U. of Lund), 90-95.

Hirschberg, J. 1995. Prosodic and other acoustic cues to speaking style in spontaneous and read speech. *Proc. XIIIth International Congress of Phonetic Sciences* (Stockholm) vol. 2, 36-43.

Horne, M., Strangert, E. and Heldner, M. 1995. Prosodic boundary strength in Swedish: final lengthening and silent interval duration. *Proc. XIIIth International Congress of Phonetic Sciences* (Stockholm) vol.1, 170-173.

Hunt, A. 1997. Training prosody-syntax recognition models without prosodic labels. In Y. Sagisaka, N. Campbell and N. Higuchi (eds), *Computing Prosody*. New York: Springer, 309-326.

Jensen, U., Moore, R., Dalsgaard, P. and Lindberg, B. 1993. Modelling of intonation contours at the sentence level using CHMMs and the 1961 O'Connor and Arnold scheme. *Proc. Eurospeech 93* (Berlin), 785-788.

Kompe, R., Batliner, A., Kießling, A., Kilian, U., Niemann, H., Nöth, E. and Regel-Brietzmann, P. 1994. Automatic classification of prosodically marked phrase boundaries in German. *Proc. International Conference on Acoustics, Speech and Signal Processing*, vol. 2, 173-176.

Kompe, R., Kießling, A., Niemann, H., Nöth, E., Schukat-Talamazzini, E., Zottmann, A. and Batliner, A. 1995. Prosodic scoring of word hypotheses graphs. *Proc. Eurospeech 95* (Madrid), vol. 2, 1333-1336.

Lari, K. and Young, S.J. 1990. The estimation of stochastic context-free grammars using the inside-outside algorithm. *Computer Speech and Language* 4, 35-56.

Macanucco, D. 1994. Automatic recognition of prosodic patterns. Unpublished Boston University course report.

Mast, M., Kompe, R., Harbeck, S., Kießling, A. Niemann, H., Nöth, E., Schukat-Talamazzini, E. and Warnke, V. 1996. Dialog act classification with the help of prosody. *Proc. International Conference on Spoken Language Processing* (Philadelphia) vol. 3, 1732-1735.

Morlec, Y., Bailly, G. and Aubergé, V. 1996. Generating intonation by superposing gestures. *Proc. International Conference on Spoken Language Processing* (Philadelphia), vol. 1, 283-286.

Nakai, M., Singer, H., Sagisaka, Y. and Shimodaira, H. 1995. Automatic prosodic segmentation by F0 clustering using superpositional modeling. *Proc. International Conference on Acoustics, Speech and Signal Processing*, vol. 1, 624-627.

Nöth, E., De Mori, R., Fischer, J., Gebhard, A., Kompe, R., Kuhn, R., Niemann, H., and Mast 1996. An integrated model of acoustics and language using semantic classification trees. *Proc. International Conference on Acoustics, Speech and Signal Processing,* vol. 1, 419-422.

Ostendorf, M., M. 1998. Linking speech recognition and language processing through prosody. *CCAI,* vol. 15, 279-303.

Ostendorf, M., Kannan, A., Austin, S., Kimball, O., Schwartz, R., and Rohlicek, J.R. 1991. Integration of diverse recognition methodologies through reevaluation of N-

Best sentence hypotheses. *Proc. DARPA Workshop on Speech and Natural Language*, 83-87.

Ostendorf, M., Wightman, C. and Veilleux, M. 1993. Parse scoring with prosodic information: An analysis/synthesis approach. *Computer Speech and Language*, 193-210.

Ostendorf, M. and Veilleux, N. 1994. A hierarchical stochastic model for automatic prediction of prosodic boundary location. *Computational Linguistics*, 20, 27-54.

Ostendorf, M., Digalakis, V. and Kimball, O. 1996. From HMMs to segment models: A unified view of stochastic modeling for speech recognition. *IEEE Trans. on Speech and Audio Proc.*, vol. 4, no. 5, 360-378.

Ostendorf, M. and Ross, K. 1997. A multi-level model for recognition of intonation labels. In Y. Sagisaka, N. Campbell and N. Higuchi (eds.) *Computing Prosody*. New York: Springer, 291-308.

Pierrehumbert, J. 1980. *The Phonetics and Phonology of English Intonation*. Ph.D. Dissertation, MIT.

Pitrelli, J., Beckman, M., and Hirschberg, J. 1994. Evaluation of prosodic labeling reliability in the ToBI framework. *Proc. International Conference on Spoken Language Processing* (Yokohama) vol.1, 123-126.

Selkirk, E. This Volume. The interaction of constraints on prosodic phrasing.

Shattuck-Hufnagel, S., Ostendorf, M. and Ross, K. 1994. Pitch accent placement within lexical items in American English. *Journal of Phonetics* 22, 357-388.

Shattuck-Hufnagel, S. This Volume. Phrase-level phonology in speech production planning. Evidence for the role of prosodic structure.

Silverman, K. Beckman, M., Pierrehumbert, J., Ostendorf, M., Wightman, C., Price, P. and Hirschberg, J. 1992. ToBI: a standard for labeling English prosody. *Proc. International Conference on Spoken Language Processing* (Banff) vol. 2, 867-870.

ten Bosch, L. 1993. On the automatic classification of pitch movements. *Proc. Eurospeech 93* (Berlin), vol. 2, 781-784.

Veilleux, N. and Ostendorf, M. 1993a. *Proc. International Conference on Acoustics, Speech and Signal Processing*, vol.II, 51-54.

Veilleux, N. and Ostendorf, M. 1993b. Prosody/parse scoring and its application in ATIS. *Proc. ARPA Workshop on Human Language Technology*, 335-340.

Wightman, C., Ostendorf, M., Price, P. and Bear, J. 1990. The use of relative duration in syntactic disambiguation. *Proc. International Conference on Spoken Language Processing*, 13-16.

Wightman, C., Shattuck-Hufnagel, S., Ostendorf, M. and Price, P. 1992. Segmental durations in the vicinity of prosodic phrase boundaries. *Journal of the Acoustical Society of America* 91, 1707-1717.

Wightman, C. and Ostendorf, M. 1994. Automatic labeling of prosodic patterns. *IEEE Trans. on Speech and Audio Proc.* 2, 469-481.

Electrical Engineering Department, University of Washington, Seattle, WA , USA

NICK CAMPBELL

TIMING IN SPEECH: A MULTI-LEVEL PROCESS

1. INTRODUCTION

The modelling of timing in speech is of particular interest to language technology because it represents an interface between cognitive and mechanical aspects in the processes of speech production. Higher-level aspects of speech timing control are related to speaker-specific and utterance-context-specific factors such that the same sequence of sounds produced by two different speakers (or by one speaker on two different occasions) will most likely have different timing characteristics. Lower level aspects that govern timing ensure on the other hand that there will also be some similarities in the productions since the vocal tract and articulatory mechanisms used to produce the individual sounds are fundamentally similar for all speakers of each language. The interaction of these two levels of influence results in patterns of timing that can be interpreted in a useful way to add an extra layer of meaning to an utterance.

We will not be concerned here to explain human mechanisms of timing perception, but rather to show how use can be made of regularities in timing characteristics both for the understanding and for the production of speech by computers. Timing variation can provide clues that help interpret meaning by marking boundaries and prominences in an utterance and by delimiting chunks of speech and showing focus relationships between them. This information can be useful for the automatic detection of simple speech-act differences as a step in the process of machine understanding of speech.

In nature, two supposedly similar speech sounds rarely have exactly the same duration, but statistical analysis of the distributional characteristics of the sounds reveals consistent patterns that form the basis of our study. A key point that will recur throughout this paper is that a measure of the *lengthening* undergone by a segment can provide much more useful information than a measure of its duration. Raw durations per se have little meaning, and simply knowing the number of milliseconds that it takes to produce a given unit of speech can be of little use unless we also know the durational distributions exhibited by similar segments and the degree to which those durations vary.

M. Horne (ed.), Prosody: Theory and Experiment, 281–334.

The following sections will examine the types of factors that influence segmental durations, and show the types of factoring that we can perform to filter out known or predictable effects in order to learn more from the weaker effects that underlie them.

We are now at a particularly interesting period with respect to timing research. Early studies (even as late as Dennis Klatt's seminal work in the mid 80's) were limited to analysing durations measured manually from paper plots of speech spectrograms or waveforms. We are now able to use automatic techniques such as hidden Markov models (Entropic Research Laboratory 1993) to rapidly segment and label large speech corpora into arbitrary-sized units. Advances in computer memory and processing speed have encouraged the development of large electronic corpora (primarily for training speech recognisers) and software for their analysis and the presentation of results. It is perhaps a good time to review past findings and determine directions for future work.

2. DURATION ACCORDING TO DENNIS H. KLATT

Science follows technology, and the limitations of the technology sometimes constrain the scope of the scientific view. The technology up to the time of Dennis Klatt required that segmental duration measurements be made from paper printouts of speech waveforms or spectrograms, so the duration of a corpus utterance was typically constrained by the size of the paper it could be conveniently plotted on. It follows that most of the data studied in early analyses were from words or phrases presented in carefully designed short 'frame' sentences. The limited size of data tokens and the prevailing 'phoneme' theory restricted early analyses to a segmental view and encouraged the notion of 'inherent' durations for vowels and consonants. Klatt made a significant contribution to our understanding of segmental timing, and we can best summarise the segment-level approach by a review of his work.

2.1 'Intrinsic' segmental duration characteristics

Peterson and Lehiste (1960) first introduced the notion of an *intrinsic duration*, defined as "the average duration of the syllable nucleus measured from minimal pairs differing in the voicing of the final consonant", from recordings of one adult male speaker (GEP) reading 1263 single words in a sentence frame "with determined stress and pitch patterns", and by comparison with data from recordings of 70 words from five different speakers "who used the same frame sentence and pitch and stress patterns as the speaker for the larger set of material''. They computed these averages for all stressed syllable nuclei for English, and found a significant effect on the durations from the nature of the consonant that follows. In particular, they found that the syllable nucleus tends to be shorter when followed by a voiceless consonant.

The short frame sentences isolated the words from each other and provided a uniform context to facilitate comparison of the effects governing their durations. However, we can assume that the pronunciation of these target words was closer to that of an isolated-word or citation-form reading than to any pronunciation they might receive in fluent continuous speech and, as we shall see later, there are considerable differences between data from isolated words and from continuous speech.

T. P. Barnwell (1971) presented an 'algorithm for the control of segmental durations in a reading machine' in which he proposed to model the duration of a vowel as

> ''a function of the word-level stress of the syllable in which the vowel appeared, the structural location of the word in which the vowel appeared, the number of syllables in the word, and the proximity of any word or syllable juncture''

From measurements of spectrograms of careful readings, by 10 subjects, of single words, and two- and three-word phrases such as ''sad fat bat'', ''night-nativist killer'', and ''tiny negative entity'', he determined the following model for vowel durations:

$$V.D. = B.D. + X \times B.D. + P.P \qquad (1)$$

where $V.D.$ represents the vowel duration, $B.D.$ is a 'basic' duration determined by the following consonants, X is a factor for stress, and $P.P$ is a vowel-specific prepausal lengthening factor in the range of 20 to 60 msec (increasing in steps of ten msec). Basic durations ($B.D.$) were adopted from House (1961) and consisted of the average durations for each vowel before a) unvoiced stop consonants, b) voiced stop consonants, c) nasals, d) unvoiced fricatives, and e) voiced fricatives. Seven structural transform values of X ranging from 0.0 to 2.1 (selected according to phrasal context and number of syllables in the word) were proposed for the above model. These were used in conjunction with five base durations for eleven vowels, with fixed durations for 19 consonants. Barnwell noted that in the case of polysyllabic words, when a syllable boundary fell between a vowel and its following consonant, the effect of that consonant on the duration of the vowel was found to be greatly reduced. He also isolated effects governing consonantal durations but these changes were small and he made do with vowel-duration control alone in his model. Consonant durations were found to vary greatly in clusters but with no measurable effect from syllable boundaries. In the case of ambisyllabic segments, where no clear syllabification decision could be made, timing effects supported the interpretation of this group as a separate acoustic category. No results were presented for the semivocalic liquids or glides.

He maintained that speech "attempts to be very rhythmic but fails because of durational constraints. Hence it may be that what is heard as rhythmic may be really the interpretation of changes in duration *in the direction of true rhythm*" (my italics). He noted that "each structural location *puts a pressure on* a word to be a certain size" (my italics), and that words with the same number of syllables tended to have similar length. However, although "an attempt was made to retain the word length" as the number of syllables increased, durational constraints such as vowel type, stress, number of syllables, postvocalic voicing, and so forth make this impossible: "All words cannot be forced into the same size but only moved in the proper direction".

Dennis Klatt summarised Barnwell's work and presented his 'Theory of Segmental Duration in English' at the 82nd Meeting of the Acoustical Society of America in 1971, in which he proposed a set of rules to model the way syntactic structure modifies durational patterns in fluent speech, and showed the usefulness of a distinction between stressed and unstressed vowels. The paper was based on a preliminary analysis of the durations of 1200 phonetic segments from several minutes of prose read by a single speaker (himself). His rules were intended to be "compatible with other rules of a complete generative phonology of English (cf. Chomsky and Halle 1968), [...] to accept as input a syntactic bracketing and a string of formatives (words, morphemes, and special markers), and transform this abstract representation of a sentence into a detailed phonetic transcription":

> "A basic duration is assigned to each segment according to whether the segment is a stressed vowel, unstressed vowel, prestressed consonant, or a consonant in non-pre-stressed position. The basic duration is then shortened by rule in multisyllabic words. Cluster rules shorten some segments such as /s/ and lengthen others such as sonorants. Many segmental rules appear, such as the modification of stressed vowel length as a function of features of the following consonant, or the extreme shortening of pre-unstressed dental consonants in intervocalic position."

Both Barnwell and Klatt employed a context-related percentage change model of segmental durations "because, in the data of Peterson and Lehiste (1960) and others, it seemed that inherently short segments were changed in duration by a smaller absolute amount of milliseconds than were inherently long segments'' (Klatt 1973). Klatt introduced a notion of incompressibility to account for the fact that two vowel shortening rules did not act independently (for polysyllabic context, and for a voiceless following consonant in data from readings of lists of 40 mono- and bi-syllabic words), since ''a vowel which is shortened by one rule becomes less compressible to other shortening influences'', and modified his model to reflect a minimum time of execution of an (assumed) articulatory program, relative to the 'inherent' duration of the vowel:

> ''An ordered sequence of duration modification rules is applied to a string of inherent segment durations in order to predict a canonical duration pattern for an English utterance. Each rule converts an input vowel duration D_i to an output vowel duration D_o according to the formula
>
> $$D_o = k\left(D_i - D_{\text{min}}\right) + D_{\text{min}} \qquad (2)$$
>
> where k is greater than zero and depends on the particular rule. The output D_o is initially set equal to the inherent duration of the vowel.''

The value of D_{min} was suggested to be about 0.45 times the 'inherent' duration of the vowel. Klatt elaborated:

> ''Inherent vowel duration has not been adequately defined in this paper, although one would like to say that it is the segmental duration to be expected in the unmarked case where minimal external constraints apply. The vowel duration measured in the frame sentence ''Say [bVb] instead'' comes close to satisfying these requirements. The important issue in our discussion is whether a different definition would change the form of [the above equation]. The answer seems to be no, there is no definition that would result in removal of the constant term D_{min}.''

The inherent duration has no special status other than as a starting point for rule application. I speculate that the reason he does not explain how to calculate inherent durations explicitly is that each value has to be determined by considerable trial and error from repeated optimisation to fit the observed data (cf. Klatt 1979:294). This means that a change to any of the rules would result in having to re-estimate all variables in the model by a further error minimisation. This, incidentally, makes it particularly difficult to adapt the model to incorporate data for different speakers or dialects.

Summarising the factors that govern segmental durations in a review of the literature, Klatt (1976) concludes that duration serves as 'a primary perceptual clue' in the distinction between 1) inherently long versus short vowels, 2) voiced versus voiceless fricatives, 3) phrase-final versus non-phrase-final syllables, 4) voiced versus voiceless postvocalic consonants, 5) stressed versus unstressed or reduced vowels, and 6) the presence or absence of emphasis. Four rules for vowel duration and four for consonant duration were shown to account well for durations in the nonsense-syllable data of Lehiste (1975) and Oller (1973). All except one used the formula in Eqn. 2 above. To model the shortening of a vowel followed by a voiceless stop, an absolute value of 45 msec was subtracted from the vowel duration. k was set to 0.6 to shorten a non-phrase-final vowel, to 0.4 for an unstressed vowel, and to 0.78 to shorten all syllables in a polysyllabic word. It was set to 0.7 to shorten non-word-initial consonants, 0.8 for an unstressed consonant, 0.7 for a word-medial consonant, and 1.6 to lengthen a phrase-final consonant. A fit accounting for 97% of the variance in vowel durations for Lehiste's data was achieved by these rules.

2.2 Connected discourse

Although Klatt was concerned from the start to work with data from fluent speech, he took a bottom-up approach that restricted him to citation-form words. This was primarily to "avoid the durational variability that can occur in either initial or final position in an utterance" but produced data that were not representative of natural speech rhythms.

> "Words to be studied are spoken in a single frame sentence at a moderate speaking rate. Thus any systematic effects of speaking rate, syntactic position (Barnwell 1971), semantic importance (Umeda and Coker 1971) or rhythmic temporal compensation (Abercrombie 1964) will not be considered. (Klatt 1974)

> "We adopt a model in which one begins by positing an inherent duration for each segment type and develop rules that modify the hypothesised basic durations. The choice reflects a bias toward attempting to account for durational changes due to the local segmental environment first, and then looking for any remaining rhythmical or semantic influences." (Klatt 1975)

On rhythm at this time, he suggested that durational changes might instead be explained by a process of stress timing (citing Abercrombie 1964), but countered a) that 'complete temporal compensation' (sic) was not found, and b) that "it would seem difficult to maintain large differences in inherent durations for different vowel types if significant rhythmic compensation were in force." He did not address the issue that inherent duration (if such exists) may not be the sole way of differentiating the different vowel types, nor that together these arguments support the weaker notion of rhythm as expounded by Barnwell, rather than the absolute isochrony assumed by Abercrombie. In

Klatt (1976) he returns again to the theme of rhythm, citing experiments by Huggins (1972) and Martin (1972, 1975). Huggins used differences in JNDs (just-noticeable differences) of phonetic segments to show that listeners are particularly sensitive to the rhythmical aspect of sentence timing. He found that if he simultaneously increased the duration of one segment and decreased the duration of an adjacent segment he sometimes had to make a larger change before subjects reported hearing a difference and that subjects were less sensitive to changes that did not disturb the temporal interval between two stressed vowels in a sentence, i.e., when the rhythm was unchanged. Martin showed that disturbances to a sentence rhythm may influence stress judgments (1972), and interfere with sentence decoding (1975). Klatt maintained however (as did Allen (1975)) that the lexical and phrase-level stress rules of English tend to produce alternating patterns of stressed and unstressed syllables and that these alone were sufficient to account for the subjective impression of equal-stress timing. No explicit control for rhythm was therefore incorporated into his model.

Klatt (1975) reports findings for vowel durations in 13 sentences of connected speech containing 236 words and 1200 phonetic segments. In contrast to earlier findings from words in short sentences, the effect of the following consonant on the length of a vowel was observed only in phrase-final syllables. Klatt summarises with:

> "A descriptive theory for the prediction of vowel duration in English discourse would have to include the following rules:
>
> - each vowel type has a different inherent duration;
> - vowels not in phrase-final syllables are shorter;
> - vowels not in word-final syllables are slightly shorter;
> - the influence of the final consonant is small except in phrase-final positions;
> - unstressed vowels are shorter in duration than stressed vowels;
>
> the duration data support the concept of vowel incompressibility: each rule cannot act independently because a vowel that has been shortened by one rule tends to resist additional shortening."

2.3 The complete Klatt model

Lindblom and Rapp (1973) proposed a set of rules for the prediction of duration based on a hierarchical relationship between the word and the phrase. Their model for the stressed vowel shows reduction in duration according to the number of stressed words in the phrase, and position of the vowel in the sequence. The relationship can be expressed by a power function:

$$D_W = \frac{D_o}{(a_W + 1)^{\alpha_W} (b_{W+1})^{\beta_W}} \tag{3}$$

where D_o is the inherent segment duration
D_W is the segment duration at the word level
α_W is the number of syllables that follow at a given point
β_W is the number of syllables that precede at a given point

This was revised by Lindblom (1974) to introduce the concept of a minimum duration that was taken up by Klatt:

$$D_W = D_{min} + \frac{D_o - D_{min}}{(a_W + 1)^{\alpha_W} (b_{W+1})^{\beta_W}} \tag{4}$$

where D_{min} is the supposed minimum duration that a certain segment can realise. At the phrase level, the duration of a stressed vowel is subject to similar shortening, independent of the structure at the word level. These rules are applied cyclically. For speech material based on nonsense utterances, the constants α_W and β_W were estimated at 0.42 and 0.12 respectively. Lyberg (1977) tested the model with data obtained from readings of 40 short meaningful phrases and estimated the constants as $\alpha_W = 0.4$, and $\beta_W \leq 0.1$. He argued that sensitivity to the number of previous stresses was only applicable phrase-finally and that shortening could be expressed by a constant term $k \cdot D_W$ elsewhere:

$$\textit{phrase finally: } D_P = \frac{D_W}{(b_P + 1)^{\beta_P}} \tag{5}$$

Here, D_P is the segment duration at the phrase level, D_W is the segment duration at the word level, b_P is the number of main stresses that precede at a given point, and β_P and k are constants. We see that a hierarchical series of effects is being modelled, with influences coming down to the segmental level from both the word and phrase.

A complete description of the final Klatt model was published in 1979 (Klatt 1979) and later summarised with minor modifications in Klatt (1987) and Allen et al. (1987). The model used 11 rules and accounted for 84% of the durational variance for his own reading of new paragraph-length text with segmental durations that differed by a standard deviation of less than 17ms. In Allen et al. (1987), Klatt acknowledges the contribution from the ''elegant rule system'' of Lindblom and Rapp, as implemented and augmented by Carlson and Granström (1976), and cites Mattingley (1968), Barnwell (1971), and Umeda (1975, 1977) as employing similar rule systems. He acknowledges that the model takes no account of the multiple stress levels conditioned by syntactic category and word frequency (as put forward by Gaitenby et al. (1972) and Coker et al. (1973)). The following is from Klatt (1987) with references from Klatt (1979). Values for the modifications to k were not given in the 1987 review. The complete model assumes that 1) each phonetic segment type has an inherent duration that is specified as one of its distinctive properties (He cites House (1961)), 2) each rule tries to effect a percentage increase or decrease in the duration of each segment, but 3) segments cannot be compressed shorter than a certain minimum duration.

The portion of the inherent duration exceeding a specified minimum is subject to multiplication by scale factors associated with phonetic context and syntactic conditions summarised by the formula:

$$DUR = MINDUR + \frac{(INHDUR - MINDUR) \times PRCNT}{100} \qquad (6)$$

where DUR is the duration of a segment, INHDUR is thc 'inherent duration' of a segment in msec, MINDUR is the minimum duration of a segment if stressed, or twice that minimum if unstressed, and PRCNT is the percentage shortening determined after cyclical application of the rules.

The value of PRCNT is initially set to 100 then modified by each applicable rule according to Eqn 7.

$$PRCNT = PRCNT \times \frac{k}{100} \qquad (7)$$

where k is the percentage change required for each of the following rules. Predicted durations can be made slightly longer or shorter to adjust for different speaking rates by setting PRCNT to an initial value other than 100, but speaking rate adjustments are said to be better made by manipulating pause durations.

1. PAUSE INSERTION. Insert a brief pause (200 msec) before each sentence-internal main clause and at other boundaries delimited by an orthographic comma. The earlier version also adds "but not before relative clauses"; it is not clear whether this change merely reflects editorial considerations of space in the later version, or whether it reflects a modification to the rules (see Goldman-Eisler 1968, Cooper, Paccia and Lapointe 1978).

2. CLAUSE-FINAL LENGTHENING: (k=140). The vowel or syllabic consonant just before a pause is lengthened. Any consonants in the rhyme (between this vowel and the pause) are also lengthened. The earlier version of this rule did not refer to the 'rhyme', but to "any consonants between this vowel and the pause" (see Gaitenby 1965, Lindblom and Rapp 1973, Oller 1973, and Klatt 1975). This difference may reflect a change from a segmental to a syllable-based view.

3. PHRASE-FINAL LENGTHENING: (k=140). Syllabic segments (vowels or syllabic consonants) are lengthened in a phrase-final syllable. Durational increases at the noun/verb-phrase boundary are more likely in a complex noun phrase or when subject-verb order is violated; durational changes are much more likely for pronouns. The lengthening is perceptually important. Note was made in the 1976 version that "A phrase-final post-vocalic liquid or nasal is lengthened", and no reference was made to "Durational increases ... etc." (see Lindblom and Rapp 1973, Klatt 1975).

4. NON-WORD-FINAL SHORTENING: (k=85). Syllabic segments are shortened slightly if not in a word-final syllable (see Lindblom and Rapp 1973, and Oller 1973).

5. POLYSYLLABIC SHORTENING: (k=80). Syllabic segments in a polysyllabic word are shortened slightly (see Lindblom and Rapp 1973, and Lehiste 1975).

6. NON-INITIAL CONSONANT SHORTENING: (k=85). Consonants in non-word-initial position are shortened (see Klatt 1974 and Umeda 1977).

7. UNSTRESSED SHORTENING: Unstressed segments are shorter and considered more compressible than stressed segments. The minimum durations for unstressed segments are halved (MINDUR = MINDUR / 2) and then stressed and secondary-stressed segments are shortened: Consonants before a stressed vowel that are in the same morpheme or form an acceptable word-initial cluster are also considered to be stressed (see Umeda 1975, 1977, Lehiste 1975).

syllabic (word-medial syll): k = 50

syllabic (others): k = 70
prevocalic liquid or glide: k = 10
others: k = 70

8. LENGTHENING FOR EMPHASIS: (k = 140).
An emphasised vowel is significantly lengthened. This lengthening can also be used to capture word frequency and discourse effects that are not otherwise incorporated in the rule system (see Bolinger 1972, Carlson and Granström 1973, Umeda 1975).

9. POSTVOCALIC CONTEXT OF VOWELS: The influence of a post-vocalic consonant (in the same word) on the duration of the vowel is such as to shorten the vowel if the consonant is voiceless. The effects are greatest at phrase and clause boundaries. No mention is made in the later version of the rules of the special case of a postvocalic sonorant-obstruent cluster, in which the obstruent determines the effect on the vowel and on the sonorant consonant (see House and Fairbanks 1953, Peterson and Lehiste 1960).

open syllable, stressed, word-final: k = 120
before a voiced fricative: k = 160
before a voiced plosive: k = 120
before an unstressed nasal: k = 85
before a voiceless plosive: k = 70
before all others: k = 100

10. SHORTENING IN CLUSTERS: Segments are shortened in consonant-consonant sequences (disregarding word boundaries, but not across phrase boundaries) (see Klatt 1973).

vowel followed by vowel: k = 120
vowel preceded by vowel: k = 70
consonant surrounded by consonants: k = 50
consonant preceded by a consonant: k = 70
consonant followed by a consonant: k = 70

11. LENGTHENING DUE TO PLOSIVE ASPIRATION:A stressed vowel or sonorant preceded by a voiceless plosive is lengthened. In contrast to all other modifications, which effect a percentage change to part of the segment's inherent duration, this is an additive modification by a fixed value of 25 msec. (see Peterson and Lehiste 1960).

2.4 The segmental view crumbles

There is by no means complete agreement in the literature on the 'facts' of segmental timing. Klatt (1975) begins to cast doubt on the applicability of results reported in earlier studies:

> "Measured durations of vowels were found to be much shorter in a connected discourse than in words spoken in isolation. The durational differences observed in these two linguistic environments cannot be described as a simple compression of the time scale because vowels become longer at phrase boundaries. These results call into question the practice of relying on published studies of acoustic-phonetic characteristics of speech that are obtained from citation forms when one wishes to quantify sentence production performance."

The fact that segmental durations only appear well-behaved in carefully prepared laboratory speech is taken up in a series of reports on segmental and syllable timing features based on analysis of two medium-sized corpora of read speech. Crystal and House (1982, 1986, 1988, 1990) and House and Crystal (1988) compared fast and slow readings of two scripts totalling about 600 words each by fourteen speakers. They included a checklist (Crystal and House 1988:Appendix B: 1571-2) of assertions in the literature for which they did *not* find support, including 21 cases of lengthening conditions for which they found no evidence, compared with only 10 claims for which they did find support and 6 for which they reported mixed findings. Many of the effects were only found pre-pausally, where lengthening is greatest. The following list details their positive findings, i.e., those that had been reported previously and were confirmed in the corpus data.

1. Vocalic durations - Tense vowels tend to be longer than lax vowels, and vowel durations vary inversely with vocalic 'height' but only for extremes of height.

2. Vocalic lengthening - Vowels are longer in phrase-final words. They are longer before voiced consonants prepausally, longer before a nasal followed by a voiced plosive, and tend to be longer before bilabials than before alveolars or velars. Vowels following [b,d,g] tend to be longer than vowels following [p,t,k]. Vowels following plosives lengthen more than those following continuants.

3. Consonantal durations - The longest word-initial consonants tend to be those which are voiceless and those with a back place of articulation. Voiced plosive releases are generally shorter than voiceless plosive

releases. Velar plosive releases are longer than labial or dental plosive releases. Prepausal plosive holds tend to be longer than non-prepausal plosive holds.

4. Consonantal lengthening - Prestressed [s] is longer than [s] before an unstressed vowel. Sounds in clusters are shorter than singletons. Although [s] and [p] are shorter in [sp] clusters than as singletons, the plosive closure times are not significantly different.

Of particular interest in these reports are the histograms of phone durations (Crystal and House 1988 (Appendix A:1570-1) which show that with the exception of long vowels and complete plosives, all groupings of segment categories though differing in their means and in the skew of the distributions show a clear modal duration close to 50ms. Crystal and House make no reference to this fact, but if the most commonly occurring duration for any phone is uniform, then we can conclude that a syllable's duration may in the default case be a simple function of the number of segments it contains. The reasons for deviance from this 'ideal' or 'base' duration will be the theme for the remaining sections of this paper.

3. RHYTHMIC REGULARITY

Several other types of model have been proposed that predict segmental durations by first taking into account the rhythmic regularity that is characteristic of fluent speech and then predicting segmental durations by accommodation into the higher-level framework. The idea of a supra-segmental level of speech control is not new. Abercrombie (1965) points out that one of the first descriptions of our language, Lodwicke's (1686) 'New Universal Alphabet', is in fact a syllabic alphabet for English. Several intermediate levels of representation are put forward in metrical phonology for timing the utterance, the intonational phrase, the foot, the syllable and the segment (see for example Pierrehumbert 1980, Selkirk 1984 or Coleman 1992).

Witten (1977) described a higher-level duration prediction algorithm for accommodating segment durations into predetermined foot timings, whereby vowel onset locations are determined by the foot duration, subject only to modification for the extreme cases of very long or very short feet. His rules were based more on Abercrombie's theory than on statistics drawn from observations of acoustic speech data, and failed to take into account other factors known to influence the segment durations. Witten acknowledged that "little is known about the strength of this tendency and the extent of deviations from it", but argued that "the designer of a prosodic synthesis strategy must needs make assumptions which cannot be properly justified" when proposing a set of rules for "imitating the rhythm of natural English speech". It is interesting to note here as an aside that European linguists seem to hold more strongly than their American counterparts to the importance of the rhythmic foot as an entity in speech timing; perhaps rhythm is one of the bigger differences between the British and American variants of the same language.

Under the Witten system, rate of talking is determined by a standard foot time (default = 480 msec), with a fast rate resulting in less isochrony as a result of a minimum syllable limit (default = 140 ms), and slower rates similarly limited by segmental expansion constraints. He uses the three basic foot duration types proposed by Abercrombie (1965): i) a short syllable followed by a long syllable, ii) two syllables of equal length, and iii) a long syllable followed by a short syllable, and posits four syllable length categories to account for salient, short-salient, word-final and proclitic positions, then adds medial, medial-short and utterance-final categories to bring the number of syllable categories up to seven. Additional specification of a silent beat is allowed for anacrusis. Durations are assigned to *beat points* of syllables, which are defined as the onset of the stressed vowel (after Allen 1972) or the preceding glide if one is present in the syllable-initial consonant cluster. Diphthongs are considered as a sequence of two vowels and triphthongs as two separate syllables - a diphthong followed by a single vowel. Phoneme durations distinguish 'intrinsic' from 'extrinsically' determined variants (after Lawrence 1974) with the former a property of the phoneme itself and the latter determined primarily by the syllable but influenced to some extent by the intrinsic (default) duration. All onset consonant durations are considered intrinsic, as are those of the first element of a diphthong and all short vowels. If the first consonant of the coda is a stop, its duration is considered extrinsic and that of any following consonants intrinsic. All other consonant durations in the coda are intrinsic except the last. Individual phoneme durations are adjusted to realise calculated syllable durations within predetermined limits on the amounts by which individual phonemes can be expanded, and under the assumption of a minimum duration for syllables. It is not clear from Witten's description how durational adjustment for different levels of sentential or emphatic stress can be implemented.

Young and Fallside (1980) described a foot-based model for predicting the timing of English speech that a) gives prominence to the stressed syllables of an utterance, b) reduces the durations of all other syllables in the utterance, and c) adjusts the durations between stress centres towards an isochronous rhythm. After first assigning stress to syllables in an earlier component of the system, the intervals between stressed syllables, or 'foot durations', are normalised. Several strategies are described; the first showing that by simply computing an average foot duration for the utterance and scaling the syllable durations between stress centres by equal amounts, function-words are given durations that sound too long, and content words too short. To resolve this, maximum and minimum limits were imposed on the extent that a syllable duration could be adjusted to fit the foot duration. However, they found that when specifying different limits for different classes of words, either the effect of rhythmic isochrony was lost or the limits became generous enough to allow the over-lengthening and shortening that they were trying to avoid. Their solution was to lengthen syllables in content words in a short foot, and shorten those in function words in a long foot. A minimum duration is specified for function words (which are typically monosyllabic) to prevent over-shortening, at which point all remaining syllables not carrying primary stress are also shortened. To maintain the rhythm, a 300 msec silent syllable is inserted at the beginning of an utterance in the case where no stressed initial syllable exists (see Halliday (1967)). Each syllable preceding a word-group boundary is lengthened by 15% (Klatt 1975, Lehiste et al. 1976), and a pause is inserted after each word-group if the last foot is shorter than average (O'Malley et al. 1973). The final foot in an utterance is further lengthened by 10% to model the paragraph-level effects noted by Lehiste et al. (1976). A variant of the Klatt model is used to determine timing events within the syllable. As with many rule systems used in speech synthesis, the speaking rate is held slower than that observed for natural speech "as an aid to achieving the highest possible intelligibility". Good results were claimed for the perceived rhythm produced by this method; and the tendency towards an isochronous rhythm is said to be achieved without the effect of content words being hurried.

Kohler (1986) proposed a top-down timing model based on foot durations for German. His cascade model computes syllable duration as a linear function of a predetermined foot duration, and assumes that segment durations are, in turn, a linear function of the syllable duration. The model for foot duration is of the form

$$R_{ijkl} = D + C_i + N_j + I_k + T_l \qquad (8)$$

where R_{ijkl} is the rhythmic foot duration, D is a duration constant for a mono-syllabic foot at a neutral pitch pattern and a medium tempo, C_i is an additive duration constant for complexity in vowels and consonants in mono-syllabic feet, N_j is an additive constant for feet with more than one syllable, I_k is an additive constant for other intonations than neutral, and T_l is an additive constant to allow for other speech rates. Syllable duration is determined by different formulae for stressed and unstressed syllables in the foot. The following equations describe the default durations for the two types of syllables:

$$S^{(S)} = b_{R_j} \times R_{ijkl} \qquad S^{(U)} = R_{ijkl} - S^{(S)} \tag{9}$$

where $S^{(S)}$ is the stressed syllable duration, $S^{(U)}$ is the unstressed syllable duration, and b_{R_j} is determined from the slope of a linear regression of stressed syllables to foot duration, factored according to the total number of syllables in the foot. For the non-default case, with feet containing syllables consisting of only a reduced vowel, the following formula is used:

$$S^{(E)} = a_{N_j} + a_{I_k} + A_{T_l} + b'_R \times R_{ijkl} \qquad S^{(St)} = R_{ijkl} - S^{(E)} \tag{10}$$

where $S^{(E)}$ is the duration of the 'ending' (or final syllable), $S^{(St)}$ is the duration of the 'stem' syllable in the foot, a_{N_j} is determined from the intercept, a_{I_k} and A_{T_l} are intercept increments depending on intonation and tempo respectively, and b'_R is the slope from a linear regression of final syllables on foot duration factored according to number of syllables in the foot. The 'syllable component' (SC), which is defined as the initial consonant (or consonant cluster) and vowel, less final consonant(s), is calculated:

$$SC = a\gamma_m + b\gamma_m \times S^{(S,U,E,St)} \tag{11}$$

where $a\gamma_m$ and $b\gamma_m$ are the intercepts and slopes of linear regressions of the syllable component on syllable duration, factored according to type of syllable component; $a\gamma_m = 0$ in C_1VC_2 syllables where C_2 = obstruent. Finally, phone duration is calculated according to

$$P = a\gamma_n + b\gamma_n \times SC \tag{12}$$

where P is the duration of the phone, and $a\gamma_n$ and $b\gamma_n$ are the intercepts and slopes of the linear regressions of phone on syllable component duration, factored according to phone type. Kohler notes that in his data there is evidence of duration adjustments between successive feet, and shows that for German there is right-to-left compensation taking place, such that later timing correction occurs to compensate undershoot or overshoot in the preceding foot. This is not incorporated in his model as described. Nor is any utterance-final lengthening. Like Witten, his model is based on the duration of the foot, but in this case the durations are determined by statistical analysis of speech data from regression coefficients in a top-down hierarchical manner.

Coker et al. (1973) proposed a model for predicting durations in an articulatory-based synthesiser that first inserts pauses or pseudo-pauses in the utterance, and then assigns *key-word* stressing, and more general lexical stress, before calculating segmental durations. Stress is considered multi-valued; they note that although people normally only "perceive" two or three levels, "the perceptual acuity for the correlates of stress (pitch, intensity, duration) is much sharper", and that in order to synthesise long passages without "obviously wrong" stress patterns, a finer distinction is required. Their definition of stress incorporates a consideration of *newness* to the text as listed in the following order of priority: New nouns > Prepositions used as complements > New infrequent verbs, adjectives, adverbs, repeated nouns > Repeated infrequent verbs > Interrogatives, quantitatives > Frequent verbs > Less frequent function words > Ordinary function words > Schwa function words. A quantised scale of three levels of stress is claimed adequate for short sentences, but six or more levels are deemed necessary for paragraph-length materials. "Words that are heavily stressed are set apart quite clearly in the stream of speech [and] weak unstressed words are run together as the syllables of a rapidly-spoken polysyllabic word." The duration for a single vowel is calculated according to:

$$T = K_1 + S(K_2 + K_3C), \qquad (13)$$

where S is a factor combining the effects of stress and situation relative to a pause, and C is a factor for the consonants following the vowel, k_1 k_2, and k_3 are approximately 80, 130 and 130 ms respectively. S and C vary in the range of $0 \sim 1$. A numerical pause-potential is assigned to the boundary of every pair of words according to a grammatical-category transition matrix. These may be realised either as actual silences in the output or as vowel elongation associated with marked pitch change when there is no actual break in the speech. For moderately rapid speech the system relies more heavily on these "pseudo pauses" than on full stops. Vowel elongation is an important feature of the pseudo-pause and is associated with either an upward or downward glide immediately preceding the pause, with a conspicuous drop in pitch on the nucleus.

Consonant durations are considered to be "a physical attribute of stress" and consonants on the boundaries of strong substantives are lengthened by 60 msec subject to an unspecified "lengthening sensitivity factor" for each consonant. Initial and final consonants are treated separately with respect to coarticulation effects as would be required to distinguish "grey tie" from "great eye" in clear disambiguating articulation. These are implemented by allophones differing in burst intensity and devoicing time after voiceless stops and by different degrees of loudness in nasals and voiced stops. They can be ordered on a scale ranging from initial consonants of word-initial stressed syllables, through medial-position allophones, to word-final ones. The rules differ as to the amount of (again unspecified) detail in timing, and to the extent that they differentiate allophonic variation inside the word to handle specific exceptions. Examples are given of the shortening of a vowel before a syllable-final nasal, shortening of voiceless stops and fricatives before nasals, and the insertion of aspiration for a voiceless stop before a weak vowel when preceded by a consonant. Although they claim that the lengthening of boundary consonants around salient words does produce a somewhat isochronous effect, the key feature in this model is not rhythm but stress. The (intuitively appealing) concept of multi-valued stress assignment will be discussed in more detail below.

3.1 Statistical modelling

The challenge of duration prediction is attractive to mathematicians as well as linguists. Riley (1990, 1992) described an application of the CART (Breiman et al. 1984) tree-based modeling procedure to segmental duration prediction. CART generates a binary prediction tree by recursively splitting subsets of the data to minimise the variance of the estimated error for each subset, allowing a combination of both real-valued and categorial factors, and so is well suited for modelling duration. However, it is vulnerable to shortages of training data and requires a well-balanced training set in order to generalise properly. If segment position were treated as a single feature, then 48 phone classes would produce 2^{48} possible combinations, so place and manner categorisation of phone type were used instead to avoid data-sparcity problems for training, in conjunction with segmental context, stress, lexical position and phrasal position as prediction factors. The model was able to reduce the 65 msec standard deviation of the original durations to 25 msec in the residuals from the prediction using a corpus of 1500 hand-labeled utterances from a single male speaker of American English. Riley compares this favourably with the 35 msec standard deviation in the residuals of a prediction using AT&T's heuristically derived duration rules with the same data, but notes that while the results look favourable on paper, they are not *noticeably* better on informal listening tests. This is explained as due to poor predictions, resulting from sparse data, standing out from the generally better predictions that the tree-based classifier is capable of when the training data is sufficient.

Pitrelli (1990) determined by similar tree-based clustering 27 linguistic properties that affect phoneme duration as a factor in speech recognition:

> For phonemes in general: identity, place of articulation, gemination, clustering, manner, place, and voicing of both left- and right-context, onset or rhyme position within the syllable, and whether or not in one of a list of 59 reducible words.
>
> For consonants: stridency and voicing of fricatives, voicing of stops and affricates, syllabicity of semi-vowels and nasals, and whether or not pre-stressed.
>
> For nasals: whether following a low vowel, whether preceded by a vowel and followed by /d/ or /dh/ (for spotting *'and the'* whether preceded by a vowel and followed by an unstressed vowel (for spotting *'in a'*).
>
> For syllables: whether sentence-final, pre-pausal, or last in one of six levels of syntactic unit, whether in a reducible word, the number of syllables in the word, and whether in word-final or word initial position, its position in the word, and the stress level of the syllable (three categories).

Such factoring methods can provide good fits when given enough training data and cope well with categorical, discrete-valued explanatory variables, but they fail to account for continuously-valued predictors such as speaking rate and stress, and offer no principled way (short of modeling a new database) of allowing for the changes in rate that are required for the prediction of timing under a variety of speaking styles and conditions. Van Santen (1992) summarised the duration prediction systems that are in current use as being of four principal classes: sequential rule systems, equation systems, table-lookup systems, and binary prediction trees. He then further simplified these by redefining binary trees as a collapsed form of lookup table. Because of the seemingly insoluble problem of finding adequate representative data for training lookup systems, he advocates instead the use of rule-based or equation systems for statistical duration modeling. If segmental duration depended on only a few factors, then a lookup table could be constructed simply by segmenting a sufficiently large body of natural speech and determining the average duration for each combination of levels and factors. However, he argues that because of the high number of factors known to be involved, the number of combinations quickly reaches several tens of thousands, so that even very large databases fail to be sufficiently representative of all possible situations. He cites data from Church and Gale (1990) arguing that the non-uniform distribution of segments in speech data exacerbates this situation.

To show that sequential rule systems can be simply reformulated as equation systems, and to challenge the essential assumption of independence between the prediction factors in such models, Van Santen rewrites the Klatt model expressed as Eqn. 14 to its equivalent, generalised version, Eqn.15. In Eqn.14, p denotes segmental identity, c_j is a level on contextual factor j (where j is a variable index that takes on the values from 1 through N), and $d(p,c_1,\ldots,c_N$ is the duration (in msec) of phoneme /p/ in the contextual condition defined by the factor levels $(c_1,\ldots,c_N)$. The (positive) coefficients k correspond to levels of the contextual factors. The difference, a re-expression in Eqn. 15 of $[D_{inh}(p)-D_{\min}(p)]$ as $f(p)$, of $D_{\min}(p)$ as $h(p)$, and of ki_{c_j} as $g_j(k_{c_j})$, highlights the importance of factor scales (Barlow et al. 1972) in such models.

$$d(p,c_1,\ldots,c_N)=k_{c_1},\ldots,k_{c_N}\left[D_{inh}(p)-D_{\min}(p)\right]+D_{\min}(p) \tag{14}$$

$$d(p,c_1,\ldots,c_N)=k_{c_1},\ldots,f(p)g_1(c_1)\ldots g_N(c_N)+h(p) \tag{15}$$

The Klatt model differs from the additive model by employing two scales, f and h, for the segmental identity factor. It employs the sum of a product of N+1 scales and a single scale, while the additive model uses simply the sum of N+1 scales. Since the amplificatory effects of some factors on certain other factors are not included in the Klatt model, and cannot be modeled in either the generalised form of that model or by generalised additive models, Van Santen proposes ordinal data analysis (Krantz et al. 1971) to cope with the sometimes complex interactions for a more rigorous duration model. He cites as an example of these interactions Umeda's (1975) finding that 'alveolar consonants tend to be shorter than velar consonants at the head of unstressed syllables but longer when at the head of stressed syllables'. This is given as a violation of single-factor independence (between 'phone type' and 'stress'), which shows a deficiency in the phone categories. It can be easily overcome, however, by relabeling flapped variants of the alveolar segments as a separate phone.

A more serious problem for equation models is illustrated with data from Luce and Charles-Luce (1985) which shows that there exists a vowel (v, voiced, phrase-medial) which is shorter than another vowel (w, unvoiced, phrase-medial), and yet *longer* than (w, unvoiced, phrase-final) when (v, voiced, phrase-final). This contradicts the joint independence of the factors 'voicing of the following consonant' and 'position in the phrase' because in phrase-medial position, the (v, voiced) combination has a shorter duration than the (w, unvoiced) combination, while in phrase-final location, the order is reversed. He shows that such data cannot be fitted by either an additive model or a generalised additive model and argues that a generalisation of the Klatt model (Eqn. 16 below) is sufficiently powerful.

$$d\left(vowel_p, voicing_c, location_l\right) = F\left[f(p)g_1(c) + g_2(b) + h(p)\right] \tag{16}$$

In Eqn.16, the effects of vowel identity are distributed over two terms, only one of which involves the phrasal position factor; i.e., splitting 'a part of the vowel factor $[h(p)]$' that is not amplified by the phrasal position factor. Thus he claims that all effects can be sufficiently modeled by sums-of-products models, and it only remains to determine the best one in each case.

4. PHONEMES OR SYLLABLES?

We can classify duration studies into two categories: those that attempt a mathematical modelling of the observed distributions by a combination of factors operating directly at a segmental level, and those that attempt to specify a higher-level framework into which the segments are accommodated; i.e., by a bottom-up versus a top-down distinction. In many of the previous studies it was implicitly assumed that phonemes have fixed stable duration characteristics, subject only to slight modification according to certain environmental factors. Little attempt has been made to model the stability of 'inherent durations' under more natural speech conditions, and they are considered as distinctive features that serve to identify the phones. An error in a model's fit is assumed to be Gaussian, random, and merely a result of natural human variation. In this section we will challenge that view, claiming that speech sounds have only minimal 'inherent' duration characteristics (if at all), and will see that under normal conditions, the distributions of segmental durations are not as well-behaved as has been traditionally assumed and that the clear citation-form data of prepared laboratory speech are less representative of real-world distributions which have considerably more variation.

4.1 Segmentation Criteria

First, a comment with respect to phonemic segmentation: It is an oversimplification to believe that the complex results of speech articulation can be adequately described by a non-overlapping sequence of phone labels. We can make no claim that clear definable boundaries exist between all phoneme-sized speech sounds, and in many cases the assignment of a label to a portion of speech can be quite arbitrary. However, for the purposes of speech synthesis at least, and much of timing research in general, the assumption of intersegmental boundaries is maintained, and durations are determined for the interboundary phoneme-sized portions of speech thus delimited. In synthesis, numbers are predicted that govern the durations of the segments in the output speech and, given that the labels used for the predictions are the same as those used in the initial description of the speech, it does not really matter what they describe as long as events in the output can be time-aligned with events in a natural speech rendition of the same utterance. We will proceed on the understanding that the segment labels described here are not to be interpreted as a statement supporting phoneme theory *per se*, but simply as a convenience for the prediction and alignment of events in speech.

The data presented in this paper are not new; they were produced by hand by a team of phoneticians using interactive speech workstations to produce non-overlapping phonetic (sub-phonemic) transcriptions and segment labels from spectrograms of directly digitised recordings according to criteria based closely on the DARPA-TIMIT segmentation guidelines (Fisher et al. 1987). These will be described below.

- **Stop consonants** were segmented into three acoustically distinct events: closure, burst and aspiration (voice onset delay), of which the closure was sub-categorised into a voiced and unvoiced portion. Stop closure onset was defined as the point at which energy in the region of F2 and the higher formants ceases to be visible on the spectrogram display. The burst marks the onset of the release phase which lasts (in voiceless stops) until the point of the first identifiable glottal pulse. In the case of stops preceding fricatives and nasals, with no identifiable release phase, only the closure was marked. Utterance-initial stops were given an arbitrarily determined 50 msec closure duration unless clear voicing was visible in the spectrogram. Only the burst, and not the pre-silence weak frication was marked in utterance-final stops. When no final release was evident, an arbitrary 50 msec duration was assigned. In all cases, both burst and aspiration segments (where transcribed) were included in the final duration of the plosive.

- **Fricative** boundaries were usually easily distinguished by the presence of high-frequency energy, but for short voiced fricatives, that were sometimes difficult to detect from the spectrogram alone, an overall drop in energy in the formants was taken as criterial.

- **Nasals** were detected from the drop in energy above 500 Hz and from a discontinuity in the formant frequencies and a lowering of formant amplitudes. Geminates (e.g. /n/-/n/ 'in nine') which showed no discernible boundary were segmented at the mid-point. Because the separate marking of nasalisation on vowels resulted in great variance in vowel segment durations, fine distinctions were disregarded.

- **Liquids and glides** presented the greatest segmentation difficulty. The formant transition was taken as criterial, using F2 for /y/, both F1 and F2 for /l/ and /w/, and F3 for /r/. The midpoint of the transition was taken as the point of segmentation in such cases.

- **Vowel and diphthong** boundaries were taken either from the abutting consonant criteria, or from the formant changes. In the example of 'do a' the unrounding of the lips in the transition from the /uu/ of 'do' to the /@/ (schwa) of 'a' was typically marked by an increase in the amplitude of the formants and/or waveform.

Inter-labeller tests of segmentation consistency (Hieronymus, Unpubl.) showed that end-points for 70% of labels were within 10 msec of each other, 50% within 5 msec, and 90% within 25 msec. Four classes of phones were tested: vowels, semi-vowels, stops, and fricatives, and no significant difference was found in the segmentation accuracy of each class. Boundary segments trailing off at the end of a sentence were noted as a major cause of disagreement.

4.2 Variability in Timing

There was considerable difference in durations observed for the same phones in identical contexts produced by four different speakers sharing the same dialect of British English. Table 1 presents summary statistics for the segmental durations of all speakers. Table 2 shows inter-speaker differences for one sentence of the corpus. The first column shows the durations predicted by averaging and illustrates the difficulty of evaluating a predictive model of speech timing; there is so much variation in the natural durations that we can not easily determine a 'right' duration for any context. And yet when listening to synthesised speech it is easy to spot a 'wrong' duration. There appears to be a consistency within the variation of the individual speakers that is not adequately captured by the segment-based mathematical models described above.

speaker:	sab	gsw	pms	jok	ave
num:	8505	8411	8426	8476	8612
mean:	79.1	76.0	71.7	70.0	71.6
sdev:	42.8	43.2	35.0	35.0	33.6
adev:	33.5	33.2	27.2	27.0	26.5
skew:	0.9	1.2	1.0	1.0	0.9
kurt:	1.3	2.3	1.7	1.9	1.1

Table 1: *Summary statistics for the segmental durations of all four speakers, for comparison with the averaged data, showing mean duration in msec, standard deviation, average deviation, skew and kurtosis.*

	ave		sab			gsw			pms			jok		
39	dh	21	dh	18	-	dh	26	-	dh	31	-	dh	21	-
39	ii	66	ii	29	-	ii	85	-	ii	106	-	i	84	-
39	=	=	=	=	-	=	=	-	=	=	-	=	=	-
39	@@	132	@@	208	1	@@	164	2	@@	141	1	@@	155	1
39	th	137	th	183	-	th	105	-	th	106	-	th	125	-
39	=	=	=	=	-	=	=	-	=	=	-	=	=	-
39	y	40	y	56	-	y	19	-	y	54	-	y	46	-
39	uu	56	uu	54	-	uu	57	-	@	40	-	uu	57	1
39	s	57	s	68	-	s	54	-	s	57	-	s	50	-
39	t	46	t	53	-	’	0	-	t	44	-	t	39	-
39	=	=	=	=	-	=	=	-	=	=	-	=	=	-
39	t	52	t	46	-	t	73	-	t	46	-	t	38	-
39	uu	17	@	18	-	@	14	-	’	0	-	@	19	-
39	=	=	=	=	-	=	=	-	=	=	-	=	=	-
39	b	57	b	65	-	b	50	-	b	30	-	b	56	-
39	ii	67	ii	63	-	ii	66	-	i	44	-	i	74	-
39	=	=	=	=	-	=	=	-	=	=	-	=	=	-
39	f	124	f	136	-	f	134	-	f	134	-	f	103	-
39	l	39	l	45	-	l	33	-	l	37	-	l	39	-
39	a	106	a	149	2	a	159	1	a	141	1	a	115	1
39	t	106	t	155	-	t	48	-	t	123	-	t	115	-
39	=	=	=	=	-	=	=	-	=	=	-	=	=	-
39	b	135	b	264	-	b	60	-	b	90	-	b	82	-
39	uh	51	@	36	--	@	63	-	@	44	-	@	54	-
39	t	53	t	71	-	t	26	-	t	50	-	t	62	-
39	=	=	=	=	-	=	=	-	=	=	-	=	=	-
39	n	51	n	36	-	n	50	-	n	37	-	n	67	-
39	au	128	au	113	1	au	160	2	au	113	-	au	111	-
39	=	=	=	=	-	=	=	-	=	=	-	=	=	-
39	i	68	i	81	-	i	52	-	i	52	-	i	72	-
39	t	33	t	13	-	t	34	-	’	0	-	t	52	-
39	s	70	s	61	-	s	84	-	s	72	-	s	66	-
39	=	=	=	=	-	=	=	-	=	=	-	=	=	-
39	@	29	@	0	-	@	21	-	@	37	-	@	38	-
39	=	=	=	=	-	=	=	-	=	=	-	=	=	-
39	s	142	s	167	-	s	143	-	s	143	-	s	116	-
39	f	81	f	95	-	f	72	-	f	57	-	f	77	-
39	i@	197	i@	260	2	i@	277	1	i@	176	1	i@	250	1

word boundaries shown by "=", elided segments by "’"

Table 2: *Aligned MRPA (Machine Readable Phonemic Alphabet) labels, durations (msec) and stress for sentence number 39: ''The earth used to be flat, but now it's a sphere.''*

Variation is observed not only between speakers but also according to speaking style, as between read and spontaneous speech. Figure 1 plots mean segmental duration against the coefficient of variance (i.e., the standard deviation expressed relative to the mean) for each phone class for four speaking styles. By plotting durational characteristics in this way, we see which phones are stable or invariant, and which are more variable, and we can differentiate those that are inherently short from those that typically tend to be long. A phone appearing in the lower left of these plots will be short and stable, one in the top left, typically short, but showing great variation in its observed durations.

The data were produced by measurement of segments in speech from four different utterance styles. For the first, 5000 words were read in isolation, with a pause between each. For the second, a subset of these words was used to make 200 meaningful sentences in which words were read in sentence order but with pauses between each. For the third, the same sentences were read as continuous text, with pauses only at the end of each sentence. The fourth set of durations are from a spontaneous monologue produced by the same speaker, but without controls on the production. The first set is 'well-behaved', and exhibits clear durational patterns, but we can see that as the production controls are reduced, the variance in duration increases considerably. In the isolated-word citation-form readings there is a good dispersion in the mean durations for each phone class, and relatively constant variance in their durations; i.e., they are well spread out along the horizontal axis and close together in the vertical axis of the plot. However, the opposite is the case for the same sequence of words read as sentences. Here the variance increases and there is considerable shortening so that the 'inherent' durations of the segments are no longer so distinct and characteristic. Separate examination of segments in word-initial and word-medial position confirmed that this increase in variance is not simply the result of more phrase-final lengthening (isolated words being produced as complete phrases in themselves), but rather that the articulation of the citation-form words was generally slower and more distinct.

When the speech contains little contextual information and the speaker is concerned to be clearly understood, then segmental durations are maximally separated, exaggerating the difference in duration between the phone types, but as the style becomes more natural and the listener can rely on prosodic phrasing to aid in the interpretation of the speech, then we find more variance in the durations and less distinction between their means; all words tend to be shorter and more varied than in the citation-form readings. The durations measured in a spontaneous monologue from the same speaker, in Figure 1, show the same trends exaggerated. We find not only that the mean durations for all segment types are low and uniform, but also that the variances are huge – note especially that the vertical scale of the plot in the bottom right of this figure is *twice* that of the other three figures. When a listener is present and the content is meaningful, the speaker can form a better understanding of the extent of their mutual knowledge and will speak more quickly through some parts of the utterance and take more time over others. There is a much greater range of variance in the natural speech than was found in the more carefully produced 'lab' speech. Unfortunately, most of the duration studies reported above were carried out on laboratory speech.

4.3 A higher-level view - the Syllable Timing Hypothesis

The syllable-timing hypothesis ((henceforth STH) was proposed by Campbell and Isard (1991) and expanded further in Campbell (1992c) to account for the interaction between higher and lower levels of timing control. It posits the syllable as a mediator and offers a way to map the effects of linguistic and semantic contexts onto the physiological constraints of speech sound

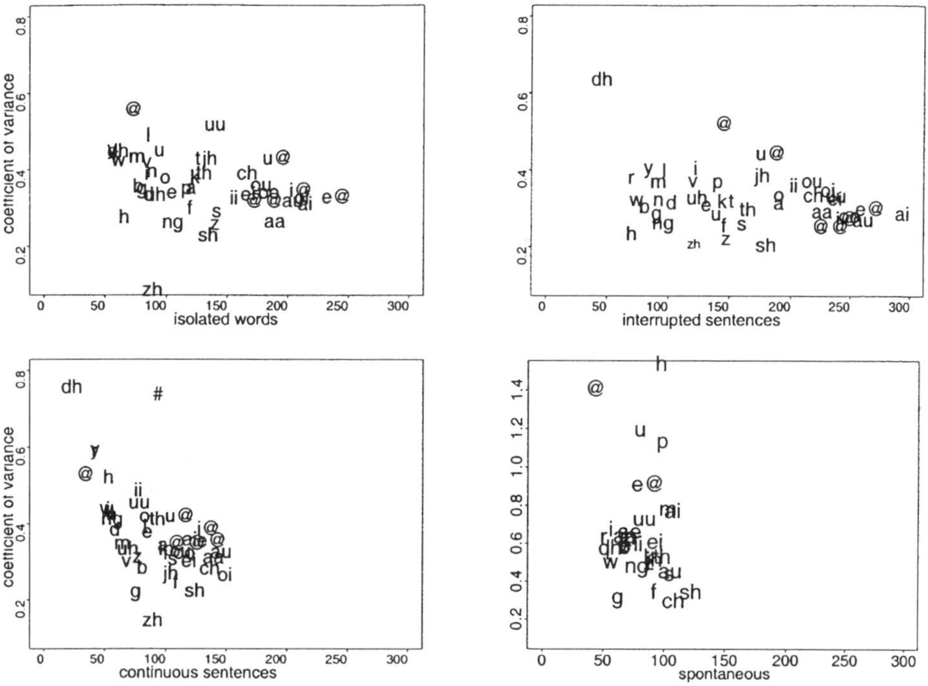

Figure 1: Dispersion of durations for four speaking styles, plotting mean duration per segment (in milliseconds) against the coefficient of variance (standard-deviation/mean)

production. By adopting a higher-level framework for duration control, it overcomes the sparcity of data problem in modelling the variability in individual phone durations. It is a fundamental principle of the STH that both consonants and vowels have similar lengthening characteristics if viewed not in percentage but in variance terms, and the STH attempts to explain how all segments in a syllable can share the available space equally according to their own elasticity. Klatt (1974) came close to this view, but did not pursue it further:

> "Prestressed consonants and stressed vowels generally appear to be of longer duration in English than unstressed consonants and unstressed vowels. That the consonant and vowel in a consonant-vowel sequence behave similarly in the duration rules provides support to the contention that the syllable is an articulatory programming unit [...] It seems likely that unstressed syllables are

> not only spoken more rapidly, but also with a reduced muscle tonus or relaxed criteria for attainment of target configurations. Thus a feature such as the presence or absence of lexical stress would seem to be equally applicable to consonants or vowels.''

i.e., that perhaps duration control would be better applied from the level of the syllable. Soon before he died he added (1987: 760) that

> ''In fact one of the unsolved problems in the development of rule systems for speech timing is the size of the unit (segment, onset/rhyme, syllable, word) best employed to capture various timing phenomena.''

Collier (1992:206) recently pointed out that

> ''The minimum domain in which [intonational phenomena] operate seems to be the syllable or word. Indeed, it is not without reason that prosody is referred to as the 'supra-segmental' level. Therefore it may come as a surprise that studies of the temporal aspects of speech usually focus on the duration of single segments. As a rule, both the acoustic analyses of databases and the synthesis rules for duration control are based on phoneme-sized units. Yet it seems intuitively more plausible that the speaker organises his articulatory behaviour in syllable-, word-, or foot-sized chunks and programs his articulatory timing at the supra-segmental level. This alternative view implies that global temporal structures determine to a large extent the duration of individual segments. If so, the actual lengths of the segments may not be very critical, as long as their combined durations add up to the required length of the suprasegmental unit.''

Kozhevnikov and Chistovich (1965) showed that when a speaker repeats the same sentence many times at the same rate of articulation then the durations of adjacent phones are strongly negatively correlated, i.e., any variance in the duration of one phone is compensated in the adjacent phones. From this, Chistovich argues that the temporal sequence of articulation must be organised, at least in part, at levels higher than the phoneme. Huggins (1968) extended this argument and showed results that ''would lead one to expect substantial compensation between the durations of any two phonemes, not necessarily adjacent, that occur between two prominence peaks.''

We see in our own data evidence to support this higher-level view in a comparison of inter-speaker durations at the two levels, where there were significantly higher correlations at the level of the syllable than at the level of the segment. Although we find considerable variation in the segment durations of different speakers in the corpus, almost all of this variability is reduced when compared at the syllable level. Table 3 shows the inter-speaker correlations at the different levels. The syllable durations of the different speakers show very high correlations ($r = 0.92$) to reveal almost 85%

agreement in durations at this level, compared with less than 60% agreement at the segment level ($r = 0.76$).

sab	1		average (spkr:spkr) r = 0.92	
gsw	0.93	1		
pms	0.92	0.92	1	
jok	0.93	0.92	0.92	1
	sab	gsw	pms	jok
sab	1		average (spkr:spkr) r = 0.76	
gsw	0.78	1		
pms	0.73	0.76	1	
jok	0.77	0.78	0.73	1
	sab	gsw	pms	jok

Table 3: *High inter-speaker correlations of durations at two levels.*

Some part of this high correlation may be due to artifacts of phone-based segmentation of the speech, where small differences in boundary determination can have relatively larger effects on segment durations than on syllable durations. However, the difference in the correlations is greater than can be explained by random segmentation error, and the higher systematicity at the higher level is in accordance with the separate experimental findings of Chistovich, Martin, and Huggins, and confirms a uniformity in inter-speaker timing at the syllable level. Because there is less speaker-specific variation in the durations at the syllable level, we can consider them to be more reliable as a basis for predictions. If the syllable-level durations can be determined independently and the segmental durations derived from them, then any errors at the segmental level will be confined to the region of a single syllable and cannot systematically percolate up to accumulate in a way that would disturb the rhythm. As long as the temporal structure of the higher-order unit is predicted correctly, then any errors at the lower segmental level will compensate for each other rather than accumulate; i.e., an overshoot in the timing of one phone will be cancelled by an undershoot of another phone in the same rhythmical unit. Martin (1972, 1979) has shown that this type of syllable-internal error is less disruptive to perception than errors which cumulate across syllables and disturb the inter-syllable balance.

4.4 Accommodation and elasticity

Segmental durations can be well predicted by fitting them into a syllable-level timing framework. Key concepts here are *elasticity* and *accommodation*; with the former describing a notional 'stretching' or 'squashing' of segment durations, and the latter the process of 'stretching' or 'squashing' in order to fit a syllable-sized frame. In the simplest case of predicting segment durations by accommodation we posit a single value of lengthening for a syllable (k) that can be thought of as the degree of stretch applied to the elasticity of its constituent segments and can be interpreted to produce appropriate segmental durations. More specifically, the value represents a quantile in a set of given distribution. If we use the mean and variance as descriptors of the distribution, then (k) would represent the excursion in standard-deviation units above or below the mean. By defining the distribution in a small number of parameters (two should be sufficient), and the position within that distribution as a single value, we are freeing the phonological specification of length from the phonetic implementation of duration.

The elasticity principle says in its strongest form that all segments in a given syllable fall at the same place in their respective distributions; i.e., they stretch or shrink together, like linked elastic bands, by different amounts that are determined by features specific to each phone, but in the simplest case all subject to the same factors controlling lengthening. Consider as an example the phonemes /æ/ (mean (æμ) = 119 msec, standard deviation (æσ) = 37ms), /t/ (mean (tμ) = 41 msec, standard deviation (tσ)= 21 msec), and /d/ (mean (dμ) = 39 msec, standard deviation (dσ) = 19 msec). If we are to combine these into the words *at* and *ad*, then by simply summing the means, the default durations for the words can be expected to total 160 msec and 158 msec respectively. Let us consider the case where the actual duration of the words is in fact longer, say 300 msec, and assume that peak and coda lengthen equivalently by a uniform amount in terms of their standard deviations. In this case a value can be found for k such that æ, μ + tμ + k(æ σ + tσ) = 300, which yields /æ/ = 208 msec and /t/ = 92 msec. A slightly different value of k is needed for æμ + dμ + k(æσ + dσ) = 300; where /æ/ is 213 msec and /d/ is 87 msec. The overall word lengths are the same, but because the /d/ is shorter than the /t/, and has a smaller variance, the vowel seems to lengthen by 5 msec to accommodate. The lengthening applied to each segment in the syllable is the same, in terms of its elasticity.

By expressing lengthening in terms of quantiles within a distribution, or as z-scores, raw millisecond durations are transformed into unitless values. With this normalised measure of length, comparison can be made not only within each type, but also between phones of different types, since a positive value of z represents lengthening, and a negative value shortening relative to the mean duration observed for all tokens of that type in the database. With normally distributed (Gaussian) data, 68% of the tokens can be expected to fall within ± 1 SD, and 99% between ± 3 SD.

$$z\ score = raw\ duration_{token} - \mu_{type} / \sigma_{type} \tag{17}$$

where μ_{type} is the mean duration observed from all tokens of that phone type, and σ_{type} is their standard deviation. The skew observed in raw segmental durations can be reduced towards Gaussian by taking the log of each duration instead of its raw value.

No speech segment is completely invariant in its duration, and the fact that a short sound (a plosive burst for example) varies less than a longer one (its closure) can be easily explained by articulatory differences that should not need to be taken into account when processing durations to interpret the structure of the speech. By these transforms we can factor out most of the articulatory-level effects and view lengthening independently from the phonetics of the message. Similarly, for any given database we need no longer be concerned with speaker characteristics such as differences in overall speaking rate and can thus model more easily the extent to which all sounds are lengthened or shortened by contextual factors.

4.5 Accommodation into a framework for timing

In English, the difference between stressed and unstressed syllable durations is large, with compensation taking place at the level of the foot causing syllables in short feet to lengthen and those in longer feet to reduce, and resulting in greater variation amongst the segmental durations of consonants and vowels alike. We can see the extent of this compression in English if we compare the histograms of word durations under two reading styles as in Figure 2. The data are from identical sentences, but in one case (isolated-word reading) the rhythm has been disturbed by insertion of a pause between each pair of words. When natural phrasing is allowed to occur (as in continuous speech) we see that a large group of words (typically function words) is considerably reduced in duration, resulting in compression to the consonants as well as to the vowels in these words. Figure 2 (bottom part) shows that there is a clear bi-modal distribution of the data in the continuous speech, which results from the reduction and de-emphasis of the words when they become functional parts of a sentence.

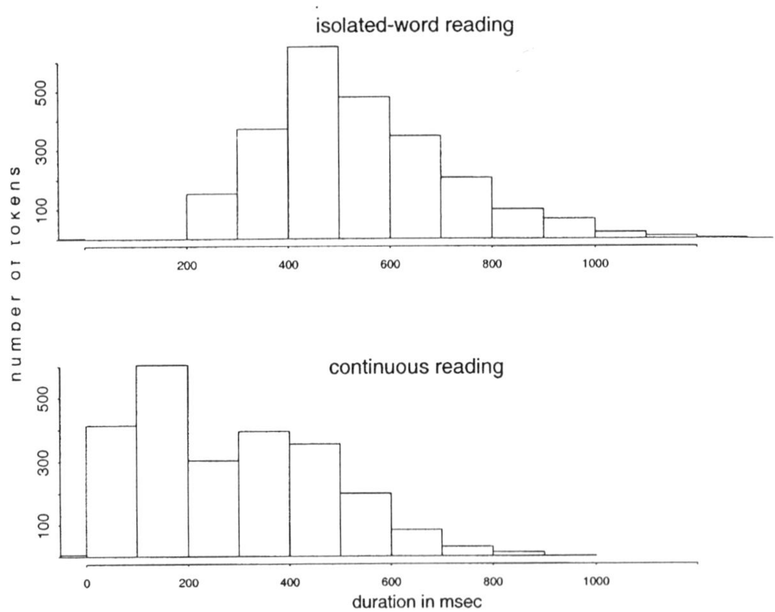

Figure 2: Histograms of word durations from two different reading styles.

To illustrate the effects of accommodation using a language with fewer phones and simpler syllable construction than English, we can take the case of Japanese: To the casual observer the timing patterns of Japanese and English sound very different; we have the impression of fast regular, almost clipped speech in the former and of 'waves of prominence' in the latter, and the two languages have been cited as polar examples of quite different stress-based and syllable-based timing structures. However, direct comparison of the segmental durations of the two languages fails to reveal much about these differences, and it is in their syntagmatic structuring, rather than in any inherent or paradigmatic characteristics that the difference can be found. Figure 3 shows the durational distributions from four male and four female Japanese speakers reading 503 phrase and sentence-length excerpts from Japanese newspapers and magazines, and compares them with equivalent durations measured from two male and two female speakers of British English (RP) reading 200 similarly constructed sentences. There appears to be surprisingly little difference at this level between the two languages. The figure plots the standard deviation of each phone as a proportion of its mean (MRPA symbols are drawn in the plot to help identify the phones but we are only concerned here with general trends) and shows that the relative variance is not uniform but rather tends to increase when the mean duration of the phones becomes shorter.

Figure 4 shows durations and equivalent z-scores for the 5 vowels and 10 consonants of Japanese. The boxes in the plot are drawn with horizontal lines indicating the 25th, 50th and 75th percentiles. Vertical lines extend above and below the boxes to one-and-a-half times the upper and lower interquartile ranges respectively. The width of each box is proportional to the log of the number of tokens in each sample, and the notches indicate significance at the 5% level in the difference of the distributions if they show no overlap. Differences in the raw durations are all significant (except perhaps between /e/ and /o/), but the z-score normalisation, showing their 'lengthening', removes such phone-specific dependencies.

We have seen that in English, when a vowel is followed by a voiced plosive in the same syllable it is generally longer than an equivalent vowel followed by an unvoiced plosive of the same place of articulation. In Japanese as can be seen from Figure 5, there is also a very clear separation of effects on vowel length according to the manner of voicing of the following plosive. By grouping together the z-scores for the vowels (which all by definition have zero mean) and displaying only values for that subset which is followed or preceeded by a plosive, further factored in the plot according to the type of plosive, we can ignore any differences in vowel type and see instead the general effects of the environment. The left-hand side of the figure shows the distribution of z-scores for all vowels in pre-plosive contexts. There are significant differences in the lengthening according to the type of following plosive but in this case the effect known for English appears to be reversed and although separation can be seen according to the voicing class of the following plosive, the vowels appear to be *shorter* relative to their mean duration when the following plosive is voiced. It is only when viewed in the context of the higher-level framework that this discrepancy is resolved. All becomes clear if we look at the distribution of lengthening of vowels when they *follow* a plosive. In Japanese, vowels show significant lengthening when they follow a voiced plosive and shortening after an unvoiced one.

Figure 5 (right-hand side) shows the difference in lengthening clearly for the two types of voicing. If we think just in terms of sequential order effects, then the two languages appear to be in contradiction. The difference, however, is in the higher-level structuring of the languages rather than in the linear segmental organisation. In Japanese, a plosive following a vowel is in a

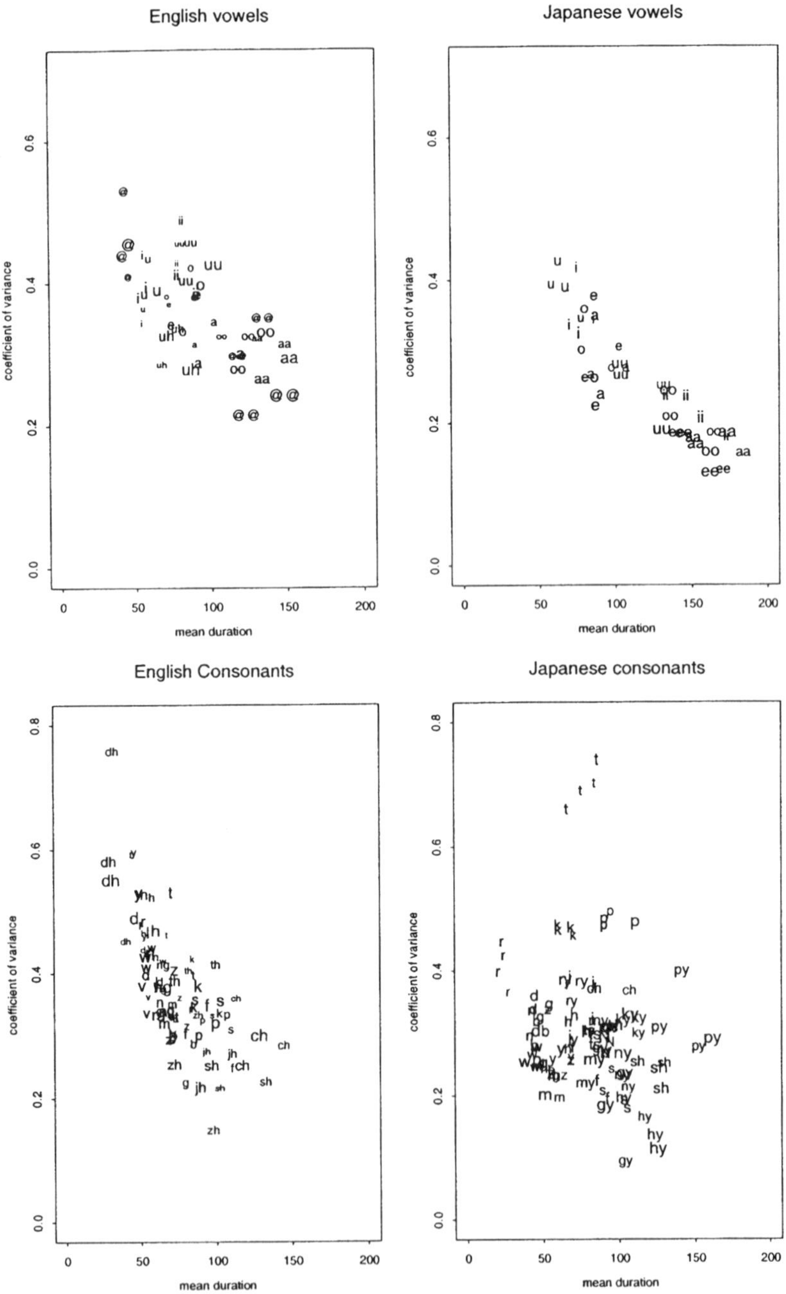

Figure 3. Japanese and English segmental durations from multiple speakers' data, showing mean durations for each phone (msec) plotted against the coefficient of variance.

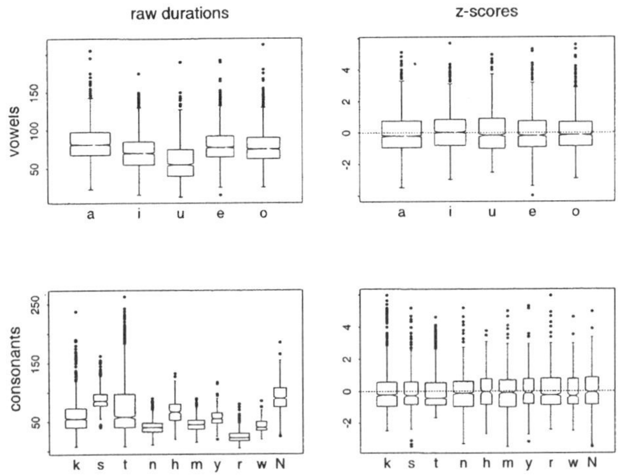

Figure 4. Segmental durations and equivalent z-scores for Japanese.

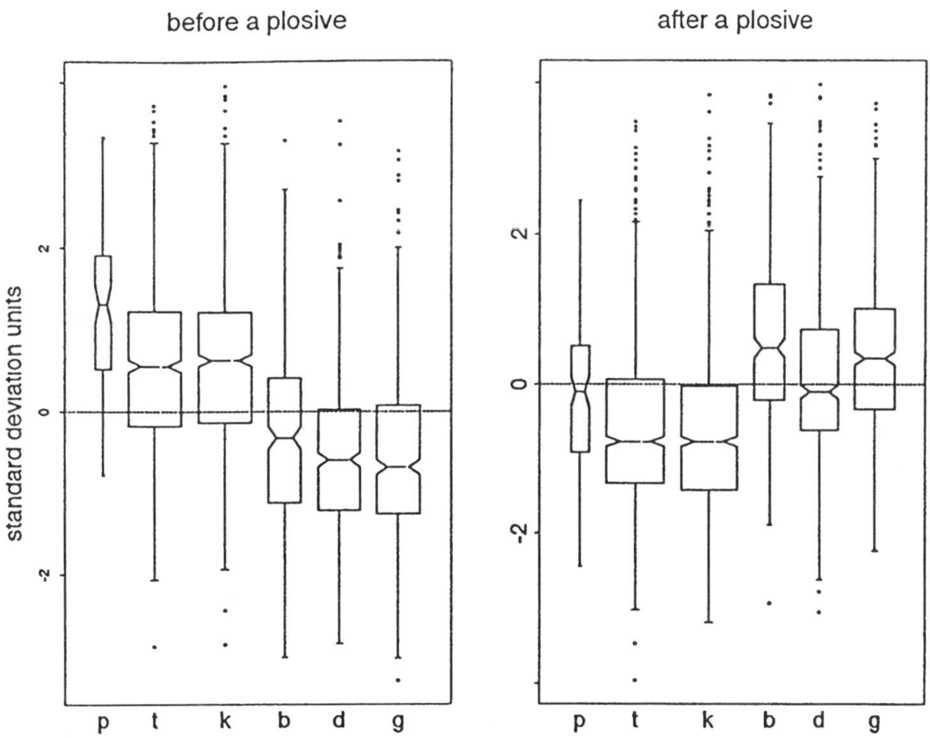

Figure 5. Vowel lengthening (expressed in z-scores) before and after plosives in Japanese.

different mora, while one that precedes the vowel is more closely affiliated in terms of moraic (CV) organisation. Seen in terms of the higher-level unit, the lengthening characteristics of vowels in both languages are consistent and, as in English, the vowel simply stretches to share the available space with the other segment(s) in the syllable. The amount of space it can have is determined by higher-level factors and the elasticity of the segments.

5. PREDICTING DURATIONS BY ACCOMMODATION

Campbell (1992a) described a two-level model that predicted segmental durations from the higher level, and employed a combination of six factors to account for 85% of the variance in the syllable durations of a twenty-minute short-story BBC radio broadcast. For comparison, the output of the MITalk duration algorithm, with duration tables and 'speech-rate' parameter optimised to fit the same corpus (Campbell 1988), accounted for only 68% of the variance. A categorial factor analysis was employed to find the six features most correlated with syllable duration (Campbell 1992c) and that could be easily determined by a text-to-speech synthesis system. Partial correlations of $r = 0.36$ were found for number of phones in the syllable, $r = 0.56$ for type of peak, $r = 0.40$ for position in the tone-group, $r = 0.15$ for type of foot, $r = 0.44$ for stress, and $r = 0.44$ for word class. The multiple correlation coefficient for the six factors combined was $r = 0.87$, indicating an ability in linear combination to account for 76% of the variance in the durations of the syllables. The difference in prediction strength can be partly explained by the use of a non-linear neural network to predict the syllable durations, but more importantly as resulting from a 'backstep' in the prediction that allowed for bottom-up influence from the segment tier, moderating the predicted syllable duration to compensate for the typical length of its components, and thereby simulating the physiological constraints on speech production that interact with the higher-level cognitive commands for global timing.

A three-layer back-propagation neural network that consisted of six analog input units with three hidden units and one output unit was trained to predict the syllable durations in an initial pass. Output from the syllable-level prediction module was a single value, limited by the sigmoid function of the neural network, that was converted to yield a number that represents the natural logarithm of the predicted syllable duration. In a second stage, the model determined the durations for each segment in the syllable by a top-down process of accommodation using a single factor of lengthening applied as a quantile to the individual segment duration distributions. The syllable-level factors are summarised here from Campbell (1992a):

1) The number of phonemes in the syllable is used as a simple approximation of its articulatory complexity.

2) The peak of the syllable is categorised into one of four classes: i) reduced vowels (schwa), ii) lax vowels, iii) tense vowels, and iv) syllabic consonants and polyphthongs.

3) Syllables are ranked according to initial, medial, or final position in major and minor intonational phrases (five categories).

4) Each syllable is labelled with a code that describes the nature of the foot it occupies and whether or not that foot is headed by a stressed syllable. All feet with three or more unstressed syllables in the tail are classed equivalently (giving eight foot categories in all). Unstressed syllables that occur immediately after a prosodic boundary and before a stressed syllable (anacrusis) are categorised separately from other unstressed syllables and typically undergo more extreme shortening.

5) Stress and accent: whether a syllable has been transcribed with a full vowel (i.e. not reduced), whether it has been flagged as stressed, and if so, whether it has also been flagged as accented. In the latter case it is further subcategorised according to whether the pitch movement is simple or complex (four classes).

6) Syllables in content words, typically information-carriers such as nouns and adjectives, are distinguished from those in closed-class function words (two classes).

Under the strongest form of the elasticity principle, segmental durations were predicted by simple accommodation, using log-transformed means and standard deviations for each phone without regard to allophonic variation. With the overall duration for the syllable determined by a neural network, the timings for each phoneme in the syllable were computed by solving Eqn. 18 for k:

$$\Delta = \sum_{i=1}^{n} \exp\left(\mu_i + k\sigma_i\right) \tag{18}$$

where Δ is the duration determined for the whole syllable and n is the number of segments in that syllable. Segment (i) is assigned the duration $\exp(\mu(i) + k\sigma(i))$. Because of the exponential in the equation, it cannot be simply solved and an iterative approach is adopted. The factor k is initially assumed to be zero valued and then grown in steps of 0.1 positive or negative until the resulting segment durations sum to fit the syllable duration. The final value of k is further reduced towards zero by a 'backstep' parameter (typically set to about 0.75, giving a granularity in the range of two to six milliseconds per segment) so that usually long segments forced into a short syllable duration will lengthen it, and usually short ones in a long syllable need not fully fill it. This 'backstep' performs an important function by compensating for the missing phone-type information and simulates a degree of interaction between the two levels. Because segments in phrase-final position are not lengthened in the same way as segments in a stressed phrase-internal syllable, a form of pre-boundary lengthening was provisionally implemented with the use of a decay variable δ to reduce the effect of any lengthening on segments further from the boundary by lengthening proportionally more towards the end of the final syllable. Thus in a final syllable, the *i*th segment was assigned a duration of $\exp\left(\mu_i + \delta^{(n-i)} k\sigma_i\right)$, where $\delta \approx 0.75$ and n the number of phones in the syllable.

5.1 Improved modelling of segmental distributions

The hierarchical approach to the modelling of speech timing described here represents both a simplification of duration prediction and an integration of the differing standpoints of existing theories. It separates prosodic lengthening effects from phonetic effects and allows statistical modelling of the durational characteristics both at the segmental and at the syllable levels. It is also capable of being extended further upwards, given sufficient training data, to allow similar modelling of the durational characteristics at the levels of the foot, the phrase, and the utterance.

The strong form of the accommodation principle is clearly too rigid an assumption though, since we see qualitatively different kinds of lengthening under the different conditions of stress, tone-group-final position, and speaking rate or rhythmic compression. At the syllable level, we can distinguish three different types of lengthening context: emphatic lengthening (type 1), pre-boundary lengthening (type 2), and rate-related lengthening (type 3). The first is governed by force of articulation, the second, a perceptually different form of lengthening marking tone-group or prosodic boundaries by decay of articulation, and the third a combination of rhythmic and semantic influences, such as the number of syllables in the foot and the given/new or function/content role of the parent word of the syllable. Since each type of lengthening has different acoustic correlates, it is preferable to consider them separately and model their effects independently. The first exerts its effects primarily on the early part of the syllable, with more aspiration in the onset and steeper spectral tilt across the vowel; the second is more noticeable in its effects towards the end of the syllable and can be accompanied by vocal fry and rapid drop in amplitude. Rhythmical timing control on the other hand is distinguished by predominently shortening the syllable, as opposed to the lengthening effects of the other two types; it can be considered uniform on all segments in the syllable (Campbell and Isard 1991). Control of speaking rate at levels above the syllable is complex, and almost certainly requires a reorganisation of the phonological structure of the utterance as well as durational modifications; it will not be considered further here.

Campbell (1993b) made detailed extensions to the two-level prediction model that took into account the different types of lengthening and also learnt the selective modifications required by coarticulatory and positional effects. A major weakness of the commonly used phonemic labelling is that it fails to differentiate between allophonic variants; in the case of /l/ (Sproat and Fujimura 1993) for example, a difference in onset or coda position can result in very different articulatory characteristics, and phones in clusters (like the /p/ in /spr/) have very different characteristics and durations when they are in isolation (if they can be considered equivalent at all). Position in the syllable and position in part of the syllable are contexts that have strong moderating effects on the speech sounds, and the durational rules must take these into account. Accordingly, it was proposed that the effects of allophonic variation and prosodic context could be modelled as a difference in the probability distribution function (pdf) of the segmental durations (Campbell 1993b). In this way, a similar accommodation computation could be performed but using context-specific pdf's instead of simple log Gaussians. Using broad-class categorisation of the phones, representing place and manner separately, and with additional dimensions to represent the prosodic contexts of stress and phrase-finality, a stochastic learning technique was employed to generalise across the features to predict the shape and scale values of a two-parameter Gamma distribution that best modelled the expected distributions for each individual phone in a given syllable/context combination. By constraining the network to generalise across the broad classes to produce a suitable value for each of the parameters we force it to learn the effects of each, and thus overcome the problem of insufficient training samples in the database by maximising the number of tokens for each training type.

Whereas the log transform brings the pdf of many phones' durations close to a Gaussian, it is only an approximation for general use and is not appropriate for allophonic variants; the Gamma distribution (see for example Levinson 1986), which does not presuppose symmetry about the mean, is better for this. The one-parameter Gamma distribution permits modelling of the 'shape' or skew of a distribution, and the introduction of a second parameter, a scaling factor, enables fitting to any range of values having that distribution shape. The bivariate Gamma probability distribution function with parameters for shape and scale can be defined as

$$\Gamma(x|p,s) = \frac{s^{-p} x^{p-1} e^{-\frac{x}{s}}}{\Gamma(p)} \tag{19}$$

where $\Gamma(p)$ is the Gamma function, p is the shape parameter, and s is the scale parameter. For high values of p the Gamma distribution closely approximates a normal distribution but exhibits increasing skew for smaller values. Campbell (1992e) gives details of the maximum likelihood estimation of these parameters. A comparison of the distributions of five speakers' English segmental durations (43 phone types, approx 200 tokens of each per speaker) with randomly-generated values from Gaussian, log and Gamma pdf's confirmed that the Gamma distribution provides a closer fit to the observed data over an average of ten trials (Campbell 1992b). Students' t tests showed the differences between the means of correlations over the ten trials to be significant, the log showing an improvement on the normal $(t_{428} = 3.9)$, and the Gamma showing a similar improvement on the log $(t_{428} = 3.2)$. Analysis of variance for speaker-specific characteristics showed no significant differences $(F_{4,210} = 1.567)$ so the mean across all speakers was taken as representative for each phoneme type. Examination of the individual results suggests that the normal distribution is failing to model the distributions of sonorant consonants and vowels, while the log distribution is failing to model the stop consonants. Significant effects were found for type of phoneme by analysis of variance in both log and Gamma distributions $(F_{47,167} = 5.89$, and 6.16 respectively), but differentiating lengthening of vowels and consonants did not show a significant effect $(F_{1,213} = 0.481$, n.s.). This is in contrast to the traditional view since in raw millisecond terms, vowels and consonants appear considerably different in their lengthening characteristics. Separation into broad phonemic class yielded significant differences at the 0.05 level with the normal ($F_{6,208} = 5.922$) and log-normal distributions ($F_{6,208} = 5.525$) but this effect was below the level of significance for the Gamma distribution ($F_{6,208} = 3.131$, n.s.), which can therefore be preferred.

To learn the shape and scale parameter values that best fitted the observed durations for phones in various prosodic and syllabic contexts, a vector of training data was prepared from the averaged aligned segmental durations of four speakers reading 200 phonemically-balanced sentences. The vector was produced by exhaustive calculation of the Gamma parameters describing the distributions of all phone types and combinations of stress, boundary, onset position, and clustering, for which the number of tokens in the database was greater than 10. The resulting scale and shape parameters were presented to a three-layer back-propagation network together with the feature vectors describing the phones and their contexts so that typical values for each phone-context combination could be learned. The problem of predicting a duration for the appropriate context is thus reduced to the sub-tasks of a) selecting the most appropriate set of (n = 32) quantised shape distributions, b) scaling it by a value estimated for that subcategory of the phone's probable distributions, and c) choosing a quantile of that scaled distribution that best represents the expected duration of the phone in that syllable context. After training, the network successfully predicted 80% of the variance in the shape parameter and 68% of the variance in the scale, modelling 75% of the variance in the segmental durations. Differentiating predicted durations from observed showed that 50% of the predictions were within 11 msec of the original, and 80% were within 21 msec. These results compare favourably with the segmental durations predicted by the simple log transform without classification, which showed less control of the variance, with 80% of the durations being only within about 40 msec. The combined model well matches the discrepancy observed in Table 3 where syllable durations between speakers showed a very high correlation (r = 0.92, almost 85% agreement) but aligned segment durations for the same utterances only matched about 60% (r = 0.76). Allowing backstep, to model the bottom-up influence of articulatory differences in segmental composition of the syllable, improved prediction at the higher level by about 5% (equivalent to raising the correlation coefficient from 0.8 to 0.9).

6. APPLICATIONS OF DIFFERENTIAL LENGTHENING

Advanced techniques exist for the automatic and semi-automatic segmentation of speech waveforms to produce a phonemic labelling of an utterance but similar techniques are not so well established for the automatic labelling of prosodic events in a speech signal. Campbell (1993a) showed with multi-speaker data of British English that significant information regarding the prosodic segmentation of an utterance can be achieved using z-score normalised duration contours from which a prominence index for boundary location, stress detection, and hierarchical ordering of focus can be obtained. Because there is a high degree of inter-speaker agreement in these contours, the events thus marked are likely to correspond to meaningful linguistic events in the speech.

6.1 Stress and focus

Durational lengthening serves as a cue to stress and focus marking in speech. Using a corpus of conference-registration dialogues that had been recorded to illustrate contrastive focus in three different speaking styles, ranging from read speech to almost spontaneous, we attempted to determine which part of each utterance had been assigned focus (Campbell 1992d). The test data consisted of syntactically and semantically identical word-strings that differed in the amount of emphasis given to each word in three or four different interpretations. Placing of emphasis was controlled by capitalisation in read texts and by (deliberate) misinterpretation to force the use of contrastive stress in dialogue.

Energy in the speech waveform was also used to assist in the location of emphasised areas of the speech by disambiguating the two types of lengthening. Since there is a strong correlation between lengthening for emphasis and increased energy in the emphasised region, and since there is also a strong correlation between boundary-related lengthening and a *decrease* in energy as the utterance trails off into the boundary, then the energy contour provides a useful layer of information. Because declination was observed in the energy contour throughout the shorter utterances, the residuals of a linear regression of the smoothed data were used. The raw energy values were first z-score normalised by phone type as was done for duration. Since the normalised energy values are scaled in the same range as those for duration, the two measures can be simply combined. Adding the energy measure to that of duration increases the strength of the values that are related to emphasis (typically having the same sign), while decreasing those associated with proximity to a boundary (duration typically positive, energy negative).

Figure 6 illustrates the combination of these two measures. A test with data of three speaking styles (sentences read in sequence, identical sentences produced in random sequence, and identical sentences elicited in conversation) showed that we were able to correctly identify more than 80% of the stressed syllables, with a false-insertion rate of less than 5% using lengthening and energy information combined, and to rank them in order of prominence to correctly identify the focussed phrase in more than 75% of the utterances. Analysis of the errors in this test showed that there were cases of sentence-medial phrase-final lengthening that were being incorrectly tagged as stressed because the energy level remained high (Campbell 1992d).

6.2 Prominences and boundaries

Whereas duration can be a fairly reliable cue to prominence, power is not. A speaker can vary the distance from microphone to mouth and the power in the recording will change accordingly. Power can also vary according to a range of external conditions that have nothing to do with the meaning carried in the speech signal. By using the onset/coda differential to distinguish lengthening due to stress or prominence from that due to finality or boundary-marking we can distinguish between the two types of lengthening in read speech from durational cues alone. Since boundary lengthening affects coda segments more strongly, and stress lengthening is strongest on onset segments, the 'slope' of lengthening through a syllable can therefore be used to distinguish between the two cases. Figure 7 shows the original differences in syllable lengthening according to prosodic context, the prediction results after factoring out syllable duration differences, and the residual from the prediction showing different slopes of lengthening for onset, peak, and coda after the overall syllable duration has been factored out.

To apply this differential, we compared the z-score of each phone with that of its neighbours to determine if the lengthening is increasing throughout the syllable (final case) or decreasing (stressed case). This 'slope' differential was used to filter out stress-lengthened syllables and indicate potential prosodic boundaries in the 200 corpus sentences for four speakers. Figure 8 plots example results from three sentences. We can see that words that are

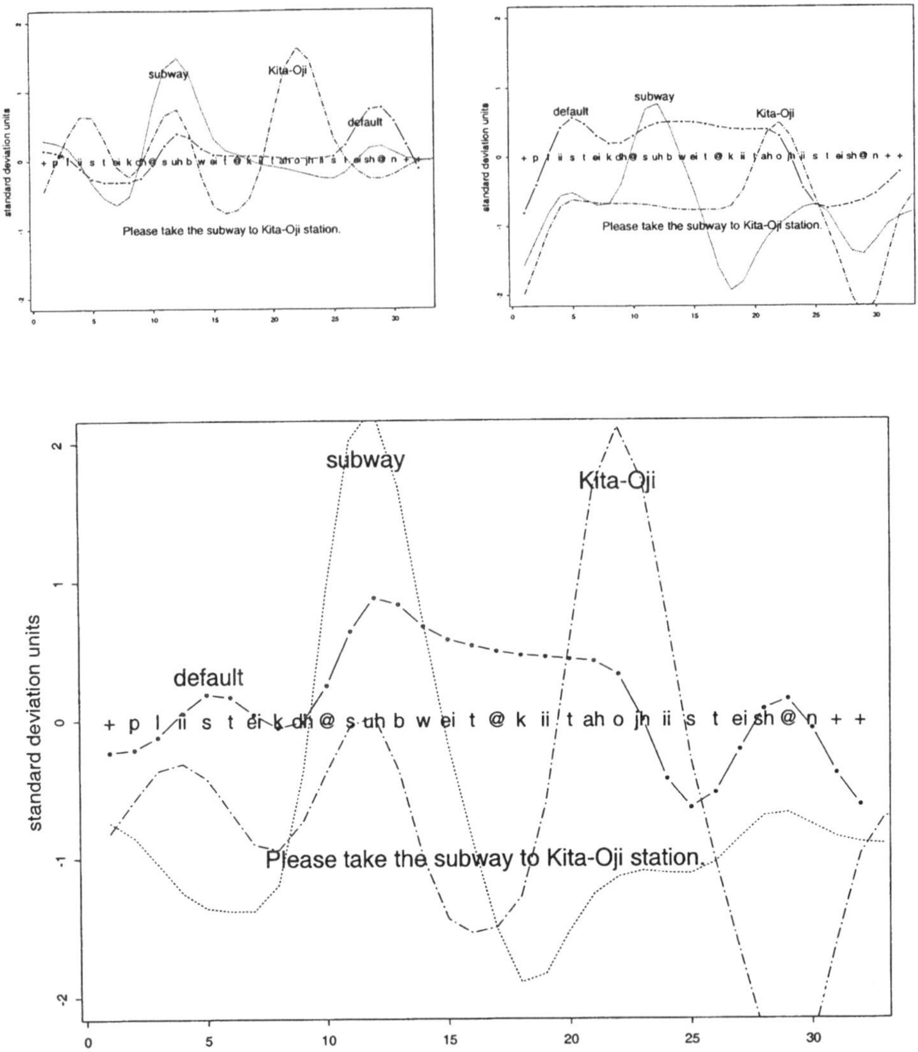

Figure 6. Three versions of the same sentence produced with focus on different words to convey different interpretations of the same word sequence. Plots of normalised and smoothed duration (top left) show prominence in the utterance but do not distinguish between segments lengthened for stress and those lengthened phrase-finally. Energy (top right) is high throughout the default reading, but drops quite sharply once the focal point has been reached. t starts lower in the marked readings. Energy shows focus well, but not which other words or parts of the utterance have been given prominence. When the two measures are combined (bottom), (here they have been added), one can see both the stress pattern and the focal prominence.

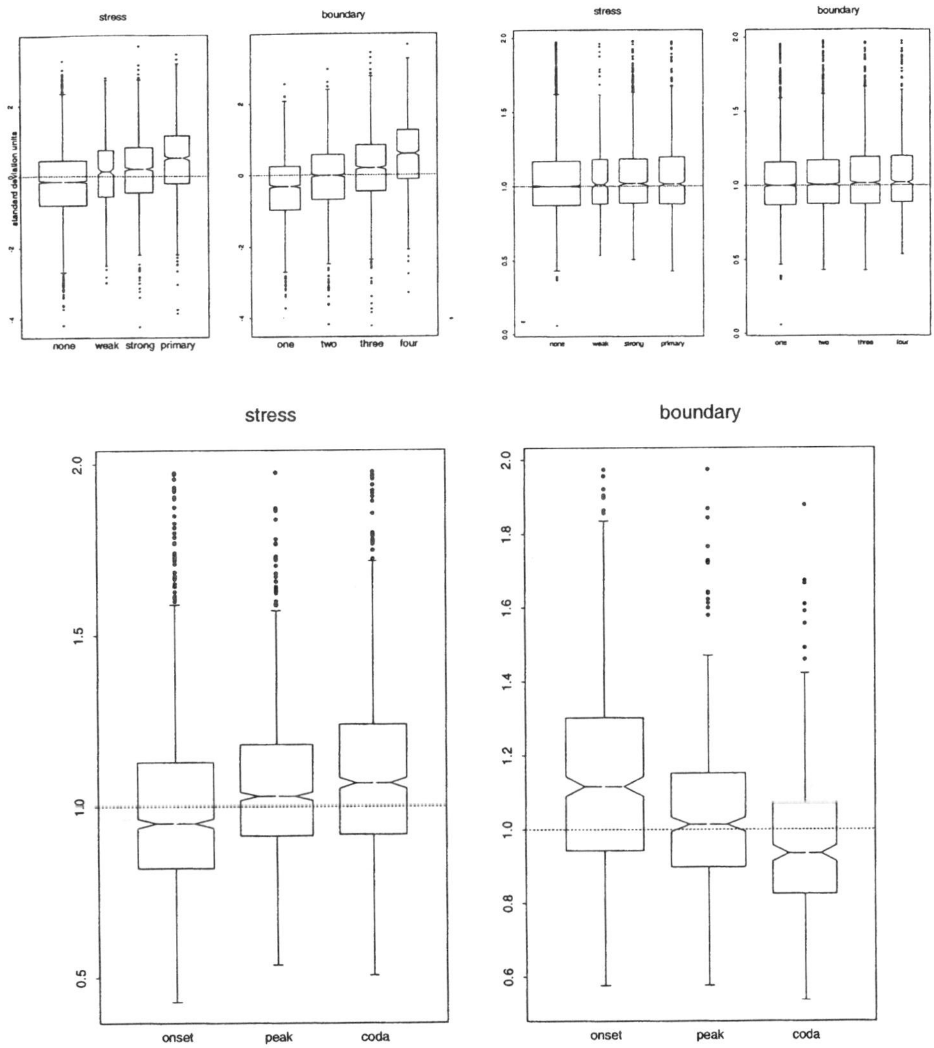

Figure 7. Top left: Four levels of stress (stress) *and four degrees of final lengthening* (boundary) *were marked in the database. The figure shows z-scores for syllables in each position. Top right: Prediction of segmental durations on the basis of syllable durations factors out most of the differences arising from contextual effects. The plot shows error* (predicted/observed) *from a prediction assuming strong elasticity. Bottom: Prediction error factored according to position within the syllable shows onset segments to be overpredicted in final syllables, and coda segments to be overpredicted in stressed syllables.*

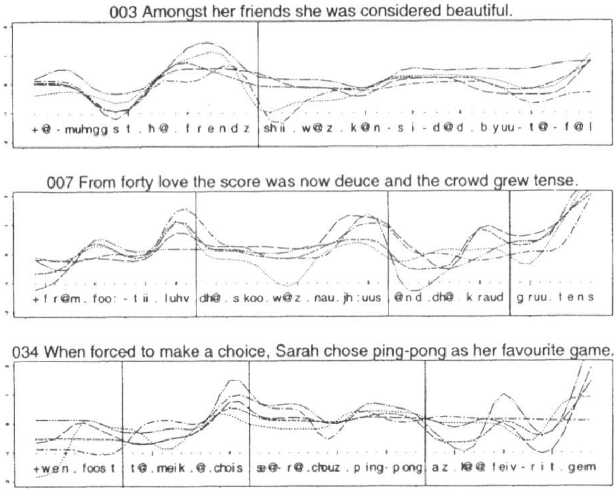

Figure 8. Plots of normalised segmental durations for three sentences and four speakers, showing a high degree of similarity in the contours. Vertical lines indicate prosodic phrase boundaries determined by the algorithm. (Raw scores have been smoothed for these plots to allow clearer comparison).

lengthened by stress as in *favourite* (bottom example) are not marked whereas those that mark the end of a prosodic unit are identified.

There is a high degree of agreement between speakers with respect to boundaries detected (73%) but speaker-specific variation reveals individual interpretations of the linguistic structures. The discussion in Campbell (1993a) focusses on the differences between speakers in placing prosodic phrase boundaries for prepositional phrases: some accordance with the syntactic boundary, others prefer a rhythmic placement, closing the previous foot and anticipating a stress on the following word or phrase.

6.3 Speech act type

More recently, we have been applying lengthening differentials to determine speech act type as a first step towards the automatic labelling of communicative function in Japanese and English dialogues (Fujio et al. 1995). There is a subset of questions that occur commonly in dialogues, such as ‘Is that so?’ and ‘Really?’ that can serve as fillers to give the speaker time to think before a more considered response. In a machine-mediated dialogue, it is essential to know whether these are functioning as real questions and so need a reply, or whether they are functioning as fillers, in which case the system should wait quietly rather than respond directly. In spoken interaction many

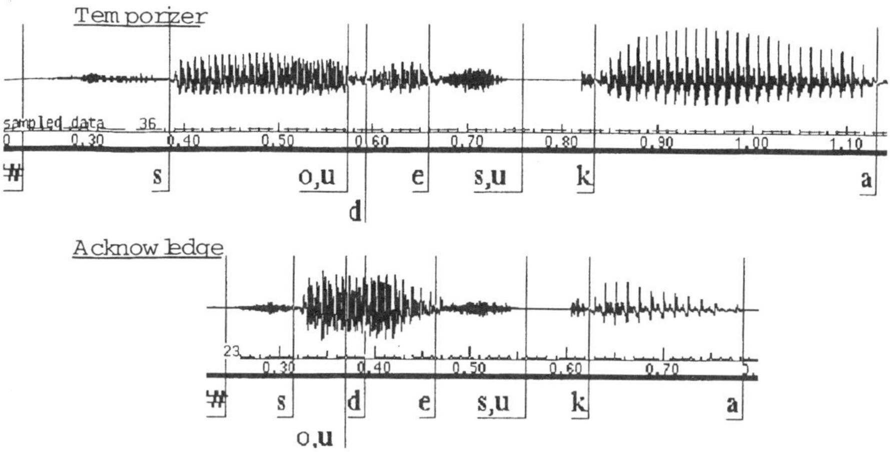

Figure 9. Two speech waves showing the difference in timing between the same words being used as a temporiser and as a genuine question. The utterance is 'Sou desu ka?' (Is that so?)

words are similarly ambiguous as to the intended meaning, and there is a large subset of communicative acts (such as Wh-questions, yes/no-questions, confirmation-questions, confirmation-questions-to-self, acknowledgements, and temporizers) that cannot be properly differentiated without prosodic information. Although some disambiguating information is provided by changes of the fundamental frequency, much can be determined from duration alone.

We examined the dialogues in the 'Speech and Language Database for Speech Translation Research' (Morimoto et al. 1994) and selected four commonly occurring phrases for a test. The question 'Sou desu ka?' (Is that so?) and the variant 'X desu ka?' (Is that X?) and the equivalent comment 'Sou desu ne' (That's right, isn't it?) and the variant 'X desu ka?' (That's X isn't it?) were frequent utterances that could not be properly interpreted from the text alone. Figure 9 shows the differences in a labelled waveform plot, where we can see that as well as differences in overall speaking rate, which can vary from case to case, there are also differences in the ratio of the vowel durations which are more consistent. We can use the ratio of normalised vowel durations to determine speech-act type in many of the cases. Figure 10 shows boxplots of ratios between z-scores for duration in two speech-act types. Table 4 shows the percentages correctly differentiated using duration in conjunction with fundamental frequency information.

Table 4. *Results for four prosodically disambiguated phrases: percentage that could not be disambiguated from text alone* (need prosody), *speech act and percentage correctly disambiguated by prosodic information* (correct).

	need prosody	speech act	correct
sou desu ka?	82.2%	Acknowledge	80.5%
		Temporizer	91.2%
x desu ka?	76.8%	Wh/YN/Conf-question	90.0%
		Conf-question-to-self	61.8%
sou desu ne	49.0%	Acknowledge/Yes	82.0%
		Temporizer	38.0%
x desu ne	76.8%	Conf-question	90.5%
		Inform	76.7%

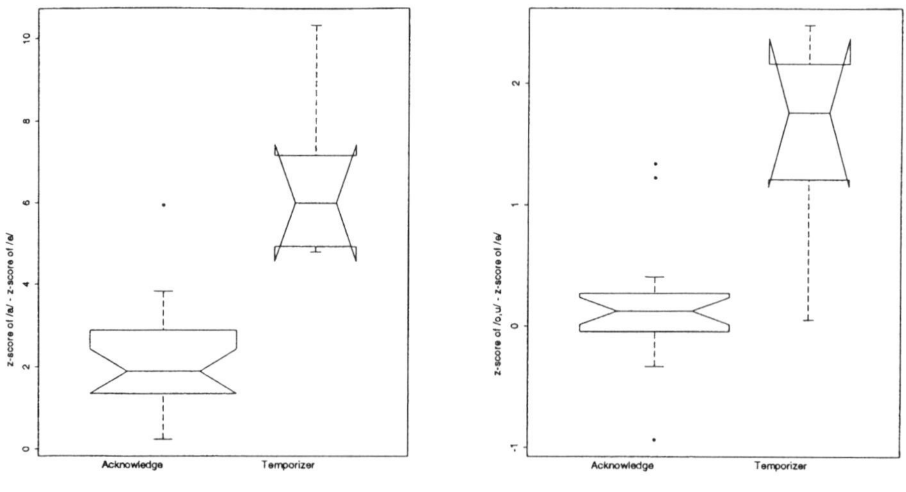

Figure 10. Analysis results for timing differentials with respect to speech-act type.

7. CONCLUSION

This paper has attempted to present an overview of the main theoretical and experimental developments of phonetic research related to speech timing from the viewpoint of practical applications in speech technology. The theoretical bias of the paper is towards a multi-tier model of speech timing that integrates physiologically-motivated 'inherent' durational characteristics into a cognitively-motivated prosodic structure at the level of the syllable. Data from both English and Japanese were presented as evidence for a higher-level of speech timing control based on rhythmic structuring, and a method was shown for predicting timing characteristics by incorporating segmental constraints into a syllable-level prosodic framework. An important feature of the model presented here is the introduction of backstep as a means to allow the bottom-up influence from differences in segmental composition while limiting prediction factors to the small number having an effect at the level of the syllable. Having shown that a good prediction of timing can be produced with limited training data for speech synthesis, we then showed how the same model can also be used for speech analysis and how slopes of lengthening within the syllable can reveal the structure of an utterance for automatic prosodic analysis.

A separate point made in several places throughout the paper is that data from different speaking styles show very different characteristics and vary even more than data taken from different speakers using a constrained speaking style. Data measured from carefully prepared laboratory speech will not necessarily be representative of natural speech, and a model that fits the durations of speaker A will not necessarily be able to match those of speaker B. In attempting to model durational characteristics we face a dilemma: high correlations can only be achieved with constrained data, and if we are to model a wide variety of styles and speakers then we need to introduce flexibility at some levels in order to find the regularity at others. The syllable durations of our English speakers matched much more closely than their segmental durations; and one can suppose a level where, for example, their phrasal durations might match much more closely than their syllable durations. This is certainly so in the case of singing for example, when different choruses have to fit the same timing framework. In order to begin to model the higher levels, we must filter out the predictable effects of the lower ones, and when a mechanical feature such as jaw movement or tongue height has an effect on the durations, then it should be modelled separately as a filter and need not be incorporated further. Similarly with the structural effects shown above; if we are to begin to model meaning through prosody, then we should not incorporate predictable elements such as position in the utterance or part-of-speech or number of syllables in the foot as part of our model – they can be applied as filters much as z-score normalisation was applied to cope with the articulatory effects, and we should work with their residuals.

The same sequence of words in the same context can have different interpretations in different utterances, and it is the structuring of information through the grouping of words and phrases with differential prominence that allows the listener to form a meaningful interpretation of what was intended from what has been said. Duration provides a clue to that structuring. We have made a start in using these cues to assist in the machine understanding of speech and in the automatic labelling of prosodic events in large corpora so that more data will be available for the further analysis of higher-level features.

REFERENCES

Abercrombie, D. 1964. Syllable quantity and enclitics in English. In D. Abercrombie, D. Fry, D. MacCarthy, N. Scott, and J. Trim (eds), *In Honour of Daniel Jones*. London: Longman, 216-222.

Abercrombie, D. 1965. *Studies in Phonetics and Linguistics*. Oxford: Oxford University Press.

Allen, G.D. 1972. The location of rhythmic stress beats in English: an experimental study. *Language and Speech* 15, 72-100 and 179-195.

Allen, G.D. 1975. Speech rhythm: Its relation to performance universals and articulatory timing. *Journal of Phonetics* 3, 75-86.

Allen, J., Hunnicutt, S. and Klatt, D. 1987. *From Text to Speech: the MITalk System*. Cambridge: Cambridge University Press.

Barlow, R.E., Bremmer, J.M. and Brunk, H.D. 1972. *Statistical Inference under Order Restrictions*. New York: Wiley.

Barnwell, T.P. 1971. An algorithm for segmental durations in a reading machine context. *Technical Report* 479, MIT, Research Laboratory of Electronics.

Bolinger, D. 1972. Accent is predictable (if you're a mind reader). *Language* 48, 633-644.

Breiman, L., Friedman, J.H., Olshen, R.A. and Stone, C.J. 1984. *Classification and Regression Trees*. Monterey, CA.: Wadsworth and Brooks/Cole Advanced Books and Software.

Campbell, W.N. 1988. Extracting speech-rate values from a real-speech database. *Proc. ICASSP-88* (New York), 683-686.

Campbell, W.N. 1992a. Syllable-based segmental duration. In G. Bailly and C. Benôit (eds), *Talking Machines*. Amsterdam: North-Holland, 211-224.

Campbell, W.N. 1992b. Gamma modelling of English segmental durations. *Journal of the Acoustical Society of Japan*, 241-242.

Campbell, W.N. 1992c. *Multi-Level Timing in Speech*. Ph.D. thesis, University of Sussex (Exp. Psychol.).

Campbell, W.N. 1992d. Prosodic encoding of English speech. *Proc. International Conference on Spoken Language Processing* (Banff), 663-666.

Campbell, W.N. 1992e. Prosodic segmentation of recorded speech. *PERILUS* XIV (Dept. of Linguistics and Phonetics, Stockholm University), 101-106.

Campbell, W.N. 1993a. Automatic detection of prosodic boundaries in speech. *Speech Communication* 13, 343-354.

Campbell, W.N. 1993b. Predicting syllable-level durations for accommodation within a syllable-level framework. *Proc. Eurospeech 93* (Berlin), 1081-1084.

Campbell, W.N. and Isard, S.D. 1991. Segment durations in a syllable frame. *Journal of Phonetics* 19, 37-47.

Carlson, R. and Granström, B. 1973. Word accent, emphatic stress, and syntax in a synthesis-by-rule scheme for Swedish. *Technical Report STL-QPSR* 2, 3, Royal Institute of Technology, Stockholm.

Carlson, R. and Granström, B. 1976. A text-to-speech system based entirely on rules. *Proc. International Conference on Acoustics, Speech, and Signal Processing*, 686-688.

Chomsky, N. and Halle, M. 1968. *The Sound Pattern of English*. New York: Harper and Row.

Church, K.W. and Gale, W.A. 1990. A comparison of the enhanced Good-Turing and deleted estimation methods for estimating probabilities of English bigrams. *Computer Speech and Language* 5, 19-54.

Coker, C., Umeda, N. and Browman, C. 1973. Automatic synthesis from ordinary English text. *IEEE Trans. Audio Electroacoust.* AU-21, 293-297.

Coleman, J.S. 1992. The phonetic interpretation of headed phonological structures containing overlapping constituents. *Phonology* 9, 1-44.

Collier, R. 1992. A comment on the prediction of prosody. In G. Bailly and C. Benôit (eds.), *Talking Machines*. Amsterdam: North-Holland, 205-208.

Cooper, W.A., Paccia, J.M. and Lapointe, S.G. 1978. Hierarchical coding in speech timing. *Cognitve Psychology* 10, 154-177.

Crystal, T.H. and House, A.S. 1982. Segmental durations in connected speech signals: Preliminary results. *Journal of the Acoustical Society of America* 72, 705-716.

Crystal, T.H. and House, A.S. 1986. Characterisation and modelling of speech-segment durations. *IEEE International Conference on Acoustics, Speech, and Signal Processing*, vol. 51.11.4, 2791-2794.

Crystal, T.H. and House, A.S. 1988. Segmental durations in connected-speech signals: Current results. *Journal of the Acoustical Society of America* 83, 1553-1573.

Crystal, T.H. and House, A.S. 1990. Articulation rate and the duration of syllables and stress groups in connected speech. *Journal of the Acoustical Society of America* 88, 101-112.

Entropic Research Laboratory, 600 Pennsylvania Avenue, Washington DC 20003. *HTK - Hidden Markov Model Toolkit*, 1993.

Fisher, W., Zue, V., Bernstein, J. and Pallett, D. 1987. An acoustic-phonetic data base. *Proc. 113th Meeting of the Acoustical Society of America*, May 1987.

Fujio, S., Campbell, W.N. and Higuchi, N. 1995. Identification of speech acts using prosody. *Proc. Acoustical Society of Japan* 1-1-1, 199-200.

Gaitenby, J. 1965. The elastic word. *Status Report on Speech Research* SR-2 (Haskins Laboratories, New Haven, CT), 3.1-3.12.

Gaitenby, J.H., Sholes, G.N. and Kuhn, G.M. 1972. Word and phrase stress by rule for a reading machine. *Status Report on Speech Research* SR-29/30 (Haskins Laboratories, New Haven, CT), 105-110.

Goldman-Eisler, F. 1968. *Psycholinguistics: Experiments in Spontaneous Speech*. New York: Academic Press.

Halliday, M.A.K. 1967. *Intonation and Grammar in British English*. The Hague: Mouton.

Hieronymus, J.L. Unpublished Report for British Telecom.

House, A. 1961. On vowel duration in English. *Journal of the Acoustical Society of America* 33, 174-1178.

House, A. and Fairbanks, G. 1953. The influence of consonant environment upon the secondary acoustical characteristics of vowels. *Journal of the Acoustical Society of America* 25, 105-113.

House, S. and Crystal, H. 1988. Segmental durations in connected speech signals: Syllabic stress. *Journal of the Acoustical Society of America* 83, 1574-1585.

Huggins, A.W.F. 1968. The perception of timing in natural speech: compensation within the syllable. *Language and Speech* 11, 1-11.

Huggins, A.W.F. 1972. Just noticeable differences for segment duration in natural speech. *Journal of the Acoustical Society of America* 4, 1270-1278.

Klatt, D.H. 1973. Voice-onset time, frication and aspiration in word-initial clusters. *Quarterly Progress Report, Research Laboratory of Electronics*, MIT 109, 124-136.

Klatt, D.H. 1974. The duration of [s] in English words. *Journal of Speech and Hearing Research* 17, 51-63.

Klatt, D.H. 1975. Vowel lengthening is syntactically determined in a connected discourse. *Journal of Phonetics* 3, 129-140.

Klatt, D.H. 1976. Linguistic uses of segmental duration in English: Acoustic and perceptual evidence. *Journal of the Acoustical Society of America* 59, 1208-1221.

Klatt, D.H. 1979. Synthesis by rule of segmental durations in English sentences. In B. Lindblom and S. Öhman (eds), *Frontiers of Speech Communications Research*. New York: Academic Press, 287-299.

Klatt, D.H. 1987. Review of text-to-speech conversion for English. *Journal of the Acoustical Society of America* 82, 737-793.

Kohler, K.J. 1986. Invariability and variability in speech timing: from utterance to segment in German. In J. Perkell and D.H. Klatt (eds), *Invariance and Variability in Speech Processes*. Hillsdale: Erlbaum, 268-298.

Kozhevnikov, V.A. and Chistovich, L.A. 1965. *Speech: Articulation and Perception*, vol. 30. Joint Publications Research Service, U.S. Department of Commerce.

Krantz, D.H., Luce, R.D., Suppes, P. and Tversky, A. 1971. *Foundations of Measurement*, Vol.1. New York: Wiley.

Lawrence, W. 1974. The phoneme, the syllable, and the parameter track. In G. Fant (ed), *Speech Communication*. Stockholm: Almqvist & Wiksell, vol. 2, 173-178.

Lehiste, I. 1973. Phonetic disambiguation of syntactic ambiguity. *Glossa* 7, 107-121.

Lehiste, I. 1975. Some factors affecting the duration of syllabic nuclei in English. In G. Drachman (ed), *Proc. First Saltzburg Conference on Linguistics*. Heidelberg: G.Narr, 81-104.

Lehiste, I., Olive, J. and Streeter, L. 1976. The role of duration in disambiguating syntactically ambiguous sentences. *Journal of the Acoustical Society of America* 60, 1199-1202.

Levinson, S.E. 1986. Continuously-variable duration hidden Markov models for automatic speech recognition. *Computer Speech and Language* 1, 29-45.

Lindblom, B. 1974. Chairman's comments on Session 2: Speech Production and Synthesis by Rules. In G. Fant (ed), *Speech Communication*. Stockholm: Almqvist & Wiksell, vol. 2, ix-xxv.

Lindblom, B. and Rapp, K. 1973. Some temporal regularities of spoken Swedish. *PILUS* 21 (Papers from the Institute of Linguistics, University of Stockholm).

Lodwick, F. 1686. An essay towards an universal alphabet. *Philosophical Transactions* of the Royal Society 16(182), ± 137.

Luce, P.A. and Charles-Luce, J. 1985. Contextual effects on vowel duration, and the consonant/vowel ratio in speech production. *Journal of the Acoustical Society of America* 78, 1949-1957.

Lyberg, B. 1977. Some observations on the timing of Swedish utterances. *Journal of Phonetics* 5, 49-59.

Martin, J.G. 1972. Rhythmic (hierarchical) versus serial structures in speech and other behaviours. *Psychological Review* 79, 487-509.

Martin, J.G. 1975. Rhythmic expectancy in continuous-speech perception. In A. Cohen and S. G. Nooteboom (eds), *Structure and Process in Speech Perception*. Berlin: Springer, 161-177.

Martin, J.G. 1979. Rhythmic and segmental perception are not independent. *Journal of the Acoustical Society of America* 65, 1286-1296.

Mattingly, I. 1968. Synthesis-by-rule of General American English. *Supplement to Status Report on Speech Research* (Haskins Laboratories, New Haven, CT), 1-223.

Morimoto, T., Uratani, N., Takezawa, T., Furuse, O., Sobashima, Y., Iida, H., Nakamura, A., Sagisaki, Y., Higuchi, N. and Yamazaki, Y. 1994. A speech and language database for speech translation research. *Proc. International Conference on Spoken Language Processing* (Yokohama), vol. 4, 1791-1794.

Oller, D. 1973. The effect of position in utterance on speech segment duration in English. *Journal of the Acoustical Society of America* 54, 1235-1247.

O'Malley, M.H., Kloker, D.R. and Dara-Abrams, B. 1973. Recovering parentheses from spoken algebraic . expressions. *IEEE transactions on Audio and Electroacoustics* AU-21, 217-220.

Peterson, G.E. and Lehiste, I. 1960. Duration of syllable nuclei in English. *Journal of the Acoustical Society of America* 32, 693-703.

Pierrehumbert, J. 1980. *The Phonology and Phonetics of English Intonation*. Ph.D. thesis, Massachusetts Institute of Technology.

Pitrelli, J. 1990. *Hierarchical Modeling of Phoneme Duration Applied to Speech Recognition*. Ph.D. thesis, MIT.

Riley, M. 1990. Statistical tree-based modeling of phonetic segmental durations. *Journal of the Acoustical Society of America* 85(S44).

Riley, M. 1992. Tree-based modeling of segmental durations. In G. Bailly and C. Benôit (eds), *Talking Machines*. Amsterdam: North-Holland, 265-273.

Selkirk, E. 1984. *Phonology and Syntax: the Relation between Sound and Structure*. Cambridge, MA: MIT Press.

Sproat, R. and Fujimura, O. 1993. Allophonic variation in English /l/ and its implications for phonetic implementation. *Journal of Phonetics* 21, 291-311.

Umeda, N. 1975. Vowel duration in American English. *Journal of the Acoustical Society of America* 58, 434-445.

Umeda, N. 1977. Consonant duration in American English. *Journal of the Acoustical Society of America* 61, 846-858.

Umeda, N. and Coker, C. 1971. Some prosodic details of American English. *Journal of the Acoustical Society of America* 49, 123 (A).

Van Santen, J.P.H. 1992. Deriving text-to-speech durations from natural speech. In G. Bailly and C. Benôit (eds), *Talking Machines*. Amsterdam: North-Holland, 275-285.

Witten, I.H. 1977. A flexible scheme for assigning timing and pitch to synthetic speech. *Language and Speech* 20, 240-260.

Young, S.D. and Fallside, F. 1980. Synthesis-by-rule of prosodic features in word concatenation systems. *International Journal of Man-Machine Studies* 12, 241-258.

ATR ITRL, Kyoto, Japan

JULIA HIRSCHBERG

A CORPUS-BASED APPROACH TO THE STUDY OF SPEAKING STYLE

1. INTRODUCTION

A corpus-based approach to the study of speaking style seeks to infer perceptually salient differences in "style" from observed differences within corpora across different conditions or between corpora produced under different speaking situations. When reliable differences are found across speakers in such corpora, we hypothesize that such differences may in fact be such that, were humans to be presented with material differing in just such ways, they might be lead to distinguish among speech that is "spontaneous" or "planned" or "conversational" or "formal". In such an approach, the question of what features distinguish one style from another is addressed by examining similar corpora for systematic differences in lexical choice, syntactic constructions, or acoustic/prosodic phenomena.

However, the very notion of speaking style is a nebulous one. Recent proposals have suggested that the traditional "read/spontaneous" distinction should be abandoned, in favor of a new characterization of speech type along multiple dimensions, making reference to the conversational participants, their number and relationship, as well as the requirements of the communicative situation itself. Such dimensions have been characterized by Eskanazi (1993) as the degree of intelligibility required by the situation, the familiarity between speaker and hearer(s), and the social status of the communicative participants. Barry (1995) proposed to extend these dimensions by adding a "listener dimension", the effect the speaker intends to have on the hearer(s), or, roughly, the communicative task at hand.

The ideal set of variables for a general investigation of the differences in speech type seem still to be determined, but these proposals are moving the debate into fruitful territory. While the corpora examined below might easily be categorized as "read" vs. "spontaneous" speech, it will also be useful to try to characterize them along speaker, hearer, and situational dimensions.

The acoustic/linguistic consequences of these variations in speech type or situation are also open to considerable debate. However, voice quality, segmental production (reduction or hyper-articulation), the presence or absence of speech disfluencies, and timing, accent prominence, intonational

M. Horne (ed.), Prosody: Theory and Experiment, 335–350.

contour, and pitch range have all been observed to vary under different speech situations (Bruce 1995, Ayers 1992, Barry 1995, Kohler 1995, Blaauw 1992, Touati 1995). This paper will investigate the importance of speech disfluencies, speech rate and pausal duration, intonational contour, and pitch range as predictors of speech type. The data will be drawn from three American English corpus studies, all using laboratory speech varying along the spontaneous/read continuum from subjects generating speech to perform a task specified for them, to subjects reading transcriptions of their own previously elicited spontaneous speech, to subjects reading sentences prepared by others. The goal will be to test potential differences over relatively large corpora of labeled speech, and to present evidence for and against the importance of these acoustic/linguistic factors as characterizing differences in speech type, insofar as different types are encompassed by the corpora compared.

2. RATE AND CONTOUR

Two phenomena widely believed to distinguish more spontaneous or casual speech from laboratory read speech are speaking rate and choice of intonational contour. An examination of 395 read and spontaneous utterances in the ARPA *atis0* training corpus, collected at Texas Instruments for use in the DARPA Air Travel Information System spoken language system evaluation task from seventeen speakers (DARPA 1990), supports these hypotheses. These speakers interacted with a simulated voice response system to make air travel plans for several different hypothetical customers. While the speaking situation differed considerably from human/human conversation, subjects did generate their own utterances within the constraints of a fairly limited vocabulary, which they learned over the course of the task. They later returned to read transcriptions of their own speech.

A simple analysis of speaking rate in this corpus indicates considerable differences in rate for the read vs. spontaneously produced utterances, as shown in Table 1.

Of the seventeen speakers represented in the corpus, the average speaking rate (in syllables per second) was faster for read sentences than for spontaneous sentences in all cases but one (Speaker 6). For the other sixteen speakers, ratios of read to spontaneous speaking rate averages ranged from 1.04 to 1.57. Whether this difference is due primarily to the presence of pauses within spontaneous utterances or simply to an overall reduction in rate remains to be examined.

With respect to choice of contour as a differentiator between spontaneous and read speech, there are similar but less clear-cut findings. It has been claimed that yes-no questions are commonly uttered with rising intonation in spontaneous speech, but not so reliably in read speech. In a small study of yes-

Speaker	Read	Spontaneous	Read/ Spontaneous
1	4.96	3.16	1.57
2	4.70	3.61	1.30
3	3.90	3.69	1.06
4	3.46	3.32	1.04
5	4.82	3.73	1.29
6	3.46	3.73	0.93
7	5.41	4.88	1.11
8	4.56	3.64	1.25
9	4.78	3.90	1.23
10	4.69	3.41	1.38
11	4.63	3.45	1.34
12	4.20	3.82	1.10
13	4.85	4.68	1.04
14	4.54	4.08	1.11
15	4.25	3.68	1.15
16	5.43	4.28	1.27
17	4.89	3.28	1.49

Table 1: *Speaking Rate (syllables per sec.) in the atis0 Corpus*

no questions in read speech (Hirschberg 1989), inverted yes-no questions and *wh*-questions from the training and test data of the speaker independent DARPA *Resource Management* (RM) database and from the TIMIT database were sampled for purposes of comparison with standard contours for those sentence types in spontaneous speech (Price et al. 1988). While only 55% of yes-no questions in the RM database were uttered with rising intonation, over 80% of yes-no questions in the TIMIT1 sample were so characterized. Production of *wh*-questions appears similar in both databases, with only 8% of the TIMIT1 *wh*-questions and 9% of the RM *wh*-questions uttered with final rise. However, this study examined only read speech, inferring the comparison with spontaneous.

Comparing spontaneous with read utterances in the *atis0* corpus, we do find apparent differences between the two styles for a broader array of contours, as shown in Table 2, where boundary types are represented in the Pierrehumbert (1980) scheme for describing intonation. 'LL%' represents final fall, the end of a standard declarative utterance in American English. 'LH%' is 'continuation rise'. 'HL%' is the high 'plateau' ending. 'HH%' is the rise of a yes-no question contour.

	LL%	LH%	HL%	HH%	?	N
declarative						288
read	268	5	3	6	6	
spon	203	31	21	31	2	
wh-question						74
read	62	1	3	3	5	
spon	46	10	6	12	0	
yes-no-question						30
read	9	3	1	15	2	
spon	13	2	11	4	0	

Table 2: *Boundary Types of Spontaneous and Read Utterances in the atis0 Corpus by Utterance Mode ('?' is used to indicate labeler uncertainty as to boundary category)*

There is a greater tendency in spontaneous speech to utter declaratives with some form of rising intonation; while fully 93.1% of read declaratives are uttered with final fall, only 70.5% of spontaneous declaratives are so uttered. Similarly, while 83.8% of read *wh*-questions are uttered with falling contours, only 62.2% of spontaneous *wh*-questions are. Although the numbers for yes-no questions are even smaller than for *wh*-questions, the observed behavior shows distinctions between read and spontaneous speech which one might not have predicted: more spontaneous yes-no questions (43.3%) are uttered with falling intonation than read yes-no questions (30%), for utterances of the same tokens.

3. DISFLUENCY

The presence of various phenomena known collectively as *disfluencies* – hesitations, self-repairs, filled pauses – is commonly thought to distinguish spontaneous speech from read speech. However, identifying what constitutes disfluency in spontaneous speech has proven a difficult topic in itself. While intuitively we may know disfluency when we hear it, the impression of disfluent speech may be described in various ways, and mapped to a large number of distinct lexical, acoustic, and prosodic realizations.

How important a cue disfluency is in identifying speech as spontaneous or not must depend in part on how frequently disfluencies occur, a question that corpus-based studies are well-suited to addressing. Blackmer and Mitton (1991) report a rate of one disfluency per 4.6 seconds for callers to a Canadian radio talk show. Studies of large recorded speech corpora have found that approximately 9-10% of spontaneous utterances contain one particular form of disfluency, *self-repairs* (Hindle 1983, Shriberg et al. 1992, Nakatani and Hirschberg 1994). Yet even where a particular type of disfluency was the subject of study, the acoustic, prosodic, and lexical phenomena which realized the disfluency varied considerably from token to token. A case in point is the study of self-repairs.

Several multi-speaker corpus-based studies of disfluent speech conducted by Nakatani and Hirschberg (1994) sought to identify the range of acoustic and phonetic features associated with one type of disfluency, self-repairs, and compared some of these features with fluent read speech. The corpus for these studies consisted of 6414 utterances from the ARPA *atis2* database (MADCOW 1992) collected at AT&T, BBN, CMU, SRI, and TI and produced by 122 speakers. The speech situation was similar to that described in Section 2: Wizard-of-Oz studies were performed in which subjects believed themselves to be interacting with a machine to procure air travel information. The speech was collected and transcribed, and subjects later returned to record readings of their own productions. In this corpus, 346 (5.4%) utterances contained at least one repair, where repair was defined as the speaker's self-correction of one or more phonemes (up to and including sequences of words) in an utterance. The speech was labeled for pitch accents and phrase boundaries following Pierrehumbert (1980). Disfluencies had previously been categorized for an earlier study of intonational phrasing (Wang and Hirschberg 1992) as 'repair' (self-correction of lexical material), 'hesitation' ("unnatural" interruption of speech flow without any following correction of lexical material), or 'other disfluency' and labeled in the corpus.

To analyze our data we proposed a model, the *repair interval model* (RIM) model, which divided repair events into three temporal intervals and identified critical time points within the intervals (Nakatani and Hirschberg 1994). A repair consists of three contiguous intervals, a *reparandum interval*, which contains the realization of the lexical material which is to be repaired); a *disfluency interval*, which runs from the termination of the fluent speech to its resumption, and may contain any number of silences, filled pauses and *cue phrases*, such as "I meant"); and a *repair interval* (the correcting material, which is intended to 'replace' the reparandum). The end of the reparandum occurs at the end of the fluent portion of the utterance, which we identified as the *interruption site*.

In our study of this single disfluency type, we found that there were many distinct indicators of self-repairs. That is, self-repairs could be realized by speakers in a number of different ways. For example, the most reliable cue to the presence of a self-correction in an utterance in our *atis2* corpus was the presence of a speech fragment in the reparandum. But such fragments occur in only 65.1% of cases in this corpus (for 95 out of 146 *simple* reparanda, or, utterances without overlapping repair intervals.) This is similar to Bear et al.'s

(1992) report of 60% (366/607) of repairs signaled by word fragments. In addition to speech fragments, some reparanda are characterized by glottalization and coarticulation at the reparanda offset, especially those ending in fragments (see also Bear et al. 1992, and Shriberg et al. 1992). So, these too may serve as cues to listeners about speaking style. Indeed, in our corpus, we found that 30.2% of reparanda offsets were marked by what we termed *interruption glottalization*, the glottalization of reparanda offsets – either during a fragment at the offset or at the end of a full word. (Note that, although interruption glottalization was usually associated with fragments, not all fragments were glottalized; in our corpus, 62% of fragments were **not** glottalized, and 9% of glottalized reparanda offsets were **not** fragments. This phenomenon appears distinct from laryngealization (creaky voice), which often occurs at the end of prosodic phrases; the latter typically extends over several syllables at the end of an intonational phrase and is associated with a decrease in energy and low fundamental frequency (Olive, Greenwood and Coleman 1993). Interruption glottalization on the other hand tends to occur only over the interrupted syllable, and does not appear to be associated with a sustained decrease in energy and fundamental frequency.) Another feature we found to characterize the end of some reparanda intervals was the presence of coarticulatory gestures preceding silence. Sonorant endings of both fragments and non-fragments in our corpus sometimes exhibited coarticulatory effects of an acoustically unrealized subsequent phoneme. A related feature was the lack of phrase-final lengthening on the last few segments in the reparandum for many cases of repairs. More generally, both of these features represented cues to disfluency in the rhythmic structure of pre-pausal segments.

Turning now to the disfluency interval, which in our RIM model (Nakatani and Hirschberg 1994) includes all cue phrases and all filled and unfilled pauses from the offset of the reparandum to the onset of the repair, we find that these phenomena, while commonly seen as good clues to spontaneous speech in general (Levelt 1983, Blackmer and Mitton 1991, Shriberg et al. 1992, and O'Shaughnessy 1992), occur but rarely in our data. In our original corpus, such phenomena occurred in conjunction with only 9.4% (36/382) of repairs. While the presence of speech fragments noted above was a reliable clue to repair phenomena, filled pauses and cue phrases were not. However, our study of this interval did suggest differences in the timing of phrase boundaries between fluent and disfluent speech. We found that the duration of silent pauses in the disfluency interval was in fact a reliable indicator of the presence of a self-repair. Overall, silent disfluency intervals are significantly shorter than fluent pauses. So, disfluency intervals in general appear to be distinct from fluent phrase boundaries. Their presence may thus also be seen as another way for listeners to distinguish among speaking styles.

One final distinction between spontaneous utterances containing self-repairs and fluent utterances was in the location of intonational phrase boundaries. Nakatani and Hirschberg (1994) tested the hypothesis that utterances containing self-repairs differ from fluent speech in terms of the location of intonational phrase boundaries, for a subset of the *atis* corpus, that collected by Texas Instruments, the TI subset. We first looked at whether phrase boundaries occurring at the repair offset differed in their pattern of occurrence from those occurring in fluent speech for the TI corpus as a whole (Wang and Hirschberg 1992). Using Wang and Hirschberg's phrase prediction procedure, with prediction trained on 478 sentences of read, fluent speech from the *atis* TI read corpus, we proceeded to investigate whether the phrasing at the repair offset in the 63 TI spontaneous utterances containing self-repairs was predictably distinct from this model of fluent phrasing. To do this, we compared the phrasing of repair utterances with the phrasing predicted for the corresponding corrected version of the utterance as identified by transcribers. For 40% of these repairs, an observed boundary occurs at the repair offset where one is predicted in fluent speech; for 33%, no boundary is observed where none is predicted; and for the remaining 27%, observed phrasing diverges from that predicted by a fluent phrasing model. In 37% of these latter cases, a boundary occurs where none is predicted, and, in the remainder, no boundary occurs when one is predicted. So, in just over a quarter of the cases, the phrasing of disfluent sentences differs from a model developed from fluent speech.

These findings comparing disfluent to fluent speech indicate that there are a large number of acoustic indicators of disfluent speech, including the presence of word fragments, distinctive glottalization and coarticulatory phenomena (whether unusually present or absent), rhythmic distinctions around the reparandum offset, the presence of filled pauses and disfluency-associated cue phrases, timing anomalies in silent disfluent intervals when compared with fluent pauses, as well as differences in intonational phrasing between disfluent and fluent speech. However, when we attempted to use these observations for the purpose of actually distinguishing between spontaneous and read speech, we found that it is hard to account for differences in read and spontaneous speaking style in terms of any of these phenomena. While a sizable fraction of spontaneous utterances contain disfluencies (about 20%, for example, of the *atis0* corpus utterances), a much larger portion do not. Even those containing a disfluent phenomenon exhibit vastly different acoustic and prosodic manifestations of that phenomenon. In a still larger sample of the *atis* corpus, only 2.8% (333/11900) of spontaneous utterances contained one or more filled pauses and 4.8% (572/11900) contained fragments, while 7.1% (844/11900) contained at least one example of either. So, while the presence of disfluencies and their various realizations, including fragments, coarticulatory and glottalization phenomena, filled pauses, cue phrases, and differences in the location and realization of intonational phrasing may variously distinguish some spontaneous utterances from non-spontaneous ones, the large majority of spontaneous utterances in our corpora contained no such cues to distinguish them, in turn, from more formal read speech. However, if in fact it is the overall occurrence of disfluency and its realization phenomena that give listeners cues to genre, over longer spans of speech, it would certainly be important to discover empirically which of these many acoustic clues provide the clearest signals to listeners that the speech they are hearing is actually spontaneous.

4. PROSODIC CORRELATES OF DISCOURSE STRUCTURE FOR SPONTANEOUS AND READ SPEECH

In a third set of studies whose results bear upon speaking style, laboratory read and spontaneous elicited speech were compared to see whether the way speakers convey discourse structural information differs from one speaking genre to the other, as previous studies have suggested (Hirschberg and Nakatani 1996). The Boston Directions Corpus, collected for a larger study of prosodic cues to discourse structure, consists of elicited monologues produced by four non-professional speakers, who were asked to give directions for increasingly complicated tasks, from simply going from one station to another on the subway, to planning a round-trip journey from Harvard Square to several Boston tourist sights. Speakers were given maps and were allowed to write notes to themselves as well as trace routes on the maps. During the recording, speakers were in visual contact only with a silent experimental partner (one of the experimenters) who traced the routes the speakers produced on her own (unseen) map. The presence of this listener, performing a task dependent upon the information the speaker conveyed, was intended to produce a more natural communicative situation, and to encourage speakers to communicate information expressively. The spontaneous elicited recordings were then transcribed, with false starts and other speech errors repaired or omitted. Subjects returned several weeks later to read their own transcribed speech. Both read and spontaneous recordings were then hand labeled for word boundaries, intonational phrasing and pitch accents, using the ToBI labeling conventions (Silverman et al. 1992; Pitrelli et al. 1994). Additional acoustic information (pausal duration between phrases, and each intermediate phrase (in Pierrehumbert 1980 terms, also equivalent to ToBI level 3 or 4 break indices) maximum F0, speaking rate, type of nuclear accent, and intonational contour) was derived automatically from the ToBI-labeled data. Results presented here are based on an analysis of the spontaneous and read speech of only one of the speakers.

The discourse segmentation for the corpus, against which the acoustic features labeled and derived were to be analyzed, was based on Grosz and Sidner's (1986) model of discourse structure. The essential goal of the analysis was to see whether distinguished features of the discourse structure of our speakers' directions, such as the beginning and ending of topic segments, could be reliably associated with any of our acoustic/prosodic features.

According to Grosz and Sidner's model, the utterances composing a discourse are grouped into discourse segments by virtue of the purposes that speakers have in uttering them. These segments in turn may be embedded relative to one another, to form a hierarchical discourse structure reflecting the hierarchical structure of the speaker's intentions. The discourse's *intentional structure* comprises the intentions or *discourse segment purposes* underlying the discourse and relations between them.

To identify this intentional structure in our read and spontaneous recordings, and to identify the differences in listener behavior that might be attributed to intonation rather than to the subject matter itself, we asked three labelers well acquainted with Grosz and Sidner's theory to label the transcriptions of read and spontaneous directions while reading the text alone (group T) and three others to label both genres while reading the transcription and listening to the speech as well (group S). Other than this difference in input modality, all labelers were given the same instructions. The text for each discourse labeling task was presented to labelers with line breaks corresponding to intermediate phrase boundaries (i.e. ToBI level 3 or higher breaks). The instructions asked labelers to analyze the phrases into discourse segments in terms of the intentions they inferred the speakers had in producing the utterances (their *discourse segment purposes*), to label these purposes for each segment, and to then specify the hierarchical relationships among segments. The results of labelers in each condition (T and S) were compared, for read and spontaneous versions of the tasks, and Consensus Labels (intermediate phrases on which all three labelers in the condition agreed) were obtained for the following model-derived categories: segment-initial (SBEG) phrase: did this intermediate phrase begin a discourse segment?; segment-final (SF): did this intermediate phrase end a discourse segment?; and segment-medial (SCONT): was this phrase neither SBEG nor SF?

Our analysis of these Consensus results by condition and by genre showed that the segmentations of group S differed significantly from those of group T. Unlike results of earlier experiments (Grosz and Hirschberg 1992; Grosz et al. 1994), Hirschberg and Nakatani (1996) found that labelers who listened to speech while segmenting produced more consensus boundaries for both read and spontaneous speech than did those who segmented from text alone. That is, labelers agreed more as to whether phrases began, ended, or merely continued a discourse segment when they could play the speech as well as read the transcription. And, curiously, although spontaneous speech is generally thought to exhibit less reliable prosodic indicators of discourse structure than read speech (cf. Ayers 1992), these results showed that our spontaneous recordings actually engendered significantly more consensus labels than did our read speech, for both groups S and T. In particular, labelers agreed more on phrases classified as SCONT – neither beginning nor ending a discourse segment.

In terms of the acoustic/prosodic features which characterized the discourse-classified intermediate phrases, we looked at the following potential cues for the intermediate phrases all labelers classified as SBEGs and SFs: F0 maximum and average F0; *rms* (root-mean-square) maximum and average (as a correlate of loudness); speaking rate; duration of preceding and subsequent pauses; nuclear accent; contour; and speaking rate. We compared segmentation labels not only for group S versus group T, but also for spontaneous versus read speech. We were particularly interested in discovering whether reliable intonational correlates for discourse structure could be found in spontaneous speech as they have been found for read speech or whether the styles are really distinct, in that speakers only vary acoustic/prosodic cues to discourse structure in more formal, planned speech.

In fact, we found strong acoustic/prosodic indicators for intermediate phrases that began (SBEG) and ended (SF) discourse segments in both spontaneous speech and read speech. (T-tests were used to test for statistical significance of differences in the means of phrases, e.g. beginning and not beginning segments. Results reported are significant at the .025 level or better.) Results on consensus SBEG labels were as follows: given group T segmentations, we found significantly higher maximum and average F0, higher maximum and average rms, and shorter subsequent pause for both spontaneous and read speech; for read speech we also found significant correlations for preceding pauses. Given group S segmentations, we found significantly higher maximum and average F0, higher maximum rms, longer preceding and shorter succeeding pauses for read and spontaneous speech; we found higher average rms as well for read speech. Results on consensus SF labels were as follows: given group T segmentations, we found significantly lower average F0 and rms maximum for both read and spontaneous speech, and lower rms average and subsequent pause in addition for read speech. Given group S segmentations, we found lower average F0, rms maximum and average, shorter preceding pause, and longer subsequent pause for both read and spontaneous speech, and in addition, lower F0 maximum for read speech. We found no statistically significant correlations for type of nuclear pitch accent or intonational contour, for phrases of any discourse category, in any condition. So, in this exercise, while we found some differences between read and spontaneous speech – in particular, additional regularities in read speech – there nonetheless seemed to be evidence that listeners used prosodic cues as indicators of discourse structure in both conditions, read and spontaneous.

When we look specifically at those features that varied from one speaking situation to another – and those that did not – some interesting additional possibilities are suggested. Table 3 compares mean values for a number of features over nine direction-giving tasks for a single speaker:

Feature	Read (N=494)	Spontaneous (N=552)
*F0 max for phrase	166 Hz	152 Hz
F0 avg for phrase	113 Hz	108 Hz
F0 max for nuclear accent	143 Hz	132 Hz
Change in F0 max for nuclear accent from prior phrase	1.05	1.04
dB change from prior phrase	-0.210	-0.210
Speaking Rate (syll/sec)	5.15	4.46
Prior pause	338 msec	460 msec
Syllables per phrase	5.84	5.45
*Difference is significant at the .001 level		

Table 3: *Comparison of Some Prosodic Features of Read vs. Elicited Spontaneous Speech*

Note that there are statistically significant differences between the speaker's spontaneous and read productions in the absolute values examined for each intermediate phrase, including maximum and average F0, maximum F0 measured within the nuclear accent (High F0 maximum), speaking rate, and duration of preceding pause. However, this does not hold for the *relative* values examined, amount of F0 change from previous phrase (calculated as the ratio between the current High F0 maximum and that of the preceding intermediate phrase) and the change in decibels from the previous intermediate phrase. In fact, these values (1.05 vs 1.04 and -.210 dB vs -.210 dB) are virtually identical in the two conditions.

Similarly, there is not any significant difference in choice of nuclear pitch accent type or choice of phrase final contour. Tables 4 and 5 show these comparisons. (Nuclear pitch accents in Table 4 are defined as the most intonationally prominent accent in the intermediate phrase, or phrase at ToBI level 3 break index. Phrase endings in Table 5 are defined in terms of tonal markings for ToBI phrases of level 3 or 4, intermediate or intonational phrases in the Pierrehumbert framework.)

Nuclear Accent Type	Read (N=494)	Spontaneous (N=552)
H*, !H*, !H*?	82%	75%
L*, L*?	8%	16%
L+H*, L+H*?, !L+H*, L+H*?	4%	5%
L*+H, L*+!H	0%	0%
H+!H*	3%	2%
*, *?	2%	0%

Table 4: *Comparison of Choice of Nuclear Pitch Accent in Read vs. Elicited Spontaneous Speech ('?' is used to indicate labeler uncertainty as to accent category)*

Ending	Read (N=494)	Spontaneous (N=552)
L-	24%	23%
H-	7%	11%
!H-	0%	0%
L-L%	34%	22%
L-H%	31%	35%
H-L%	2%	7%
!H-L%	0%	1%
H-H%	1%	0%

Table 5: *Comparison of Choice of Phrase Ending in Read vs. Elicited Spontaneous Speech*

As Table 4 shows, while there are minor differences in the proportion of high (H*) vs. low (L*) nuclear accents, the overwhelming majority of nuclear accents in both conditions are high, with only a smattering of other accent types. Similarly, while there are some differences in phrase endings, particularly in the proportion of falling (L-L%) endings between the read and spontaneous conditions, with more variety of phrase endings in the spontaneous condition, the overall patterns are not significantly different from one another.

While these results from the Hirschberg and Nakatani (1996) study are very preliminary, holding only for a single speaker and two variants of a limited laboratory task, they do suggest some interesting possibilities: First, it may be that, while global parameters such as range and timing vary significantly from one speaking situation to the next for a given speaker, the **local** variables that allow for more immediate comparison between one phrase and its predecessor remain remarkably constant. It is entirely plausible, though not yet demonstrated, that it is not the global variables – overall range or timing – that convey to listeners the information they appear to acquire about the structure of the discourse they are listening to. Rather, it is to the local variables, the amount of change from one phrase to the next, that we must look to find the meaningful correlates of discourse structure, and this information is conveyed independently of speaking situation. The similar lack of difference found in choice of contour across the two conditions may be attributable to the nearly identical semantic content of the material in the two conditions, with speakers first generating directions and later returning to read transcriptions of their own prior efforts. If variation in intonational contour does in fact contribute to the semantico-pragmatic content of the message being conveyed, we would not be surprised to find little variation of contour in two such "semantically similar" situations, independent of the differences in speech type. So, the variation observed in global parameters, in overall range and timing, appears to be a relevant area for further exploration in studies of prosodic variation due to speaking style, according to this study.

5. DISCUSSION

Data from several large corpora of American English read and spontaneous speech do provide some comparative information on issues of speech type and raise some interesting additional questions for future analysis.

A study of corpora collected for several DARPA tasks supports the hypotheses that read speech is more rapid than spontaneous speech and that choice of intonational contour does indeed appear to be different for at least some sentence types in read vs. spontaneous utterances. However, a precise characterization of how differences observed in speaking rate between the faster read and slower spontaneous *atis* utterances remains to be determined. And, while an examination of read speech found that indeed only 55–80% of yes-no questions were read with rising "yes-no question" contours, thus confirming the hypothesis that read speech does not produce "natural" intonational contours, a comparison of read and spontaneous versions of sentences in the *atis* corpus shows, paradoxically, that over 43% of spontaneous yes-no questions were uttered with **falling** intonation, compared with only 30% of read yes-no questions. Thus, the notion of a "natural" association between sentence type and intonational contour for yes-no questions in spontaneous speech may need to be re-examined. Similarly, in spontaneous speech, about 30% of declarative sentences were uttered with some type of rising intonation, while over 93% of read declaratives were uttered with a final fall. And *wh*-questions, commonly believed to be normally uttered with falling intonation in English, were only so uttered about 62% of the time in spontaneous speech, while in read speech about 84% of such questions were uttered with falling contours. One may speculate that the association between contour type and sentence type may be something that speakers adhere to more consistently when they are asked to read, particularly longer texts, such as the *atis* dialogues, rather than when they produce the same dialogues naturally.

Other questions about comparative speaking style are raised by studies of disfluency in spontaneous speech. While hesitations and self-repairs may occur in collections of read speech, they are rarely preserved; and their presence in speech presented to listeners does seem to provide a useful cue to listeners that the speech they have been presented with is spontaneous. However, do these phenomena, which are quite difficult to define precisely, actually occur often enough to provide reliable cues in general as to speaking style? In the studies of disfluency discussed above, only about 20% of spontaneous utterances contained any form of disfluency – from filled pauses to self-repairs to labeler-observed "hesitations". And the auditory cues that marked such disfluencies themselves range widely, from "abnormal" phrasings to speech fragments to glottalization and coarticulatory deviations from fluent speech. Determining which of these cues are primary indicators of disfluent speech is clearly important, but only speech fragments and filled pauses appear to occur with any great frequency, even for the 20% of spontaneous utterances that contain disfluencies. So, it is clearly important to search for other cues to speaking style variation in addition to acoustically observable indicators of disfluency.

Finally, it has been speculated that a difference between read and spontaneous speaking styles might be inferred from the fact that regularities observable in read speech in the use of prosodic cues to discourse structure, such as pitch range, speaking rate, and pausal duration, have **not** also been observed for spontaneous speech. However, the preliminary results discussed above from the Boston Directions Corpus suggest that this speculation may be incorrect. Statistically reliable associations between prosodic variation and subject-labeled discourse structure have indeed been found for spontaneous as well as read productions of speakers giving directions. However, while these differences are interesting, the similarities found in the two conditions – in local features such as change in F0 and loudness over the previous phrase, and in intonational contours – may be even more intriguing. If these patterns hold over other speakers, we may be closer to separating out the functions of global features such as overall pitch range and timing and local features involving amount of change in those features and contours, with global features reflecting differences in speech type and local features conveying structural and semantico-pragmatic information in similar ways, independent of speech type. Since it would be odd to find that speakers normally adopt radically different strategies to convey similar information in different speaking situations, and harder still to imagine that speakers simply fail to convey certain types of information in one speech situation that they reliably convey in another, such a hypothesis seems eminently plausible.

REFERENCES

Ayers, G.M. 1992. Discourse functions of pitch range in spontaneous and read speech. Presented at the Linguistic Society of America Annual Meeting.

Barry, W.J. 1995. Phonetics and phonology of speaking styles. *Proc. XIIIth International Congress of Phonetic Sciences* (Stockholm), vol. 2, 4-10.

Bear, J., Dowding, J. and Shriberg. E. 1992. Integrating multiple knowledge sources for detection and correction of repairs in human-computer dialog. *Proc. 30th Annual Meeting,* Association for Computational Linguistics, Newark, DE., 56-63.

Blaauw, E. 1992. Phonetic differences between read and spontaneous speech. *Proc. International Conference on Spoken Language Processing* (Banff), vol. 1, 751-758.

Blackmer, E.R. and Mitton, J.L. 1991. Theories of monitoring and the timing of repairs in spontaneous speech. *Cognition*, 39 173-194.

Bruce, G. 1995. Modelling Swedish intonation for read and spontaneous speech. *Proc. XIIIth International Congress of Phonetic Sciences* (Stockholm), vol. 2, 28-35.

Cooper, W.E. and Sorenson, J.M. 1977. Fundamental frequency contours at syntactic boundaries. *Journal of the Acoustical Society of America* 62, 683-692.

DARPA. 1990. *Proc. DARPA Speech and Natural Language Processing Workshop*, Hidden Valley PA, June. San Francisco:Morgan Kaufmann.

Eskanazi, M. 1993. Trends in speaking styles research. *Proc. Eurospeech 93* (Berlin), 501-509.

Gee, J.P. and Grosjean, F. 1983. Performance structure: A psycholinguistic and linguistic appraisal. *Cognitive Psychology* 15, 411-458.

Grosz, B. and Hirschberg, J. 1992. Some intonational characteristics of discourse structure. *Proc. International Conference on Spoken Language Processing* (Banff), 429-432.

Grosz, B., Hirschberg, J. and Nakatani, C. 1994. A study of intonation and discourse structure and relations. *Proc. Workshop on the Integration of Natural Language and Speech Processing.* AAAI, August.

Grosz, B and Sidner, C. 1986. Attention, intentions, and the structure of discourse. *Computational Linguistics*, 12, 175-204.

Hindle, D. 1983. Deterministic parsing of syntactic non-fluencies. *Proc. 21st Annual Meeting, Association for Computational Linguistics* (Cambridge, MA), 123-128.

Hirschberg, J. 1989. Distinguishing questions by contour in speech recognition tasks. *Proc. DARPA Speech and Natural Language Processing Workshop*, Cape Cod MA, October. San Francisco: Morgan Kaufmann.

Hirschberg, J. and Nakatani, C. 1996. A prosodic analysis of discourse segments in direction-giving monologues. *Proc. 34th Annual Meeting, Association for Computational Linguistics*, Santa Cruz.

Kohler, K.J. 1995. Articulatory reduction in different speaking styles. *Proc. XIIIth International Congress of Phonetic Sciences* (Stockholm) vol. 2, 12-19.

Levelt, W. 1983. Monitoring and self-repair in speech. *Cognition* 14, 41-104.

MADCOW. 1992. Multi-site data collection for a spoken language corpus. *Proc. DARPA Speech and Natural Language Processing Workshop,* 7-14, Harriman NY, Fĕbruary. San Francisco: Morgan Kaufmann.

Nakatani, C. and Hirschberg, J. 1994. A corpus-based study of repair cues in spontaneous speech. *Journal of the Acoustical Society of America* 95, 1603-1616.

Olive, J., Greenwood, A. and Coleman, J. 1993. *The Acoustics of American English Speech: A Dynamic Approach.* New York: Springer.

O'Shaughnessy, D. 1992. Analysis of false starts in spontaneous speech. *Proc. International Conference on Spoken Language Processing* (Banff), 931-934.

Pierrehumbert, J.B. 1980. *The Phonology and Phonetics of English Intonation.* Ph.D. thesis, M.I.T. Distributed by the Indiana University Linguistics Club.

Pitrelli, J., Beckman, M. and Hirschberg, J. 1994. Evaluation of prosodic transcription labeling reliability in the ToBI framework. *Proc. International Conference on Spoken Language Processing* (Yokohama) vol. 2, 123-126.

Price, P., Fisher, W.M., Bernstein, J. and Pallett, D.S. 1988. The DARPA 1000-word Resource Management Database for continuous speech recognition. *Proc. International Conference on Acoustic, Speech and Signal Processing* (New York) vol. 1, 651-654.

Selkirk, E.O. 1984. *Phonology and Syntax: The Relation between Sound and Structure.* Cambridge, MA.: MIT Press.

Shriberg, E., Bear, J. and Dowding, J. 1992. Automatic detection and correction of repairs in human-computer dialog. *Proc. DARPA Speech and Natural Language Processing Workshop*, 419-424, Harriman, NY. San Francisco: Morgan Kaufmann.

Silverman, K., Beckman, M., Pierrehumbert, J., Ostendorf, M., Wightman, C., Price, P. and Hirschberg, J. 1992. ToBI: A standard scheme for labeling prosody. *Proc. Second International Conference on Spoken Language Processing* (Banff), 867-870.

Touati, P. 1995. Pitch range and register in French political speech. *Proc. XIIIth International Congress of Phonetic Sciences* (Stockholm) vol. 4, 244-247.

Wang, M. and Hirschberg, J. 1992. Automatic classification of intonational phrase boundaries. *Computer Speech and Language* 6, 175-196.

AT&T Labs-Research, Florham Park, NJ, USA

INDEX

Text, Speech and Language Technology

1. H. Bunt and M. Tomita (eds.): *Recent Advances in Parsing Technology.* 1996 ISBN 0-7923-4152-X
2. S. Young and G. Bloothooft (eds.): *Corpus-Based Methods in Language and Speech Processing.* 1997 ISBN 0-7923-4463-4
3. T. Dutoit: *An Introduction to Text-to-Speech Synthesis.* 1997 ISBN 0-7923-4498-7
4. L. Lebart, A. Salem and L. Berry: *Exploring Textual Data.* 1998 ISBN 0-7923-4840-0
5. J. Carson-Berndsen, *Time Map Phonology.* 1998 ISBN 0-7923-4883-4
6. P. Saint-Dizier (ed.): *Predicative Forms in Natural Language and in Lexical Knowledge Bases.* 1999 ISBN 0-7923-5499-0
7. T. Strzalkowski (ed.): *Natural Language Information Retrieval.* 1999 ISBN 0-7923-5685-3
8. J. Harrington and S. Cassiday: *Techniques in Speech Acoustics.* 1999 ISBN 0-7923-5731-0
9. H. van Halteren (ed.): *Syntactic Wordclass Tagging.* 1999 ISBN 0-7923-5896-1
10. E. Viegas (ed.): *Breadth and Depth of Semantic Lexicons.* 1999 ISBN 0-7923-6039-7
11. S. Armstrong, K. Church, P. Isabelle, S. Nanzi, E. Tzoukermann and D. Yarowsky (eds.): *Natural Language Processing Using Very Large Corpora.* 1999 ISBN 0-7923-6055-9
12. F. Van Eynde and D. Gibbon (eds.): *Lexicon Development for Speech and Language Processing.* 2000 ISBN 0-7923-6368-X; Pb: 07923-6369-8
13. J. Véronis (ed.): *Parallel Text Processing.* Alignment and Use of Translation Corpora. 2000 ISBN 0-7923-6546-1
14. M. Horne (ed.): *Prosody: Theory and Experiment.* Studies Presented to Gösta Bruce. 2000 ISBN 0-7923-6579-8
15. A. Botinis (ed.): *Intonation.* Analysis, Modelling and Technology. 2000 ISBN 0-7923-6605-0
16. H. Bunt and A. Nijholt (eds.): *Advances in Probabilistic and Other Parsing Technologies.* 2000 ISBN 0-7923-6616-6

KLUWER ACADEMIC PUBLISHERS – DORDRECHT / BOSTON / LONDON